# Lecture Notes in Mathematics

A collection of informal reports and seminars
Edited by A. Dold, Heidelberg and B. Eckmann, Zürich

Series: Department of Mathematics, University of Maryland, College Park
Adviser: J. K. Goldhaber

266

---

# Conference on Harmonic Analysis

## College Park, Maryland, 1971

Edited by Denny Gulick and Ronald L. Lipsman,
University of Maryland, MD/USA

---

Springer-Verlag
Berlin · Heidelberg · New York 1972

AMS Subject Classifications (1970): 22-02, 42-02, 43-02, 46-02

ISBN 3-540-05856-7 Springer-Verlag Berlin · Heidelberg · New York
ISBN 0-387-05856-7 Springer-Verlag New York · Heidelberg · Berlin



Offsetdruck: Julius Beltz, Hemsbach/Bergstr.

# PREFACE

During the five-day period November 8-12, 1971, an International Conference on Harmonic Analysis took place on the College Park campus of the University of Maryland, highlighting a special year of functional analysis in the Department of Mathematics. The participants focused their attention on two aspects of harmonic analysis -- group representations and spectral synthesis. On the one hand, these two fields have many roots in common; on the other hand, their paths have diverged during recent decades. As a result, one aim of the conference was to reintroduce the two fields to each other. A second aim was to report recent work within each area, so as to generate continued interest and vigor within harmonic analysis.

The enclosed papers represent either outlines or detailed expositions of all lectures presented at the conference except those given by F. Bruhat, B. Kostant, C. C. Moore, and R. Spector, and except for the pair of special expository lectures, given by J.-P. Kahane and Y. Katznelson. It is of course impossible to reproduce the individual and group discussions that took place during that week. However, we hope that this volume will serve both those who attended and those who could not attend, and that this permanent record will add to the stimulation of the conference itself.

Nearly all the papers were typed at the institutions of the respective speakers. Nevertheless, we have proofread the manuscripts, and bear responsibility for any inaccuracies they contain.

Denny Gulick

Ron Lipsman

## ACKNOWLEDGMENTS

We thank our colleague Robert L. Ellis for his help on the conference organizing committee, Jacob Goldhaber for his assistance as chairman of the Mathematics Department, Betty Vanderslice and Patricia Berg for their assistance in the editing of manuscripts. We also extend our thanks to the Institutes and Conferences Division of University College at the University of Maryland, and in particular Edwin Crispin and his staff, for their role in setting up the conference. Moreover, we appreciate greatly the generous support granted the conference by the National Science Foundation. Finally, we are grateful to Springer-Verlag for its prompt and efficient publication of these proceedings.

CONTENTS

* For papers with more than one author, an asterisk follows the name of the author who delivered the lecture.

# ANALYSE HARMONIQUE DES MESURES NON BORNEES SUR LES GROUPES ABELIENS LOCALEMENT COMPACTS

par

Loren N. Argabright, University of Nebraska

et

Jesús Gil de Lamadrid, University of Minnesota et Université de Rennes

## 1. Introduction

Ce travail contient un résumé des principales notions et résultats (sans preuves) d'une théorie de l'analyse harmonique des mesures sur un groupe abélien localement compact $G$. Un manuscrit [2] (voir aussi [1] ), à paraître ailleurs, contient les détails de la première étape de cette théorie qui est une étude des propriétés fondamentales de la transformation de Fourier des mesures transformables. Un autre travail, à paraître ultérieurement, traitera les détails de la théorie des mesures presque périodiques. Le problème qu'on se pose dans cet exposé est de placer dans le cadre des mesures sur $G$ l'analyse harmonique classique des fonctions. C'est une sorte de théorie des " distributions tempérées " dans les groupes topologiques, mais

---

Au cours de ce travail, les auteurs ont bénéficié de subventions de la (U.S.A.) National Science Foundation.

la théorie exposée ici est assez différente de la théorie de L. Schwartz [7] et de celle qu'on obtient en développant dans cette direction les idées de Bruhat [4] sur les distributions dans les groupes topologiques.

L'outil principal dans ce travail est d'employer la convolution par les fonctions " d'essai ", appartenant à des classes diverses, pour ramener une notion ou un résultat, connu dans le domaine des fonctions, au domaine des mesures. Cette idée a son origine dans le travail de L. Schwartz sur les distributions tempérées, où les fonctions d'essai sont des fonctions indéfiniment différentiables sur $R^n$. Dans notre cas, faute de la notion de dérivée, les espaces de fonctions d'essai sont le plus souvent l'espace $K(G)$ de toutes les fonctions complexes continues sur $G$ à support compact, et des espaces obtenus à partir de $K(G)$ au moyen des opérations élémentaires de l'analyse harmonique.

En appliquant cette idée générale, on obtient diverses classes de mesures importantes pour ce travail. Par exemple, on dit qu'une mesure $\mu$ sur $G$ est à <u>translatées bornées</u>, si, quel que soit $f \in K(G)$, la convolée $f * \mu(x) = \int_G f(xy^{-1}) d\mu(y)$, (toujours bien définie, toujours fonction continue de x), est une fonction bornée. La classe $M_B(G)$ de mesures sur $G$ à translatées bornées contient la plupart des autres classes de mesures étudiées ici. C'est pour les éléments de $M_B(G)$ qu'on peut établir l'existence des convolutions étudiées dans la suite.

Par le même principe, on obtient les mesures fortement (et faiblement) presque périodiques, et les mesures transformables. Pour une mesure appartenant à cette dernière classe, on obtient une mesure $\hat{\mu}$ sur le groupe dual $\Gamma$ de $G$, sa transformée de Fourier, qui généralise la théorie classique. Le même principe de convolution mène à la caractérisation de certaines autres classes de mesures. Par exemple, par ce moyen on caractérise les mesures sur $\Gamma$ qui sont transformées de Fourier de mesures sur $G$, et les mesures sur $G$ ayant pour transformées de Fourier des mesures qui sont absolument continues.

La théorie des mesures presque périodique est analogue et généralise la théorie classique de Bohr - Eberlein ; seulement, dans le cas de mesures, de nouveaux phénomènes se produisent.

## 2. Généralités sur les convolutions et les mesures à translatées bornées

On note $M(G)$ la classe de toutes les mesures (complexes, régulières) sur $G$, et $C(G)$ l'espace de toutes les fonctions complexes continues bornées sur $G$. Quel que soit $\mu \in M(G)$, $|\mu|$ désignera la (mesure) valeur absolue de $\mu$. Comme Bourbaki [3], on dira que les deux mesures $\mu_1$, $\mu_2 \in M(G)$ sont convolables, si, quel que soit $f \in K(G)$, la fonction $F(x,y) = f(xy)$ appartient à $L_1(\mu_1 \otimes \mu_2)$. Dans ce cas là, la convolée $\mu_1 * \mu_2$ est donnée par la formule

$$(2.1) \qquad \int_G f(x)d\mu_1 * \mu_2(x) = \int_G \int_G f(xy)d\mu_1(x)d\mu_2(y).$$

On ne connaît pas de conditions très générales garantissant l'existence de la convolée de deux mesures. On sait, par exemple, que si l'une des mesures est à support compact, la convolée existe. On obtient maintenant deux caractérisations des mesures à translatées bornées. On note $A_x$ la translatée d'une partie $A$ de $G$ par l'élément $x \in G$.

Théorème 2.1

*Soit* $\mu \in M(G)$ . *Alors* $\mu$ *est à translatées bornées si et seulement si, quel que soit* $A$ , *partie compacte de* $G$ , *la fonction* $t(x) = |\mu|(A_x)$ *de* $x$ *est bornée*.

Ce théorème explique la terminologie *translatées bornées*.

Théorème 2.2

*Soit* $\mu \in M(G)$ . *Alors* $\mu$ *est à translatées bornées si et seulement si* $\mu$ *est convolable avec toute mesure bornée*.

On note $M_B(G)$ l'ensemble de toutes les mesures sur $G$ à translatées bornées, et $M_F(G)$ l'espace de toutes les mesures complexes bornées (finies) sur $G$ .

Théorème 2.3

*L'espace* $M_B(G)$ *est stable par la convolution par les mesures de* $M_F(G)$ . *D'ailleurs, quels que soient* $\mu \in M_B(G)$ *et*

$\Theta_1, \Theta_2 \in M_F(G)$ on a la relation d'associativité

(2.2) $\mu * (\Theta_1 * \Theta_2) = (\mu_1 * \Theta_1) * \Theta_2$.

## 3. Les mesures transformables

Soient $B(G)$ l'espace des transformées de Fourier-Stieltjes de mesures finies sur $\Gamma$, et $K_2(G)$ l'espace vectoriel engendré par les fonctions de la forme $g*h$, $g$, $h \in K(G)$. Une mesure $\mu$ sur $G$ est dite transformable si, quel que soit $f \in K_2(G)$, $f * \mu \in B(G)$. Le théorème suivant justifie la terminologie. On note $M_T(G)$ l'espace de toutes les mesures transformables sur $G$. On emploiera la notation $\check{f}$, quel que soit $f \in L_1(G)$, pour désigner la transformée inverse de Fourier, à savoir, $\check{f}(\gamma) = \hat{f}(\gamma^{-1})$, $\gamma \in \Gamma$,

Théorème 3.1

A toute mesure $\mu \in M_T(G)$ on peut faire correspondre une mesure $\hat{\mu}$ sur $\Gamma$, et une seule, telle que, quel que soit $f \in K(G)$, $\check{f} \in L_2(\hat{\mu})$ et

(3.1) $$\int_G f * f^*(x) d\mu(x) = \int_\Gamma |\check{f}(\gamma)|^2 d\hat{\mu}(\gamma).$$

L'application $\mu \to \hat{\mu}$ de $M_T(G)$ dans $M(\Gamma)$ est injective.

On appellera $\hat{\mu}$ la <u>transformée de Fourier</u> de $\mu$ . Le théorème suivant justifie le nom. On identifiera désormais toute fonction localement intégrable $f$ sur $G$ avec la mesure $f(x)dx$.

Théorème 3.2

*L'espace $M_T(G)$ de mesures transformables contient tous les espaces $L_p(G)$ , $1 \leq p \leq 2$ , $M_F(G)$ et l'espace $M_P(G)$ de toutes les combinaisons linéaires de mesures de type positif. Dans tous les cas, $\hat{\mu}$ coïncide avec la transformée de Fourier connue pour ces mesures.*

Pour $L_p(G)$ , la théorie est classique. Pour $M_F(G)$ il s'agit de la transformée de Fourier - Stieltjes. Pour $\mu$ de type positif, $\hat{\mu}$ coïncide (à part que notre notation ne se conforme pas à la sienne) avec la transformée de Fourier de Godement [6] . En particulier, si $\mu$ est une fonction continue de type positif, $\hat{\mu}$ est la mesure bornée sur $\Gamma$ que lui fait correspondre le théorème de Bochner. Il se trouve que la transformée de Fourier de la mesure de Dirac en l'élément neutre (corresp. la mesure de Haar) de $G$ est la mesure de Haar (corresp. , la mesure de Dirac en l'élément neutre) de $\Gamma$ , ce qui montre que, en général, ni la mesure $\mu \in M_T(G)$, ni sa transformée de Fourier $\hat{\mu}$ n'est ni une mesure bornée ni une fonction.

On note $M_{\hat{T}}(\Gamma)$ l'espace des mesures sur $\Gamma$ qui sont transformées de Fourier de mesures dans $M_T(G)$ , et $A(G)$ le sous - espace de $B(G)$ formé par les transformées de Fourier des fonctions de $L_1(\Gamma)$.

Théorème 3.3

Toute mesure $\hat{\mu} \in M_{\hat{T}}(\Gamma)$ est à translatées bornées.

Par contre, il n'est pas vrai que toute $\mu \in M_T(G)$ est à translatées bornées, comme le montre un exemple inédit de J.I. Richards d'une mesure de type positif sur $\mathbb{R}$ qui n'est pas à translatées bornées. Dans les deux théorèmes suivants, on se sert de la convolution par divers types de " fonctions d'essai " pour donner des caractérisations de diverses classes de mesures. Le premier caractérise les mesures $\mu \in M_T(G)$ telles que les $\hat{\mu}$ sont des fonctions. L'exemple élémentaire d'une telle mesure est une $\mu \in M_F(G)$.

Théorème 3.4

Soit $\mu$ une mesure transformable sur $G$ . Alors, $\hat{\mu}$ est absolument continue si et seulement si $h * \mu \in A(G)$ , quel que soit $h \in K_2(G)$.

Théorème 3.5

Soit $\nu \in M(\Gamma)$ . Alors, il existe $\mu \in M_T(G)$ telle que $\nu = \hat{\mu}$ , si et seulement si $\nu$ est à translatées bornées et $\hat{h} * \nu \in B(\Gamma)$ , quel que soit $h \in K_2(G)$ . En outre, $\mu$ est absolument continue si et seulement si $\hat{h} * \nu \in A(\Gamma)$ , quel que soit $h \in K_2(G)$.

4 - L'inversion de la transformée de Fourier et la formule sommatoire de Poisson.

Quel que soit $\mu \in M_T(G)$ , on note $\check{\mu}$ la mesure sur $\Gamma$ définie par la formule

$$(4.1) \qquad \int_\Gamma \phi(\gamma)d\check{\mu}(\gamma) = \int_\Gamma \phi(\gamma^{-1})d\hat{\mu}(\gamma)$$

quel que soit $\phi \in K(\Gamma)$. On pourrait s'attendre à la formule d'inversion

$$(4.2) \qquad \mu = \check{\hat{\mu}}$$

Malheureusement, en général, $\hat{\mu}$ n'appartient pas nécessairement à $M_T(\Gamma)$ et l'expression à droite de (4.2), en général, n'a pas de sens. On note $I(G)$ l'espace de toutes les $\mu \in M_T(G)$ telles que $\hat{\mu} \in M_T(\Gamma)$ . Or, on a le théorème suivant.

Théorème 4.1

Toute $\mu \in I(G)$ satisfait à l'identité (4.2).

En revenant à la condition de transformabilité énoncée dans le théorème (3.1), on voit que l'identité (3.1) peut être formulée comme suit :

$$(4.3) \qquad \int_G g(x)d\mu(x) = \int_\Gamma \check{g}(\gamma)d\hat{\mu}(\gamma) \ ,$$

quel que soit $g \in K_2(G)$ , où $\check{g}(\gamma) = \hat{g}(\gamma^{-1})$ , pour $\gamma \in \Gamma$ . On obtient une forme généralisée de la formule sommatoire de Poisson en cherchant (mais, malheureusement, ne trouvant jamais) la classe la plus large des fonctions $g$ satisfaisant à (4.3) , étant donnée une mesure $\mu \in \mathcal{M}_T(G)$ .

Théorème 4.2 (Formule sommatoire de Poisson généralisée)

Supposons que $\mu$ soit une mesure transformable sur $G$ et que $f$ soit une fonction complexe, soit continue, soit supportée par un ensemble, réunion dénombrable de compacts. On suppose de plus que $f$ soit intégrable (c'est à dire, $f \in L_1(G)$ ) et convolable avec $\mu$ et que $\hat{f} \in L_1(\hat{\mu})$ . Alors, localement, pour presque tout $x \in G$,

$$(4.4) \qquad f * \mu\ (x) = \int_G f(xy^{-1})d\mu(y) = \int_\Gamma \gamma(x)\hat{f}(\gamma)d\hat{\mu}(\gamma).$$

D'ailleurs, si la première intégrale dans (4.4) est bien définie et représente une fonction continue de $x$ au voisinage d'un point $u \in G$ , la formule est exacte pour $x = u$ . En particulier, pour $u = e$ , l'élément neutre de $G$ , on a

$$(4.5) \qquad \int_G f(y^{-1})d\mu(y) = \int_\Gamma \hat{f}(\gamma)d\hat{\mu}(\gamma).$$

Dans le théorème, on doit interpréter l'écriture de chaque intégrale comme contenant le fait démontrable de l'intégrabilité de la fonction en question. Si $\mu$ est une mesure de Haar d'un sous - groupe fermé $H$ de $G$, considérée comme mesure sur $G$, alors $\mu$ est de type positif, donc transformable. On peut montrer dans ce cas que $\hat{\mu}$ est une mesure de Haar du groupe orthogonal $H^{\perp}$ de $H$. Dans ce cas, on retrouve, à partir de (4.5) la formule classique de Poisson

$$(4.6) \qquad \int_H f(y)dy = \int_{H^{\perp}} \hat{f}(\gamma)d\gamma .$$

## 5. Les mesures transformables et les distributions

On peut considérer la théorie de mesures transformables comme une sorte de théorie de distributions tempérées dans les groupes abéliens localement compacts, dans le cas spécial où toutes les distributions envisagées sont des mesures. Bruhat [4] a introduit dans les groupes localement compacts une théorie des distributions tempérées, généralisation directe de la théorie de L. Schwartz. A partir du travail de Bruhat, on peut développer la transformation de Fourier des distributions tempérées dans les groupes abéliens, de façon analogue à la transformation de Fourier-Schwartz. Dans ce cadre d'idées, on a le théorème suivant.

Théorème 5.1

Toute mesure transformable $\mu$ est une distribution tempérée, et sa transformée de Fourier, en tant que distribution, coïncide avec la mesure $\hat{\mu}$.

La démonstration du théorème 5.1 dépend du théorème suivant.

Théorème 5.2

Toute mesure à translatées bornées est à croissance lente.

On devra distinguer entre les mesures transformables sur G et les mesures tempérées sur G dont les transformées de Fourier (en tant que distributions) sont elles - mêmes des mesures, puisque la transformée de Fourier $\hat{\mu}$ d'une mesure transformable $\mu$ sur G est une distribution tempérée sur $\Gamma$, dont la transformée de Fourier (en tant que distribution) est une mesure $\mu'$, définie par la formule $\mu'(f) = \int_G f(x^{-1})d\mu(x)$, $f \in K(G)$. D'autre part $\hat{\mu}$ n'est nécessairement pas transformable. Autrement (théorème 3.3), $\mu$ serait à translatées bornées, ce qui, on l'a vu, n'est pas toujours vrai.

## 6. Les mesures presque périodiques

Soit $\mu \in M(G)$. On dit que $\mu$ est fortement (resp. faiblement) presque périodique si, quel que soit $f \in K(G)$, $f * \mu$ est une fonction fortement (resp. faiblement) presque périodique. Evidemment, toute mesure fortement presque périodique est faiblement presque périodique.

Théorème 6.1

Toute mesure faiblement presque périodique est à translatées bornées.

Il convient d'introduire la topologie suivante sur $M_B(G)$ . Pour chaque $f \in K(G)$ , on définit la semi - norme $||\mu||_f = ||f * \mu||_\infty$ , $\mu \in M_B(G)$ . L'expression $||\mu||_f$ définit bien un nombre fini en raison de ce que $\mu$ est à translatées bornées. Toutes les idées topologiques employées ici à l'égard de $M_B(G)$ , doivent être entendues par rapport à la topologie localement convexe de $M_B(G)$ définie par les semi - normes $||\ ||_f$ , $f \in K(G)$ . Dans ce sens, on parlera de la topologie faible (affaiblie). Quel que soit $x \in G$ , on appellera la convolution $\delta_x * \mu$ la translatée de $\mu$ par $x$ .

Théorème 6.2

Soit $\mu \in M_B(G)$ . Alors, une condition nécessaire et suffisante pour que $\mu$ soit fortement (resp. faiblement) presque périodique est que l'ensemble $\{\delta_x * \mu\}_{x \in G}$ des translatées de $\mu$ soit relativement compact pour la topologie donnée (affaiblie) de $M_B(G)$ .

On note $SAP(G)$ (resp. , $WAP(G)$ ) l'espace des mesures fortement (faiblement) presque périodiques sur $G$ . On a, donc, $WAP(G) \supset SAP(G) \supset M_B(G)$ . En particulier, le théorème 6.2 montre que la mesure de Haar (notée $\lambda$ ) est faiblement presque périodique. La relation entre la presque périodicité des mesures et celle des fonctions peut être décrite par le théorème suivant.

Théorème 6.3

Soit $f$ une fonction continue sur $G$. Alors $f$ est une fonction fortement (faiblement) presque périodique si et seulement si $f$ est uniformément continue et est une mesure fortement (faiblement) presque périodique.

Donc, les espaces $\mathcal{SAP}(G)$ et $\mathcal{WAP}(G)$ contiennent les espaces correspondants SAP(G) et WAP(G) de fonctions presque périodiques. On voit sans peine, à partir du théorème 6.2, que les espaces $\mathcal{SAP}(G)$ et $\mathcal{WAP}(G)$ sont stables pour les translations par les éléments de $G$.

Théorème 6.4

Il existe sur $\mathcal{WAP}(G)$ une forme linéaire continue $M$, et une seule, qui reste invariante par les translatées par les éléments de $G$, telle que $M(\lambda) = 1$. Cette forme linéaire prolonge à $\mathcal{WAP}(G)$ la forme linéaire sur WAP(G) définie par la moyenne ordinaire. Quel que soit $\mu \in \mathcal{WAP}(G)$, la mesure $M(\mu)\lambda$ est la seule mesure sur $G$, invariante pour les translations et contenue dans l'enveloppe convexe fermée de l'ensemble $\{\delta_x * \mu\}_{x \in G}$ des translatées de $\mu$.

Les théorèmes 6.2 et 6.4 sont à la base de l'extension aux mesures presque périodiques de la théorie classique de fonctions presque périodiques. Quelques résultats suffisent pour en donner une idée. On connaît l'isomorphisme canonique de l'algèbre stellaire SAP(G) sur l'algèbre stellaire $C(G_b)$ de toutes les fonctions complexes continues sur le compactifié de Bohr $G_b$ de G.

Théorème 6.5

*Il existe un homomorphisme* $\mu \to \mu_b$ *de* $\mathcal{WAP}(G)$ *dans* $M(G_b)$ *tel que, quels que soient* $\mu \in \mathcal{WAP}(G)$ , $f \in SAP(G)$ , *on ait* $f\mu \in \mathcal{WAP}(G)$ *et*

$$(6.1) \qquad \int_{G_b} f(x_b)d\mu_b(x_b) = M(f\mu)$$

*La restriction à* $\mathcal{SAP}(G)$ *de cet homomorphisme est injective et prolonge l'isomorphisme canonique de* SAP(G) *sur* $C(G_b)$.

On appelle mesure faiblement presque périodique *nulle* une mesure $\mu \in \mathcal{WAP}(G)$ telle que $\mu_b = 0$.

Théorème 6.6 (*Décomposition d'Eberlein généralisée*).

*Toute mesure* $\mu \in \mathcal{WAP}(G)$ *possède une décomposition unique de la forme*

(6.2) $$\mu = \mu_a + \mu_0 ,$$

où $\mu_a$ est une mesure fortement presque périodique et $\mu_0$ est une mesure faiblement presque périodique nulle.

Le rapport entre les mesures presque périodiques et les mesures transformables est explicité par le théorème suivant.

Théorème 6.7

Quel que soit $\mu \in M_T(G)$ , $\hat{\mu}$ est faiblement presque périodique. La mesure $\hat{\mu}$ est fortement presque périodique si et seulement si $\mu$ est discrète.

Théroème 6.8

Etant donnée une mesure transformable $\mu$ sur $G$ , sa partie discrète $\mu_d$ et sa partie continue $\mu_c$ sont transformables. En outre, la décomposition d'Eberlein généralisée de $\hat{\mu}$ correspond à la décomposition $\mu = \mu_d + \mu_c$ de $\mu$ , dans le sens que $(\mu_d)\hat{} = (\hat{\mu})_a$ et $(\mu_c)\hat{} = (\hat{\mu})_0$ .

## REFERENCES


1. L. N. Argabright and J. Gil de Lamadrid

   " Fourier transforms of unbounded measures "

   Bull. Amer. Math. Soc., Vol. 77 (1971) , PP. 355-359

2. L. N. Argabright and J. Gil de Lamadrid

   " Fourier analysis of unbounded measures on locally compact abelian groups "

   A paraître

3. N. Bourbaki

   " Eléments de mathématique . Intégration "

   Chapitres 7 - 8 - Hermann - Paris - 1963

4. F. Bruhat

   " Distributions sur un groupe localement compact et applications à l'étude des représentations des groupes p - adiques "

   Bull. Soc. Math. France - T. 89(1961) - PP. 43 à 75

5. J. Glimm

   " Lectures on harmonic analysis "

   Courant Institute Notes - New York University - 1965

6. R. Godement

   " Introduction aux travaux de Selberg "

   Séminaire Bourbaki - 1956-1957 - Exposé n° 144
   Secrétariat de Mathématiques - Paris - 1959 -

7. L. Schwartz

   " Théorie des distributions "

   T. II - nouvelle édition - Hermann - Paris - 1966

# FIBER BUNDLE STRUCTURES AND HARMONIC ANALYSIS OF COMPACT HEISENBERG MANIFOLDS

by

Louis Auslander
Graduate Center, City University of New York, and
Institute for Advanced Study, Princeton, N. J.

and

Jonathan Brezin*
University of Minnesota, Minneapolis, Minnesota

1. Introduction: We are going to be working with the three-dimensional nilpotent Lie group $\mathbb{N}$, whose underlying topological space is 3-dimensional Euclidean space $\mathbb{R}^3$ and whose group operation is given by $(x,y,z)(x',y',z') = (x+x',y+y',z+z'+xy')$. This group is commonly called the (3-dimensional) Heisenberg group. The subset $\Gamma$ of $\mathbb{N}$ consisting of those points $(a,b,c)$ with $a,b,c \in \mathbb{Z}$ is a subgroup of $\mathbb{N}$, and the quotient $\mathbb{N}/\Gamma$ is compact. For technical reasons, it is convenient to work with cosets of the form $\Gamma g$, $g \in \mathbb{N}$ -- thus, despite writing $\mathbb{N}/\Gamma$, we will mean cosets of the form $\Gamma g$. The manifold $\mathbb{N}/\Gamma$ supports a unique translation-invariant probability measure, which we shall denote by $\nu$. Our main concern will be the analysis of the regular representation $R$ of $\mathbb{N}$ on $L^2(\mathbb{N}/\Gamma,\nu)$, which is defined by

$$(R_g\varphi)(\Gamma h) = \varphi(\Gamma hg)$$

for all $\varphi \in L^2$, and all $g,h \in \mathbb{N}$.

Let $\mathbb{C}$ be the subgroup of $\mathbb{N}$ consisting of all elements of the form $(a,b,z)$, where $a$ and $b$ lie in $\mathbb{Z}$ and $z$ lies in $\mathbb{R}$: $\mathbb{C}$ is the subgroup spanned by $\Gamma$ together with the center $\mathbb{Z}=z\mathbb{N}$ of $\mathbb{N}$, $z\mathbb{N}$ being $\{(0,0,z) : z \in \mathbb{R}\}$. The subgroup $C$ is normal in $\mathbb{N}$, and $\mathbb{N}/C$ is the 2-dimensional torus $\mathbb{T}^2$. Since $\mathbb{Z}/\Gamma \cap \mathbb{Z} = \mathbb{T}$, we have that $\mathbb{N}/\Gamma$ is a principal circle bundle over $\mathbb{T}^2$:

---

*Both authors are partially supported by grants from The National Science Foundation. The first author is a John Simon Guggenheim Fellow and the second is an Alfred P. Sloan Fellow.

$$\begin{array}{ccc} \mathbb{T} & \longrightarrow & \mathbb{N}/\Gamma \\ & & \downarrow p \\ & & \mathbb{T}^2 \end{array} \qquad (*)$$

When one passes to fundamental groups, the diagram (*) becomes a short exact sequence

$$1 \longrightarrow \mathbb{Z} \longrightarrow \Gamma \longrightarrow \mathbb{Z} \times \mathbb{Z} \longrightarrow 1.$$

Returning to the diagram (*), we see that functions on $\mathbb{T}^2$ can be viewed (by composing with p) as functions on $\mathbb{N}/\Gamma$. Thus, $C^\infty(\mathbb{T}^2)$ operates on $L^2(\mathbb{N}/\Gamma, \nu)$ via pointwise multiplication. The structure of $L^2(\mathbb{N}/\Gamma, \nu)$ as a module over $C^\infty(\mathbb{T}^2)$ can be used to show that the direct-sum decomposition of $L^2(\mathbb{N}/\Gamma, \nu)$ into irreducible R-invariant subspaces exhibited by Brezin [2] and Richardson [4] is "essentially" unique. How this comes about will be indicated in this note.

2. Presenting $\mathbb{N}$: The presentation of $\mathbb{N}$ as the middle term of the group extension $1 \longrightarrow Z \longrightarrow \mathbb{N} \longrightarrow \mathbb{R}^2 \longrightarrow 1$ is analogous to the presentation of $\mathbb{N}/\Gamma$ in (*). For what follows it is convenient to look at $\mathbb{N}$ from a somewhat different point of view. Let $\mathbb{A}$ denote the subgroup $\{(0,y,z) : y,z \in \mathbb{R}\}$ of $\mathbb{N}$, and let $\mathbb{P} = \{(x,0,0) : x \in \mathbb{R}\}$. We then have a split exact sequence $1 \longrightarrow \mathbb{P} \longrightarrow \mathbb{N} \longrightarrow \mathbb{A} \longrightarrow 1$ and $\mathbb{N}$ is the semi-direct product $\mathbb{A}\cdot\mathbb{P}$. Our choice of $\mathbb{P}$ and $\mathbb{A}$ is compatible with $\Gamma$ in the sense that if $\Gamma_{\mathbb{A}} = \Gamma \cap \mathbb{A}$ and $\Gamma_{\mathbb{P}} = \Gamma \cap \mathbb{P}$, then $\Gamma = \Gamma_{\mathbb{A}} \cdot \Gamma_{\mathbb{P}}$. (It is worth observing that we have indeed made a choice of $\mathbb{A}$ and $\mathbb{P}$ -- neither $\mathbb{A}$ nor $\mathbb{P}$ is unique; they are merely the natural choices, coordinates in $\mathbb{N}$ being given.)

Let $\nu^*$ be the unique translation-invariant measure on $\mathbb{N}/\Gamma_{\mathbb{A}}$, and let $R^*$ denote the regular representation of $\mathbb{N}$ on $L^2(\mathbb{N}/\Gamma_{\mathbb{A}}, \nu^*)$. It turns out that, to analyze R, it is helpful first to analyze $R^*$. The analysis of $R^*$ is essentially done in the Stone-von Neumann theorem:

Let n and m be two integers. Given $f \in L^2(\mathbb{R})$, we set $(S_{n,m}f)(x,y,z) = e(nz + my)f(x)$, where $e(\theta) = e^{2\pi i\theta}$ for $\theta \in \mathbb{R}$. It is not hard to see that $S_{n,m}f$ is constant on right $\Gamma_{\mathbb{A}}$ cosets and hence defines a function on $\mathbb{N}/\Gamma_{\mathbb{A}}$. Furthermore, $S_{n,m}$ is an isometry onto its image in $L^2(\mathbb{N}/\Gamma_{\mathbb{A}}, \nu^*)$. Let $H^*(n,m)$ denote the image of $S_{n,m}$. It is not hard to see that

$L^2(\mathbb{N}/\Gamma_{\mathbb{A}},\nu) = \Sigma_{n=-\infty}^{\infty}\Sigma_{m=-\infty}^{\infty} \oplus H^*(n,m)$. The interesting part of the sum is that where $n \neq 0$, because when $n = 0$, the functions in $H^*(n,m)$ are functions on $\mathbb{N}/\Gamma_{\mathbb{A}} Z = \mathbb{R} \times \mathbb{T}$ and hence submit to classical Fourier analysis.

Theorem 1: *Assume $n \neq 0$. Each $H^*(n,m)$ is then an irreducible $R^*$-invariant subspace of $L^2(\mathbb{N}/\Gamma_{\mathbb{A}},\nu^*)$. Furthermore, given any two integers $m$ and $m'$, we have that the restrictions $R^*|H^*(n,m)$ and $R^*|H^*(n,m')$ are unitarily equivalent.*

Proof: For each $(x,y,z) \in \mathbb{N}$ and each $\varphi \in L^2(\mathbb{R})$, define

$$(U(n,m)_{(x,y,z)}\varphi)(t) = \mathbb{e}\,(n(z + ty) + my)\varphi(t + x),$$

for all $t \in \mathbb{R}$. It is easy to see that $U(n,m)$ is an irreducible unitary representation of $\mathbb{N}$ on $L^2(\mathbb{R})$. Also, direct computation reveals that if $g \in \mathbb{N}$ and $\varphi \in L^2(\mathbb{R})$, then

$$S_{n,m}U(n,m)_g\varphi = R^*_g S_{m,n}\varphi.$$

It follows that $H^*(n,m)$ is an irreducible $R^*$-invariant subspace.

That $R^*|H^*(n,m)$ and $R^*|H^*(n,m')$ are unitarily equivalent is the Stone-von Neumann theorem.

Q.E.D.

Consider now the covering space $\pi : \mathbb{N}/\Gamma_{\mathbb{A}} \longrightarrow \mathbb{N}/\Gamma$. The group of covering transformations is a copy of $\mathbb{Z}$ and the transformations can be written $C(j) : \Gamma_{\mathbb{A}}g \longmapsto \Gamma_{\mathbb{A}}(j,0,0)g$ for all $j \in \mathbb{Z}$. Suppose now that $\varphi \in H^*(n,m)$. Where is $\varphi \circ C(j)$? Writing $\varphi = S_{n,m}\psi$, we have

$$\begin{aligned}(\varphi \circ C(j))(x,y,z) &= \varphi(x + j,\ y,\ z + jy)\\ &= \mathbb{e}\,(n(z + jy) + my)\psi(x + j)\\ &= \mathbb{e}\,(nz + (m + j)y)\psi_j(x)\\ &= (S_{n,m+nj}\psi_j)(x,y,z).\end{aligned}$$

Hence, composition with $C(j)$ sends $H^*(n,m)$ to $H^*(n,m+nj)$.

Let $\varphi$ be a function in $H^*(n,m)$ having compact support. We can then form $J\varphi = \Sigma_{j=-\infty}^{\infty}\varphi \circ C(j)$. It is easy to see that $J\varphi$ (viewed as a function of $(x,y,z)$) is constant not only on right $\Gamma_{\mathbb{A}}$ cosets, but on right $\Gamma$ cosets as well. Let us set $H(n,m)$ equal to the closure in $L^2(\mathbb{N}/\Gamma,\nu)$ of family of functions $J\varphi$, where $\varphi$ traces the functions in $H^*(n,m)$ having compact support. One can show (cf.[3],section 3) that $J$ extends to an isometry from $H^*(n,m)$ onto $H(n,m)$, and that $R_g J = J(R_g^*|H^*(n,m))$.

From the fact that composing with $C(j)$ sends $H^*(n,m)$ to $H^*(n,m+jn)$, one can see that

(1) $H(n,m) = H(n,m')$ if, and only if, $m \equiv m'$ modulo n.

(2) $H(n,m) \perp H(n,m')$ if $m \not\equiv m'$ modulo n.

(3) $L^2(\mathbb{N}/\Gamma,\nu) = H(0) \oplus \Sigma_{n\neq 0}\Sigma_{m=0}^{|n|-1} \oplus H(n,m)$, where by $H(0)$ we mean those functions in $L^2(\mathbb{N}/\Gamma,\nu)$ constant on cosets of $\Gamma\mathbb{Z}$.

The decomposition of $L^2(\mathbb{N}/\Gamma,\nu)$ that appears in (3) is, except (obviously) for the summand $H(0)$, a decomposition into irreducible $R$-invariant subspaces. This decomposition is unique in the following sense:

Let $\varphi \in C^\infty(\mathbb{N}/\mathbb{A}\Gamma)$, and let $\psi \in H(n,m)$. Then, multiplying $\varphi$ and $\psi$ pointwise ($\varphi$ being viewed as defined on $\mathbb{N}/\Gamma$), we get another function $\varphi\psi$ in $H(n,m)$ -- in fact, given $\theta \in H^*(n,m)$, we have $\varphi J\theta = J\varphi\theta$. Hence, if we view $L^2(\mathbb{N}/\Gamma,\nu)$ as a module over $C^\infty(\mathbb{N}/\mathbb{A}\Gamma)$, then each $H(n,m)$ will be a $C^\infty(\mathbb{N}/\mathbb{A}\Gamma)$-submodule.

Theorem 2: The decomposition $\Sigma_{n\neq 0}\Sigma_{m=0}^{|n|-1} \oplus H(n,m)$ of $H(0)^\perp$ is the unique decomposition of $H(0)^\perp$ into irreducible $R$-invariant subspaces all of which are $C^\infty(\mathbb{N}/\mathbb{A}\Gamma)$ submodules.

This theorem is a special case of the results announced in [1]. The proof is a bit involved, and so will be omitted.

3. More on $L^2(\mathbb{N}/\Gamma, \nu)$ as a $C^\infty(\mathbb{N}/\mathbb{Z}\Gamma)$-module: We have already seen that if $C^\infty(\mathbb{N}/\mathbb{Z}\Gamma)$ is viewed as operating on $L^2(\mathbb{N}/\Gamma, \nu)$ by pointwise multiplication, then the subspaces $H(n,m)$ are essentially characterized by being multiplied into themselves by the functions in $C^\infty(\mathbb{N}/\mathbb{Z}\Gamma)$ constant on $\mathbb{A}\Gamma$ cosets. One is led by this phenomenon to ask: if $K$ is an infinite-dimensional irreducible $R$-invariant subspace of $L^2(\mathbb{N}/\Gamma,\nu)$, what can one say about the subalgebra $\{\varphi \in C^\infty(\mathbb{N}/\mathbb{Z}\Gamma) : \varphi K \subseteq K\} = A(K)$ of $C^\infty(\mathbb{N}/\mathbb{Z}\Gamma)$?

As a first result, we have:

*Theorem* 3: *Set* $H(n) = \Sigma_{m=0}^{|n|-1} \oplus H(n,m)$. *If* $K$ *is an irreducible* R-*invariant subspace of* $H(n)$, *then* $e(jnx + kny) \in A(K)$ *for all* $j$ *and* $k \in \mathbb{Z}$.

This is easily seen by pulling the assertion back to $\mathbb{N}/\Gamma_{\mathbb{A}}$ via the map J. The theorem has a striking corollary: $C^{\infty}(\mathbb{N}/\mathbb{Z}\Gamma)$ is a finitely-generated module over $A(K)$. More is true:

*Theorem* 4: $C^{\infty}(\mathbb{N}/\mathbb{Z}\Gamma)$ *is a free module over* $A(K)$ *whose dimension divides* $n^2$ *and is divisible by* $n$.

The idea behind theorem 4 is quite simple: Let $G = \{(j,k) \in \mathbb{Z}^2 : e(jx + ky) \in A(K)\}$. Then $G$ is a subgroup of $\mathbb{Z}^2$ that contains $(n\mathbb{Z})^2$ (by theorem 3). $A(K)$ consists precisely of the functions whose only non-zero Fourier coefficients are those in G. Hence $C^{\infty}(\mathbb{N}/\mathbb{Z}\Gamma)$ is a free module over $A(K)$ whose dimension is the index in $\mathbb{Z}^2$ of G.

Our final result is a generalization of theorem 2:

*Theorem* 5: *Let* $G$ *be a subgroup of* $\mathbb{Z}^2$ *that contains* $(n\mathbb{Z})^2$ *and has index* $n$ *in* $\mathbb{Z}^2$. *Further, let* $A$ *be the smallest closed (in the topology of* $C^{\infty}$ *convergence) subalgebra of* $C^{\infty}(\mathbb{N}/\mathbb{Z}\Gamma)$ *that contains* $\{e(jx + ky) : (j,k) \in G\}$. *Then there exist precisely* $n$ *distinct irreducible* R-*invariant subspaces* $K_1,\ldots,K_n$ *in* $H(n)$ *such that* $A(K_i) = A$. *These subspaces also satisfy:*

(1) $K_i \perp K_j$ if $i \neq j$.

(2) $H(n) = K_1 \oplus \cdots \oplus K_n$.

## REFERENCES

1. Auslander, L., and Brezin, J. Invariant Subspace Theory for Three-dimensional Nilmanifolds. To appear, Bulletin A.M.S.

2. Brezin, J. Harmonic analysis on nilmanifolds. TAMS, 190, 611-618 (1970).

3. ———— Function theory on metabelian solvmanifolds. To appear, J. of Funct. Analysis.

4. Richardson, L. Thesis. Yale University, 1970.

# UNIONS OF SETS OF INTERPOLATION

by

S.W. Drury

Saint John's College, University of Cambridge.

1. Introduction. Our object here is to examine the convolution device introduced in [1] and to discuss several applications it has found. We shall use throughout the standard notations of harmonic analysis as found in [2]. In particular when G is a locally compact abelian group A(G) and B(G) denote the Banach algebras of absolutely convergent Fourier series on G and Fourier-Stieltjes transforms on G respectively.

Let $\mathcal{B}$ be a Banach algebra realised as an algebra of functions on its maximal ideal space X. Let Y be a closed subspace of X and let $\mathcal{B}|_Y$ denote the Banach algebra of restrictions with the quotient norm. In the case where $\mathcal{B}|_Y$ is isomorphic to $C_o(Y)$ we say that Y is a set of interpolation for $\mathcal{B}$ and we are assured the existence of a constant $\alpha$ $(0<\alpha\leq 1)$ such that

$$\alpha\, ||f||_{\mathcal{B}|_Y} \leq ||f||_\infty \leq ||f||_{\mathcal{B}|_Y} \quad \forall f \in C_o(Y),$$

by virtue of the closed graph theorem. A set Y satisfying this condition is called an $I_\alpha$ set. Alternatively, by the Hahn-Banach theorem the condition has the dual formulation

$$||\mu||_{(\mathcal{B}|_Y)'} \geq \alpha\, ||\mu||_M \qquad \forall \mu \in M(Y).$$

From this it may be seen that Y is an $I_\alpha$ set if and only if all the totally disconnected subsets of Y are $I_\alpha$ sets. If Y and Z are interpolation sets it is natural to ask whether $Y \cup Z$ is. We may assume that Y and Z are disjoint and totally disconnected in the sense that it suffices to find $\alpha > 0$ such that $Y' \cup Z'$ is an $I_\alpha$ set whenever Y' and Z' are disjoint totally disconnected subsets of Y and Z respectively. It is easy to see that $Y \cup Z$ is an interpolation set if and only if there exists a separating function

$s \in \mathcal{B}$ such that

$$|s(y) - 1| \leq 1/_3 \quad \forall y \in Y, \ |s(z)| \leq 1/_3 \quad \forall z \in Z.$$

In the case $Y \cap Z \neq \emptyset$ it will be necessary to construct s for each $Y'$ and $Z'$ with uniform control of norm. In the case where $\mathcal{B}$ has no identity it suffices to find s in the multiplier algebra of $\mathcal{B}$.

2. The convolution device. Let K be a compact $I_\alpha$ set and let $\Omega$ be a set of continuous functions of unit modulus on K which form a group under pointwise multiplication. For all $\delta > 0$ there exist functions $f_\omega$ $(\omega \in \Omega)$ in $\mathcal{B}$ such that

1) $\quad f_\omega(x) = \omega(x) \qquad \forall x \in K, \forall \omega \in \Omega.$

2) $\quad \| f_\omega \|_{\mathcal{B}} \leq \alpha^{-1} (1+\delta) \qquad \forall \omega \in \Omega.$

If we consider f as a function on $X \times \Omega$ and denote $f_x(\omega) = f_\omega(x)$ the most we can say about $f_x$ is

3) $\quad \| f_x \|_{C(\Omega)} \leq \alpha^{-1}(1+\delta) \qquad \forall x \in X,$

although for $x \in K$ we note that $f_x$ is an algebraic character on $\Omega$.

Restricting attention for the moment to the case where $\Omega$ is finite we consider the convolution

$$g(x,\omega) = \int f(x, \omega\lambda^{-1})\, f(x,\lambda)\, d\eta(\lambda),$$

where $\eta$ is the normalized translation invariant measure on $\Omega$. Clearly we have

1') $\quad g_\omega(x) = \int (\omega\lambda^{-1})(x) . \lambda(x)\, d\eta(\lambda) = \omega(x) \qquad \forall x \in K, \forall \omega \in \Omega.$

2') $\quad \|g_\omega\|_{\mathcal{B}} \leq \sup_{\lambda \in \Omega} \|f_\lambda\|^2_{\mathcal{B}} \leq \alpha^{-2} (1+\delta)^2 \quad \forall \omega \in \Omega.$

For a fixed element $x \in X$ we have

3') $\quad \|g_x\|_{A(\Omega)} = \|f_x * f_x\|_{A(\Omega)} = \|f_x\|^2_{L^2(\Omega)}$

$$\leq \|f_x\|^2_\infty \leq \alpha^{-2}(1+\delta)^2.$$

Thus (3') represents a real advance over (3), and (2') only a trivial loss relative to (2).

The above device has a topological version. Suppose for instance that $\Omega$ is an infinite group compact for the topology of uniform convergence on K that K is metrizable and that $\mathcal{B}$ is separable. We may choose f according to the borel section theorem [3] and find the same conclusion.

3. Interpolation sets for group algebras. Let G be a discrete abelian group. The sets of interpolation for A(G) are called Sidon sets. To show that the union of two Sidon sets is again Sidon it suffices to prove the following theorem.

Theorem. Let G be a discrete abelian group and K a finite $I_\alpha$ subset of G for A(G). Then for all $\alpha$ $(0<\alpha<1)$ there exists a separating function $s \in A(G)$ such that

a) $\quad s(x) = 1 \qquad \forall x \in K$

b) $\quad ||s||_{A(G)} \leq C(\alpha,\varepsilon)$

c) $\quad |s(x)| \leq \varepsilon \qquad \forall x \in G \setminus K$

where $C(\alpha, \varepsilon)$ is a constant depending only on $\alpha$ and $\varepsilon$.

Proof. We set

$$\Omega = \{\phi \ ; \ \phi : K \to \{\pm 1\}\},$$

which definition conforms with the situation of §2. The dual group H of $\Omega$ is the free abelian group of order two generated by K and there is a canonical inclusion $j : K \to H$. Since j(K) is independent in H the set

$$\text{graph } (j) = \{(x, j(x)); \ x \in K\}$$

is independent in $G \times H$. Thus, by means of a Riesz product, we may construct a function $p \in A(G \times H)$ with the properties a') $p(x,j(x)) = 1 \qquad \forall x \in K$

b') $||p||_{A(GxH)} \leq 8\,\varepsilon^{-1}\,\alpha^{-2}$

c') $|p(x,y)| \leq \frac{1}{2}\,\varepsilon\,\alpha^2 \qquad \forall x \in G\backslash K, \forall y \in H$

The spectral analysis of p in terms of the characters of H reads

$$p(x,y) = \sum_{\omega\in\Omega} \hat{p}(x,\omega)\ \omega(y),$$

and we write

$$s(x) = \sum_{\omega\in\Omega} \hat{p}(x,\omega)\ g\ (x,\omega),$$

where g is the function of §2.

By (1') and (a')

a) $s(x) = \sum_{\omega\in\Omega} \hat{p}(x,\omega)\ \omega_o\, j(x) = p(x,j(x)) = 1 \quad \forall x \in K,$

by (2') and (b')

b) $||s||_{A(G)} \leq \{\sum_{\omega\in\Omega} ||\hat{p}_\omega||_{A(G)}\}\{\sup_{\omega\in\Omega}||g_\omega||_{A(G)}\} \leq C(\alpha,\varepsilon)$

and by (3') and (c')

c) $|s(x)| \leq ||p_x||_\infty ||g_x||_{A(\Omega)} \leq \varepsilon \qquad \forall x \in G\backslash K$

as required.

The corresponding result for topological Sidon sets is proved by Déschamps-Gondim in [4]. In the case where G is a general locally compact abelian group, sets of interpolation are called Helson sets. The union problem for Helson sets was first solved by Varopoulos in [5] but subsequently a short elegant solution was found by Herz [6]. We give a brief account of Herz′ method below in the case where G is compact. By remarks already made in §1, it suffices to show the following:

Theorem. Let G be a compact abelian group and K a compact totally disconnected $I_\alpha$ subset of G for A(G). Then for all $\varepsilon$ $(0 < \varepsilon < 1)$ and every open set U containing K there exists a function $s \in A(G)$ such that

a) $$s(x) = 1 \qquad \forall x \in K$$

b) $$||s||_{A(G)} \leq C(\alpha, \varepsilon)$$

c) $$|s(x)| \leq \varepsilon \qquad \forall x \in G\backslash U.$$

Before proceeding to the proof we need some machinery for a lemma. Let

$$\Omega = \{\phi;\ \phi: K \to \mathbb{T},\ \phi \text{ continuous}\},$$

a discrete abelian group, and let H denote the dual group of $\Omega$, that is, the compact free abelian group generated by K. There is a topological embedding $j : K \to H$, and $j(K)$ is a Kronecker set of a very particular type in H.

Lemma (Herz). The theorem is true in the particular case of $j(K)$ in H and $\alpha = 1$. The proof is non-trivial and is not given here.

Next we introduce the canonical mapping $\pi : H \to G$ which arises as the dual of the inclusion $\hat{G} \subsetneq \Omega$, and the 'change of variables' mapping $\sigma : G \times H \to G \times H$ given by

$$\sigma(x,y) = (x - \pi(y), y)$$

and which takes graph (j) bijectively onto $\{0\} \times j(K)$. We may take U to have the form $K + 2V$, where V is a neighbourhood of $0_G$. By the lemma we have a separating function $p^* \in A(H)$ with the properties

a) $$p^* \circ j(x) = 1 \qquad \forall x \in K$$

b) $$||p^*||_{A(H)} \leq C(\alpha, \varepsilon)$$

c) $|p^*(y)| \leq \frac{1}{2}\varepsilon\alpha^2 \qquad \forall y \in H\setminus(j(K) + \pi^{-1}(V))$

It is easy to construct a function $q \in A(G)$ such that

a) $q(0_G) = 1$

b) $||q||_{A(G)} = 1$

c) $q(x) = 0 \qquad \forall x \in G \setminus V.$

All this work is to produce the function $p = (q \otimes p^*)_o\sigma \in A(G \times H)$

enjoying the properties

a) $p(x,j(x)) = 1 \qquad \forall x \in K$

b) $||p||_{A(GxH)} \leq C(\alpha, \varepsilon)$

c) $|p(x,y)| \leq \frac{1}{2}\varepsilon\alpha^2 \qquad \forall x \in G \setminus U, \ \forall y \in H,$

and analogous to the function p introduced previously.

At this point Herz introduces the generalized Fejér kernel. It is well known (c.f.[2] §2.6.7) that there exists a nested family $\lambda$ of finite symmetric subsets of $\Omega$ such that

$$\lim_{\Lambda \to} \frac{\text{card}((\omega+\Lambda) \cap \Lambda)}{\text{card}(\Lambda)} \to 1 \qquad \forall \omega \in \Omega.$$

Since the generalized Fejér kernel $K_\Lambda \in L^1(H)$

$$K_\Lambda = \{\text{card } \Lambda\}^{-1} \{\hat{k}_\Lambda\}^2$$

($k_\Lambda$ denotes the characteristic function of $\Lambda$) satisfies

$$\hat{K}_\Lambda(\omega) \underset{\Lambda \to}{\to} 1 \qquad \forall \omega \in \Omega,$$

for a given $\delta > 0$ there exists $\Lambda \in \lambda$ such that

$$\sum_{\omega\in\Omega} |1 - \hat{K}_\Lambda(\omega)| \ ||\hat{p}_\omega||_{A(G)} \leq \delta.$$

The convolution device becomes

$$g_x = \{\text{card } \Lambda\}^{-1} (k_\Lambda . f_x) * (k_\Lambda . f_x) \qquad \forall x \in G,$$

and has the properties

a) $\quad g(x,\omega) = \hat{K}_\Lambda(\omega) . \omega(x) \qquad \forall x \in K, \forall \omega \in \Omega$

b) $\quad ||g_\omega||_{A(G)} \leq (1+\delta)^2 \alpha^{-2} \qquad \forall \omega \in \Omega$

c) $\quad ||g_x||_{A(\Omega)} \leq (1+\delta)^2 \alpha^{-2} \qquad \forall x \in G.$

Proceeding as before we see that the function $s^* \in A(G)$

$$s^*(x) = \sum_{\omega\in\Omega} \hat{p}_x(\omega)\ g(x,\omega)$$

has the properties

a) $\quad |1 - s^*(x)| \leq \sum_{\omega\in\Omega} |1 - \hat{K}_\Lambda(\omega)|\ ||\hat{p}_\omega||_{A(G)} \leq \delta \qquad \forall x \in K$

b) $\quad ||s^*||_{A(G)} \leq (1+\delta)^2\ C(\alpha, \varepsilon)$

c) $\quad |s^*(x)| \leq \frac{1}{2}\ \varepsilon (1+\delta)^2 \qquad \forall x \in G \setminus U.$

Using again the condition that K is a set of interpolation we correct $s^*$ to a function $s \in A(G)$ which satisfies the required conditions when $\delta$ is small enough.

4. Interpolation sets for uniform algebras.

A uniform algebra $A$ is a Banach algebra with identity the norm of which coincides with the uniform norm on its maximal ideal space X. For the sake of simplicity we shall assume that X is metrizable. A compact subset K of X is said to be peak if and only if there exists a function $p \in A$ such that

$$p(x) = 1 \quad \forall x \in K, \quad |p(x)| < 1 \quad \forall x \in X \setminus K.$$

Two necessary and sufficient conditions for K to be peak are the following [7]

1) (Bishop). There exists $\varepsilon$ $(0<\varepsilon<1)$ and C $(C\geq 1)$ such that for all open U containing K there exists $p \in A$ satisfying

a) $p(x) = 1$ $\quad \forall x \in K$

b) $||p||_\infty \leq C$

c) $|p(x)| \leq \varepsilon$ $\quad \forall x \in X \setminus U$.

2) (Glicksberg). $k_K \cdot \mu$ belongs to $A^o$ whenever $\mu$ does. ($A^o$ denotes the annihilator of $A$ in M(X)).

One consequence of condition (2) is that if all the totally disconnected subsets of K are peak then so is K.

We have the following key lemma

Lemma (Varopoulos). Let K be a totally disconnected compact $I_\alpha$ subset of X and let L be a compact subset of X disjoint from K. Suppose that there exist constants C $(C\geq 1)$ and $\varepsilon$ $(\varepsilon>0)$ such that for all $x \in K$ there exists a function $f_x \in A$ with the properties

a) $f_x(x) = 1$ $\quad \forall x \in K$

b) $||f_x||_\infty \leq C$ $\quad \forall x \in K$

c) $|f_x(y)| \leq \varepsilon$ $\quad \forall x \in K,\ \forall y \in L$.

Then for all $\delta > 0$ there exists $f \in A$ such that

a) $|1 - f(x)| \leq \delta$ $\quad \forall x \in K$

b) $||f||_\infty \leq C\alpha^{-2}$

c) $|f(y)| \leq \varepsilon\alpha^{-2}$ $\quad \forall y \in L$.

Proof. By an elementary compactness argument there exists a topological partition of K

$$K = \bigcup_{j=1}^{J} K_j,\ K_j \text{ open and closed in } K\ (1\leq j\leq J),\ K_{j_1} \cap K_{j_2} = \emptyset$$

$(1\leq j_1<j_2\leq J)$, and functions $f_j \in A$ with the properties

a) $$|f_j(x) - 1| \leq \delta \qquad \forall x \in K_j$$

b) $$||f_j||_\infty \leq C$$

c) $$|f_j(y)| \leq \varepsilon \qquad \forall y \in L.$$

We now identify the set $\{j;\ 1 \leq j \leq J\}$ with a finite abelian group G in any manner whatsoever. For instance the obvious identification with $\mathbb{Z}(J)$ will do. Then each element $\omega \in \Omega = \hat{G}$ is really a function of unit modulus on K

$$\omega(x) = \langle \omega, j \rangle \quad \text{if } x \in K_j.$$

The spectral analysis of f reads

$$f(x,j) = \sum_{\omega\in\Omega} \hat{f}(x,\omega) \langle \omega, j \rangle,$$

and we set

$$f(x) = \sum_{\omega\in\Omega} \hat{f}(x,\omega)\ g\ (x,\omega).$$

Since $\Omega$ is finite, $f \in A$. When $x \in K_j$ we have $f(x) = f(x,j)$. Thus

a) $$|1 - f(x)| \leq \delta \qquad \forall x \in K$$

b) $$||f||_\infty \leq \alpha^{-2} \sup_{\substack{x\in X \\ 1\leq j\leq J}} |f(x,j)| \leq C\alpha^{-2}$$

c) $$|f(y)| \leq \alpha^{-2} \sup_{1\leq j\leq J} |f(y,j)| \leq \varepsilon\alpha^{-2} \qquad \forall y \in L.$$

We point out that in this case condition (2') on g is subordinate to (3'). It is condition (3') that is used in the estimate (b). This completes the lemma but if we wish we may go one step further and correct f, thus obtaining $f^* \in A$ with the properties

a) $$f^*(x) = 1 \qquad \forall x \in K$$

b) $$||f^*||_\infty \le C\alpha^{-2} + \delta(1+\delta)\alpha^{-1}$$

c) $$|f^*(y)| \le \varepsilon\alpha^{-2} + \delta(1+\delta)\alpha^{-1} \qquad \forall y \in L.$$

Corollary 1. Let K be a compact $I_\alpha$ set every point of which is peak. Then K is peak.

Proof. By Glicksberg's criterion we may suppose that K is totally disconnected. The result follows from Bishop's criterion and the lemma with C = 1, $\varepsilon$ and $\delta$ small enough, and L arbitrary.

Let P and Q be compact $I_\alpha$ and $I_\beta$ sets respectively. A necessary condition for $P \cup Q$ to be of interpolation is that for all $x \in P$, $y \in Q$, there exists $f \in A$ with $f(x) = 1$, $f(y) = 0$ and $||f||_\infty \le C$.

Corollary 2. The above condition is sufficient.

Proof. Without loss of generality we may assume that P and Q are disjoint and totally disconnected. By hypothesis there exists $f_{x,y} \in A$ such that

a) $$f_{x,y}(x) = 1 \qquad \forall x \in P, \ \forall y \in Q$$

b) $$||f_{x,y}||_\infty \le C$$

c) $$f_{x,y}(y) = 0 \qquad \forall x \in P, \ \forall y \in Q.$$

We apply the lemma for the first time with K = P and L = {y}, where y is an arbitrary element of Q. After a suitable correction the function $h_y = 1 - f_y^*$ satisfies

a) $h_y(y) = 1$ $\quad\forall y \in Q$

b) $||h_y||_\infty \leq (1+\delta)\, C\alpha^{-2}$

c) $h_y(x) = 0$ $\quad\forall x \in P,\ \forall y \in Q.$

Applying the lemma for the second time with $K = Q$ and $L = P$, we have a separating function of the required type. Both the corollaries are due to Varopoulos [8] [9]. The reader may also like to consult Bernard [10].

## 5. Bibliography


[1] S.W. Drury, Sur les ensembles de Sidon, C.R. Acad. Sci. Paris, t 271 (1970) p.162.

[2] W. Rudin, Fourier Analysis on Groups, Interscience tract No.12.

[3] N. Bourbaki, Topologie Générale, Ch.9, §6.1 and §6.8, Hermann.

[4] M. Déschamps-Gondim, Sur les ensembles de Sidon topologiques, C.R. Acad. Sci. Paris t 271 (1970) p.1247.

[5] N. Th. Varopoulos, Groups of continuous functions in Harmonic Analysis, Acta Math., v 125 (1970) p.109.

[6] C. Herz, Drury's lemma and Helson sets (to appear).

[7] A. Browder, Introduction to function algebras, §2.3 and §2.4, Benjamin.

[8] N. Th. Varopoulos, Ensembles pics et ensembles d'interpolation pour les algèbres uniformes, C.R. Acad. Sci. Paris, t 272 (1971) p.866.

[9] N. Th. Varopoulos, Sur la réunion de deux ensembles d'interpolation d'une algèbre uniforme, C.R. Acad. Sci. Paris, t 272 (1971) p.950.

[10] A. Bernard, Algèbres quotients d'algèbres uniformes, C.R. Acad. Sci. Paris, t 272 (1971) p.1101.

# LACUNARITY FOR COMPACT GROUPS, III

by

Robert E. Edwards, Edwin Hewitt[(1)], and Kenneth A. Ross[(2)]

Australian National University, University of Washington and University of Oregon

## 1. INTRODUCTION AND NOTATION

The origin of the present paper lies in a famous theorem about trigonometric series on the circle group which admit Hadamard gaps. Consider such a series, say

$$\Sigma_{k\varepsilon Z} c_k \exp(in_k x),$$

where the $c_k$ are complex numbers and the $n_k$ integers,

$$0 \leqq n_1 < n_2 < n_3 < \cdots,$$

$$c_{-k} = \overline{c_k}, \quad n_{-k} = -n_k,$$

$$\inf_{k>0} n_{k+1}/n_k = q > 1.$$

Suppose that the (real-valued) symmetric partial sums

$$s_n(x) = \Sigma^n_{k=-n} c_k \exp(in_k x)$$

satisfy the condition

$$\sup_{n\geqq 1} s_n(x) < \infty$$

for every $x$ in some nonvoid open interval; the conclusion is that

$$\Sigma_{k\varepsilon Z}|c_k| < \infty .$$

We call this the Fatou-Zygmund theorem. This theorem goes back to Fatou [4], who announced the result without proof

(1) Supported by National Science Foundation Grant GP-28513.
(2) Supported by National Science Foundation Grant GP-8382.

for $q > 2$, $\mathrm{Re}(c_k) = 0$, and the stronger hypothesis that $s_n(x)$ converges for all $x$ in some nonvoid open interval. The full theorem is due to Zygmund [10]; see also [11], Vol. I, Th. (6·3). Gapoškin [5] has extended the theorem to a much wider class of lacunary sets in $Z$.

Our aim is to study the species of lacunarity related to each of a wide variety of implications similar to that expressed by the Fatou-Zygmund theorem, and to do this by a uniform method based on functional analytic arguments. In this announcement, we describe briefly only that variation which is closest to the original. Proofs and other variations will appear elsewhere.

NOTATION In general we use the notation of [6]. In particular, $G$ will denote a compact Abelian Hausdorff topological group with normalised Haar measure $\lambda$ and (discrete) character group $X$. If $S$ is a subset of $G$, $M(S)$ will denote the set of complex Radon measures $\mu$ on $G$ satisfying $\mathrm{Supp}\,\mu \subset S$. If $E$ stands for any one of the usual spaces of functions or measures on $G$, $E_r$ stands for the set of real elements of $E$ and $E_+$ stands for the set of nonnegative elements of $E$. If further $P$ is a subset of $X$, $E_P$ consists of the P-spectral elements of $E$, i.e.

$$E_P = \{f \in E : \hat{f}(\chi) = 0 \text{ for every } \chi \in X \setminus P\}.$$

The set of trigonometric polynomials on $G$ is denoted by $T(G)$.

If $P$ is a symmetric subset of $X$, $F_h(P)$ denotes the set of Hermitian symmetric complex-valued functions on $P$: a complex-valued function $\phi$ on $P$ belongs to $F_h(P)$ if and only if $\phi(\chi^{-1}) = \overline{\phi(\chi)}$ for all $\chi \in P$. Also $B_h(P)$ denotes the set of bounded elements of $F_h(P)$, and $D_h(P)$ the set of elements of $B_h(P)$ which assume values of absolute value one at every point of $P$.

In order to introduce suitable finite partial sums, we suppose given an increasing sequence $S = (X_n)_{n=1}^{\infty}$ of finite symmetric subsets of $X$ and write $X_{\infty} = \cup_{n=1}^{\infty} X_n$. (It turns out later that the choice of $S$ is not really significant, but it is needed to begin with.) We consider symmetric subsets $P$ of $X_{\infty}$, and assign to each $u$ in $F_h(P)$ the partial sums

$$s_n u = \Sigma_{\chi \varepsilon P_n} u(\chi)\cdot\chi \ \varepsilon\ T_{P,r}(G),$$

where $P_n = P \cap X_n$. We write $s_n^+ u = (s_n u)^+ = \max(s_n u, 0)$. We also write

$$D_n = \Sigma_{\chi \varepsilon X_n} \chi,$$

the Dirichlet kernel corresponding to $S$, and note that if $f \ \varepsilon\ L^1_{P,r}(G)$ and $u = \hat{f}|P$, then $s_n u = D_n * f$.

Finally, $W$ will throughout denote a nonvoid $\lambda$-measurable subset of $G$ satisfying $W \subset ((\mathrm{int} W))^-$, and $\xi_W$ will denote its characteristic function.

## 2. FATOU-ZYGMUND PROPERTIES AND THE BASIC INEQUALITY

We denote by $FZ(P,S,W)$ the statement: If $u \ \varepsilon\ F_h(P)$ and

$$\sup_{n \geq 1} \| \xi_W s_n^+ u \|_u < \infty,$$

then

$$u \ \varepsilon\ \ell^1(P).$$

If $FZ(P,S,W)$ holds, we say that $P$ has the $FZ(S,W)$-property, and that $P$ is an $FZ(S,W)$-set. As will appear, the $FZ(S,W)$-property is a species of lacunarity at least as strong as

Sidonicity; we have been unable to decide whether it is in fact equivalent to Sidonicity, though all countable symmetric Sidon sets known to us possess the FZ(S,G)-property. An analogous but apparently not identical property has been studied by Déchamps-Gondim [1] for connected compact Abelian groups $G$.

A functional analytic argument leads to a basic inequality characterising FZ(S,W)-sets analogous to that available for Sidon sets ([6], (37.2)).

THEOREM I. *If* $P$ *possesses the* FZ(S,W)-*property, then to every positive integer* $n_o$ *there correspond a finite subset* $\Sigma$ *of* $P$ *and a positive real number* $\varkappa$ *(both possibly depending on* $n_o$*) such that*

$$(1)\quad \|\hat{f}\|_1 \leqq \varkappa \max\{\max_{\chi\varepsilon\Sigma} |\hat{f}(\chi)|,\ \sup_{n\geqq n_o} \|\xi_W (D_n * f)^+\|_u\}$$

*for every* $f \varepsilon T_{P,r}(G)$. *Conversely, if the inequality (1) holds for at least one positive integer* $n_o$ *and every* $f$ *in* $T_{P,r}(G)$, *then* $P$ *possesses the* FZ(S,W)-*property.*

Combining Theorem I with (3.9) of [3], we deduce

COROLLARY II. *If* $P$ *possesses the* FZ(S,W)-*property for some* $W$, *then* $P$ *is a Sidon set.*

## 3. MATCHING THEOREMS AND EQUIVALENCES

THEOREM III (*Main matching theorem*). *In order that* $P$ *possess the* FZ(S,W)-*property, it is necessary and sufficient that every* $\beta$ *in* $B_h(P)$ *admit an expression*

$$\text{(i)}\quad \beta = (\alpha + \hat{\nu}_\infty + \Sigma_{n=1}^{\infty} D_n\hat{\nu}_n)|P,$$

*where* $\alpha \varepsilon B_h(P)$ *has a finite support,* $\nu_\infty \varepsilon M_+((W^-)^{-1})$, $\nu_n \varepsilon M_+((W^-)^{-1})$ $(n = 1,2,\ldots)$, *and*

$$\Sigma_{n=1}^{\infty} \|\nu_n\| < \infty .$$

If this condition is fulfilled, there exist a finite subset $\Sigma$ of $P$ and a positive real number $\varkappa$ (both possibly depending on $P$ and $W$) such that every $\beta$ in $B_h(P)$ admits an expression (i) in which $\operatorname{Supp}\alpha \subset \Sigma$ and

$$\|\alpha\|_\infty + \|\nu_\infty\| + \Sigma_{n=1}^{\infty}\|\nu_n\| \leqq \varkappa\|\beta\|_\infty .$$

THEOREM IV. _The following statements are equivalent:_

(a) $P$ _possesses the_ $FZ(S,W)$-_property_;

(b) _there exists a positive real number_ $\varkappa$ _such that to every_ $\beta \in B_h(P)$ _there corresponds_ $\mu \in M_+((W^-)^{-1})$ _satisfying_

$$\|\mu\| \leqq \varkappa\|\beta\|_\infty ;$$

_and_

(i) $\lim_{m\to\infty}[\sup_{\chi\in P\setminus P_m}\{|\beta(\chi) - \hat{\mu}(\chi)|\}] = 0$;

(c) _to every_ $\beta \in D_h(P)$ _there corresponds_ $\mu \in M_+((W^-)^{-1})$ _satisfying_

(ii) $\limsup_{m\to\infty}[\sup_{\chi\in P\setminus P_m}\{|\beta(\chi) - \hat{\mu}(\chi)|\}] < 1$.

Noting that (b) trivially implies (c), and that the left hand members of (i) and (ii) are each equal to

$$\inf_\Sigma \sup_{\chi\in P\setminus\Sigma}\{|\beta(\chi) - \hat{\mu}(\chi)|\},$$

$\Sigma$ ranging over all finite subsets of $P$, we see that, if $S = (X_n)_{n=1}^{\infty}$ and $S^\dagger = (X_n^\dagger)_{n=1}^{\infty}$ are two increasing sequences of finite symmetric subsets of $X$, and if $P$ is a symmetric subset of the union of each of them, then $P$ possesses the $FZ(S,W)$-property if and only if it possess the $FZ(S^\dagger,W)$-property. Accordingly, if $P$ is any countable symmetric subset of $X$, we may (and will) say that $P$ possesses the $FZ(W)$-_property_, and that $P$ is an $FZ(W)$-_set_, if and only if it possesses the $FZ(S,W)$-property for some one (and hence for every) $S$ whose union contains P.

The most obvious choice of W for closer study is W = G; here we have further equivalences.

THEOREM V. Let P be a countable symmetric subset of X not containing 1. The following statements are equivalent:

(a) P is an FZ(G)-set;

(b) every $\beta \in B_h(P)$ has the form $\hat{\mu}|P$ for some $\mu \in M_+(G)$;

(c) every $\beta \in B_h(P) \cap c_o(P)$ has the form $\hat{f}|P$ for some $f \in L^1_+(G)$;

(d) there is a positive real number $\kappa$ such that $\|\hat{f}\|_1 \leqq \kappa \max(f)$ for every $f \in T_{P,r}(G)$;

(e) $\|\hat{\mu}\|_1 < \infty$ for every $\mu \in M_{P,r}(G)$ such that $\mu \leqq c\lambda$ for some ($\mu$-dependent) real number c;

(f) $\|\hat{f}\|_1 < \infty$ for every $f \in L^1_{P,r}(G)$ such that ess sup(f) < $\infty$;

(g) P is a Sidon set and there is a real number $\kappa > 1$ such that $\kappa^{-1} \leqq \max(f) \leqq \kappa$ for every f in $T_{P,r}(G)$ such that min(f) = -1.

## 4. DRURY'S THEOREM FOR FZ(G)-SETS

Drury [2] showed that the union of two Sidon sets is again a Sidon set. There is an exact analogue for FZ(G)-sets.

THEOREM VI. The union of two FZ(G)-sets is again an FZ(G)-set.

The proof is an adaptation of that of Drury's theorem.

## 5. STEČKIN SETS

If $\Delta$ is a subset of X and t a positive integer, write $S(\Delta,t)$ for the set of functions a from $\Delta$ into {1,0,-1} such that

$$\Sigma_{\chi\varepsilon\Delta}|a(\chi)| = t.$$

For $\psi \in X$, denote by $S(\Delta,t,\psi)$ the set of $a$ in $S(\Delta,t)$ satisfying

$$\prod_{\chi\varepsilon\Delta} \chi^{a(\chi)} = \psi.$$

As in [9], $\Delta$ is said to possess property (R) if there is a real number $A \geq 1$ such that

$$\text{card}(S(\Delta,t,\psi)) \leq A^t$$

for every $\psi \in X$ and every integer $t > 1$. Further, $\Lambda$ is termed a Stečkin set if it is a finite union of sets each possessing property (R).

By adapting a known argument ([8], 5.7.5; see also [7]), we prove

THEOREM VII. *If* $\Delta \subset X$ *is a countable Stečkin set, then* $\Delta \cup \Delta^{-1}$ *is an* FZ(G)-*set.*

BIBLIOGRAPHY

[1] Déchamps-Gondim, Myriam. Compacts associés à un ensemble de Sidon. C. R. Acad. Sci. Paris, Série A, 271 (1970), 590-592.

[2] Drury, Stephen W. Sur les ensembles de Sidon. C. R. Acad. Sci. Paris, Série A, 271 (1970), 162-163.

[3] Edwards, Robert E., Edwin Hewitt and Kenneth A. Ross. Lacunarity for compact groups, II. To appear Pacific J. Math.

[4] Fatou, Paul. Séries trigonométriques et séries de Taylor. Acta Math. 30 (1906), 335-400.

[5] Gapoškin, V. F. On the question of absolute convergence of lacunary series. Izv. Akad. Nauk SSSR, Ser. Mat. 31 (1967), 1271-1288.

[6] Hewitt, Edwin and Kenneth A. Ross. Abstract Harmonic Analysis. 2 vols. Berlin-Heidelberg-New York: Springer-Verlag, 1963, 1970.

[7] Rider, Daniel. Gap series on groups and spheres. Canad. J. Math. 18 (1966), 389-398.

[8] Rudin, Walter. Fourier Analysis on Groups. New York, N.Y.: Interscience Publishers, 1962.

[9] Stečkin, S. B. On the absolute convergence of Fourier series (third communication). Izv. Akad. Nauk SSSR, Ser. Mat. 20 (1956), 385-412.

[10] Zygmund, Antoni. Quelques théorèmes sur les séries trigonométriques et celles de puissances. Studia Math. 3 (1931), 77-91.

[11] Zygmund, Antoni. Trigonometric Series, 2nd edition. 2 vols. Cambridge, England: Cambridge University Press, 1959, reprinted 1968.

# A NEW LOOK AT MACKEY'S IMPRIMITIVITY THEOREM

by

J. M. G. Fell
University of Pennsylvania

1. Introduction. Let $G$ be a separable locally compact group and $H$ a closed subgroup of $G$. Generalizing Frobenius' construction for finite groups, Mackey [7] showed how, starting from a unitary representation $U$ of $H$, one can construct an induced unitary representation $T = \mathrm{Ind}(U)$ of $G$. Now the construction of $T$ also yields a projection-valued measure $P$ on the coset space $G/H$ which, together with $T$, forms a so-called system of imprimitivity for $G$ based on $G/H$. We refer to the pair $T,P$ as the system of imprimitivity induced by $U$. In [6] Mackey proved his well-known Imprimitivity Theorem asserting the converse: Every system of imprimitivity for $G$ based on $G/H$ is equivalent to one induced by a unitary representation $U$ of $H$; and $U$ is unique up to equivalence.

Now it was observed by Glimm [4] that from any locally compact Hausdorff topological G-space $M$ one can construct a certain Banach *-algebra $L_1(G,M)$, called the $G$, $M$ transformation group algebra, whose non-degenerate *-representations are in natural one-to-one correspondence with the systems of imprimitivity for $G$ based on $M$. Consequently, the Imprimitivity Theorem asserts a natural one-to-one correspondence between the equivalence classes of non-degenerate *-representations of two Banach *-algebras, namely $L_1(G,G/H)$ and the $L_1$ group algebra $L_1(H)$ of $H$. This correspondence is functorial, that is, it is accompanied by a correspondence of intertwining operators. So we may say that $L_1(G,G/H)$ and $L_1(H)$ have isomorphic *-representation theories.

---

This research was carried out while the author was partially supported by the National Science Foundation grant NSF-GP-20644.

This suggests that the Imprimitivity Theorem ought to be a special case of a more abstract theorem asserting that two *-algebras A and B related by some appropriate structure have isomorphic *-representation theories. Our purpose here is to show that this is indeed the case. One striking feature will be the complete symmetry between A and B in the statement of the result (Theorem 2), in spite of the very unsymmetrical appearance of their prototypes $L_1(G,G/H)$ and $L_1(H)$ in the classical theorem. This entitles us to regard the generalized Imprimitivity Theorem as a duality theorem, like the Pontryagin Duality Theorem for locally compact Abelian groups.

The work reported here was done in collaboration with Professor Marc Rieffel; he will publish his version of these results separately, in a somewhat different context from ours. Besides Mackey's fundamental papers, two earlier works provided special impetus and inspiration for our work: One is the purely algebraic paper [8] of K. Morita; the other is Loomis' proof [5] of the Imprimitivity Theorem.

In this paper we shall not need any assumptions of separability.

All linear spaces are over the complex field.

2. The Abstract Inducing Process. Our first step must be to formulate a more abstract version of the inducing process.

Let A and B be two *-algebras, and E a linear space which is both a left A-module and a right B-module. Suppose that we are also given a conjugate-bilinear B-form on E, that is, a map $s,t \to [s,t]$ of $E \times E$ into B such that $[s,t]$ is linear in s and conjugate-linear in t. We make the following assumptions:

(1) $$[s,t]^* = [t,s]$$

(2) $$[s,t]b = [sb,t]$$

(3) $$[as,t] = [s,a^*t]$$

$(s,t \in E;\ a \in A;\ b \in B)$. It follows from (1) and (2) that

(4) $$b[s,t] = [s,tb^*].$$

Now take a $*$-representation $S$ of $B$ in a Hilbert space $X$. We say that $S$ is <u>completely positive</u> (<u>with respect to</u> $E$ <u>and</u> $[\ ,\ ]$) if $S_{[t,t]} \geq 0$ (i.e., is a positive operator) for all $t$ in $E$. Assuming this, we construct a new Hilbert space $Y$ as follows: Introduce into $M = E \otimes X$ the conjugate-bilinear form $(\ ,\ )_o$ as follows:

$$(s \otimes \xi\ ,\ t \otimes \eta)_o = (S_{[s,t]}\xi\ ,\ \eta).$$

It follows from (2), (4), and complete positivity that $(\ ,\ )_o$ is positive, that is, $(u,u)_o \geq 0$ for all $u$ in $M$. Factoring out from $M$ the null space of $(\ ,\ )_o$, we obtain a pre-Hilbert space $Y'$ whose completion with respect to $(\ ,\ )_o$ is the promised Hilbert space $Y$.

Now an easy calculation based on (3) shows that the natural A-module structure of $M$ lifts to an A-module structure of $Y'$ given by:

$$a(s \tilde{\otimes} \xi) = as \tilde{\otimes} \xi$$

$(a \in A;\ s \in E;\ \xi \in X)$, where $s \tilde{\otimes} \xi$ is the image of $s \otimes \xi$ in $Y'$; and we have

$$(au,v)_o = (u,a^*v)_o \qquad (a \in A;\ u,v \in Y').$$

Hence, provided that for each $a$ in $A$ the operation $u \to au$ is continuous on $Y'$ with respect to $(\ ,\ )_o$, there will be a unique $*$-representation $T$ of $A$ on $Y$ satisfying

(5) $$T_a(s \tilde{\otimes} \xi) = as \tilde{\otimes} \xi \qquad (a \in A;\ s \in E;\ \xi \in X).$$

If this is the case we say that $S$ is <u>inducible to</u> $A$ (<u>with respect to</u> $E$ and $[\ ,\ ]$), and denote $T$ by $\mathrm{Ind}_{B\uparrow A}(S)$, the $*$-representation of $A$ <u>induced</u> by $S$.

3. Induced Representations of Groups. The classical unitary inducing process on groups is a special case of the abstract process described above. To see this, take a locally compact group $G$ with unit $e$, left Haar measure $\lambda$, and modular function $\Delta$; and let $H$ be a closed subgroup of $G$, with left Haar measure $\nu$ and modular function $\delta$. As usual $L(G)$ is the convolution $*$-algebra of all continuous complex functions on $G$ with compact support, with product and involution given by

$$(\phi * \psi)(x) = \int \phi(y)\psi(y^{-1}x)d\lambda y,$$
$$\phi^*(x) = \Delta(x^{-1})\phi(x^{-1})^-$$

($\phi,\psi \in L(G)$; $x \in G$). Similarly for $L(H)$ ($\lambda,\Delta$ being replaced by $\nu,\delta$). We shall consider $L(G)$ as a right $L(H)$-module by means of the formula:

$$(6) \quad (\phi b)(x) = \int_H \phi(xh^{-1})b(h)(\delta(h)\Delta(h))^{-\frac{1}{2}}d\nu h \qquad (\phi \in L(G);\ b \in L(H)).$$

(Formula (6) differs from the usual convolution $\phi * b$ by the factor $\Delta(h)^{\frac{1}{2}}\delta(h)^{-\frac{1}{2}}$. The purpose of this discrepancy is to secure property (10) below.) We verify the property:

$$(7) \qquad (\phi * \psi)b = \phi * (\psi b) \qquad (\phi,\psi \in L(G);\ b \in L(H)).$$

Next we define a conjugate-bilinear $L(H)$-form on $L(G)$ as follows:

$$[\phi,\psi](h) = \Delta(h)^{\frac{1}{2}}\delta(h)^{-\frac{1}{2}}(\psi^* * \phi)(h)$$

($\phi,\psi \in L(G)$; $h \in H$). As a matter of fact, $[\ ,\ ]$ is obtained by the formula

$$(8) \qquad [\phi,\psi] = p(\psi^* * \phi)$$

from the linear map $p\colon L(G) \to L(H)$ given by

$$p(\phi)(h) = \Delta(h)^{\frac{1}{2}}\delta(h)^{-\frac{1}{2}}\phi(h).$$

Notice that $p$ has the properties:

(9) $$p(\phi^*) = (p(\phi))^*,$$

(10) $$p(\phi b) = p(\phi) * b$$

$(\phi \in L(G);\ b \in L(H))$. From (7), (8), (9), (10) we obtain for [ , ]:

(11) $$[\phi,\psi]^* = [\psi,\phi],$$

(12) $$[\phi,\psi] * b = [\phi b,\psi]$$

$(\phi,\psi \in L(G);\ b \in L(H))$.

Now $L(G)$, being a *-algebra, is a left $L(G)$-module under left multiplication, and we conclude from (8):

(13) $$[\chi * \phi , \psi] = [\phi , \chi^* * \psi] \qquad (\chi,\phi,\psi \in L(G)).$$

Comparing (11), (12), (13) with (1), (2), (3), we see that we have a special case of the abstract inducing context of §2, with $L(G)$, $L(H)$, $L(G)$ playing the roles of $A$, $B$, $E$ respectively. The induced representations of §2 become, in our present context, the classical induced representations of Mackey (as generalized to the non-separable situation by Blattner [1]). More precisely we have:

Theorem 1. *Let $U$ be any (strongly continuous) unitary representation of $H$ and $S$ its integrated form on $L(H)$. Then $S$ is completely positive and inducible to $L(G)$ (with respect to $L(G)$ and [ , ]); and $T = \operatorname{Ind}_{L(H)\uparrow L(G)}(S)$ is the integrated form of a unique unitary representation $V$ of $G$. This $V$ is exactly the unitary representation of $G$ induced by $U$ in the Mackey-Blattner sense.*

The complete positivity of $S$ with respect to [ , ] was essentially proved by Blattner in [2].

We shall write down explicitly the unitary equivalence between our abstract inducing process and the Mackey-Blattner construction in the group context. As above, let $U$ be a unitary representation of $H$, acting in the Hilbert space $X$. The Hilbert space $Z$ of the Mackey-Blattner induced representation is obtained as follows: Let $Z'$ be the linear space of all those norm-continuous functions $f: G \to X$ which satisfy:

(i) $f(xh) = \delta(h)^{\frac{1}{2}}\Delta(h)^{-\frac{1}{2}}U_{h^{-1}}(f(x))$  $(x \in G;\ h \in H)$,

(ii) $f$ has compact support in $G/H$ (i.e., vanishes outside $CH$ for some compact subset $C$ of $G$).

By (i), if $f$ and $g$ are in $Z'$, the complex-valued function $\rho: x \to (f(x),g(x))$ on $G$ is a "rho-function" in the sense of [7], and so gives rise (by Lemma 1.5 of [7]) to a regular complex Borel measure $\rho^o$ with compact support on $G/H$. We denote the total mass of $\rho^o$ (that is, $\rho^o(G/H)$) by $(f,g)_Z$. Then $Z'$ is a pre-Hilbert space with $(\ ,\ )_Z$ as inner product. Its completion is the Hilbert space $Z$. If $x \in G$, the operation $f \to xf$ of left translation on $Z'$ (where $(xf)(y) = f(x^{-1}y)$) extends to a unitary operator $V_x$ on $Z$; and $V: x \to V_x$ is the unitary representation of $G$ induced by $U$ in the Mackey-Blattner sense.

Let $F: L(G) \otimes X \to Z'$ be the linear map given by

$$[F(\phi\otimes\xi)](x) = \int_H \Delta(h)^{\frac{1}{2}}\delta(h)^{-\frac{1}{2}}\phi(xh)U_h\xi d\nu h \qquad (\phi \in L(G);\ \xi \in X;\ x \in G).$$

It is easy to see that the range of $F$ is indeed contained in $Z'$. If $S$ is the integrated form of $U$ one calculates that

$$(F(\phi\otimes\xi),F(\psi\otimes\eta))_Z = (S_{[\phi,\psi]}\xi,\eta) \qquad (\phi,\psi \in L(G);\ \xi,\eta \in X).$$

Hence the formula $F_o(\phi \tilde{\otimes} \xi) = F(\phi \otimes \xi)$ $(\phi \in L(G);\ \xi \in X)$ defines a linear isometry $F_o: Y \to Z$, where $Y$ is the space of $T = \operatorname{Ind}_{L(H)\uparrow L(G)}(S)$. It turns out that $F_o$ is onto $Z$. Furthermore, $F$ commutes with left translation; so $F_o$ intertwines $T$ with the integrated form of the Mackey-Blattner induced representation $V$. Therefore $F_o$ is the desired equivalence between the abstract inducing process and the Mackey-Blattner construction in the group case.

As a matter of fact, the inducing process of §2 embraces not only the classical induced representation but also the induced system of imprimitivity. Indeed, let $M$ denote $G/H$, and $C_o(M)$ the $C^*$-algebra of all continuous complex functions on $M$ vanishing at infinity. We recall from Stone's Theorem that regular projection-valued Borel measures on $M$ are essentially the same things as non-degenerate $*$-representations of $C_o(M)$. Now $L(G)$ is a left $C_o(M)$-module satisfying (3) under the natural definition

$$(f\phi)(x) = f(xH)\phi(x) \qquad (f \in C_o(M);\ \phi \in L(G);\ x \in G). \tag{14}$$

If $U$ is a unitary representation of $H$ whose integrated form on $L(H)$ is $S$, it turns out that $S$ is inducible to $C_o(M)$ with respect to [ , ] and the left action (14) of $C_o(M)$ on $L(G)$, and that the induced representation $\operatorname{Ind}_{L(H)\uparrow C_o(M)}(S)$ corresponds by Stone's Theorem to the projection-valued measure on $M$ associated in Mackey's sense with $\operatorname{Ind}(U)$.

We can go further. Let us denote by $L(G,M)$ the convolution $*$-algebra whose underlying linear space consists of all continuous complex functions on $G \times M$ with compact support, the operations and norm being as follows:

$$(f * g)(x,m) = \int_G f(y,m)g(y^{-1}x, y^{-1}m)\,d\lambda y,$$

$$f^*(x,m) = \Delta(x^{-1})f(x^{-1}, x^{-1}m)^-,$$

$$\| f \| = \int_G \sup_{m \in G/H} |f(x,m)| d\lambda x.$$

The completion of the normed *-algebra L(G,M) is the transformation group algebra $L_1(G,M)$ referred to in the Introduction. If U is a unitary representation of H, we pointed out in the Introduction that the system of imprimitivity induced by U corresponds naturally to a certain *-representation W of $L_1(G,M)$. I claim that W can be obtained directly by the inducing process of §2. To see this we must first make L(G) into a left L(G,M)-module. If $f \in L(G,M)$ and $\phi \in L(G)$, put

$$(15) \qquad (f\phi)(x) = \int_G f(y,xH)\phi(y^{-1}x)d\lambda y \qquad (x \in G).$$

One verifies that $f\phi \in L(G)$, and that (15) defines a left L(G,M)-module structure for L(G) satisfying condition (3). Now if S is the integrated form of U on L(H), it can be shown that S is inducible to L(G,M) with respect to [ , ] and the left action (15), and that the resulting induced representation $\operatorname{Ind}_{L(H)\uparrow L(G,M)}(S)$ is just the restriction to L(G,M) of the W described above.

4. Imprimitivity Bimodules. In Loomis' proof [5] of the Imprimitivity Theorem, a key role is played by a certain conjugate-bilinear L(G,G/H)-form on L(G), to be defined in (18). On examining this form we notice two interesting facts: First, the properties of (18) are highly symmetrical with those of [ , ]. Secondly, the process adopted in Loomis' proof for recovering the inducing representation U of H from the induced system of imprimitivity (or, more precisely, from the corresponding *-representation of $L_1(G,G/H)$) is a special case of the abstract inducing process of §2, provided we replace [ , ] by (18). With this in mind, we shall define an abstract

structure called an imprimitivity bimodule, containing <u>two</u> conjugate-bilinear forms with the symmetrical properties suggested by [ , ] and (18).

Let A and B be any two fixed *-algebras.

<u>Definition</u>. An <u>imprimitivity bimodule for</u> A, B is a system $\mathcal{E} = (E, [\ ,\ ]_A, [\ ,\ ]_B)$, where:

i) E is a linear space which is a left A-module and a right B-module,

ii) $[\ ,\ ]_A$ is a conjugate-bilinear A-form on E; $[\ ,\ ]_B$ is a conjugate-bilinear B-form on E,

iii) for r, s, t in E, a in A, and b in B we have:

$$[r,s]_A^* = [s,r]_A \ , \qquad [r,s]_B^* = [s,r]_B$$

$$a[r,s]_A = [ar,s]_A \ , \qquad [r,s]_B b = [rb,s]_B$$

$$[rb,s]_A = [r,sb^*]_A \ , \qquad [ar,s]_B = [r,a^*s]_B$$

$$[r,s]_A t = r[t,s]_B$$

iv) the linear spans of $\{[r,s]_A \mid r,s \in E\}$ and $\{[r,s]_B \mid r,s \in E\}$ are A and B respectively.

<u>Remarks</u>. (I) These postulates imply the associative law:

$$(ar)b = a(rb) \qquad (a \in A;\ b \in B;\ r \in E).$$

(II) Even without Postulate (iv), the identities (iii) imply that the linear spans mentioned in Postulate (iv) are two-sided *-ideals of A and B respectively.

(III) Suppose we are given only "half an imprimitivity bimodule"; that is to say, we have a *-algebra B, a right B-module E, and a conjugate-bilinear B-form $[\ ,\ ]_B$ on E satisfying (1), (2) of §2, and such that the range of $[\ ,\ ]_B$ linearly spans all of B. Then one can canonically construct another *-algebra A and a conjugate-bilinear A-form $[\ ,\ ]_A$ on E, and also make E into a left A-module, so that E, $[\ ,\ ]_A$, $[\ ,\ ]_B$ becomes an imprimitivity bimodule for A, B.

Now fix an imprimitivity bimodule $\mathcal{E} = (E, [\ ,\ ]_A, [\ ,\ ]_B)$ for A, B.

The symmetry with which A and B occur in $\mathcal{E}$ is very important. To make it precise, let $\overline{E}$ be the complex-conjugate space of E (differing from E only in its scalar multiplication: $(cr)_{\overline{E}} = (\bar{c}r)_E$). Then $\overline{E}$ becomes a left B-module and a right A-module under the actions : of B and A given by

$$(16) \qquad b:r = rb^*, \qquad r:a = a^*r \qquad (a \in A;\ b \in B;\ r \in E).$$

If we now define

$$[r,s]_A' = [s,r]_A \ , \quad [r,s]_B' = [s,r]_B \ ,$$

we verify, by the symmetry of the above definition, that

$$\overline{\mathcal{E}} = (\overline{E}, [\ ,\ ]_B', [\ ,\ ]_A')$$

is an imprimitivity bimodule for B, A ($\overline{E}$ being of course equipped with the module-structures (16)). We refer to $\overline{\mathcal{E}}$ as the complex-conjugate of $\mathcal{E}$.

As in §2, a *-representation S of B is completely positive if $S_{[t,t]_B} \geq 0$ for all t in E. In this case we can try to form the induced representation $\operatorname{Ind}_{B\uparrow A}(S)$ of A. Similarly, a *-representation T of A is completely positive if $T_{[t,t]_A} \geq 0$ for all t in E. In this case we can try to form the induced representation $\operatorname{Ind}_{A\uparrow B}(T)$ of B with respect to $\overline{E}$ and $[\ ,\ ]_A'$ (the objects that result from replacing $\mathcal{E}$ by $\overline{\mathcal{E}}$). The fundamental fact about these induced representations is the following duality theorem:

Theorem 2. *If* S *is any completely positive non-degenerate *-representation of* B, *then* $\operatorname{Ind}_{B\uparrow A}(S)$ *exists and is a completely positive non-degenerate *-representation of* A. *Furthermore,* $\operatorname{Ind}_{B\uparrow A}(S)$ *is inducible to* B, *and*

$$\operatorname{Ind}_{A\uparrow B}(\operatorname{Ind}_{B\uparrow A}(S)) \cong S.$$

*By symmetry, the same holds on interchanging* A *and* B. *Thus the map*

$$S \to \operatorname{Ind}_{B\uparrow A}(S) \tag{17}$$

*is a bijection from the family* $\mathcal{S}$ *of all equivalence classes of completely positive non-degenerate *-representations of* B *onto the family* $\mathcal{T}$ *of all equivalence classes of completely positive non-degenerate *-representations of* A; *and its inverse is*

$$T \to \operatorname{Ind}_{A\uparrow B}(T).$$

As an important adjunct to this Theorem we have:

Theorem 3. *If* S *and* T *correspond under the map (17), the commuting algebras of* S *and* T *are *-isomorphic. In particular,* S *is topologically irreducible if and only if* T *is.*

From Theorem 3 we deduce that the spaces of intertwining operators between corresponding pairs of representations in $\mathcal{S}$ and $\mathcal{T}$ are isomorphic; and that this isomorphism is functorial in the obvious sense. Thus (17) is an isomorphism of the "completely positive *-representation theories" of A and B.

Here is another useful fact. A *-representation S of B is *compact* if $S_b$ is a compact operator for every b in B; and similarly for *-representations of A.

*Theorem* 4. *If* S *and* T *correspond under (17), then* S *is compact if and only if* T *is compact*.

Theorem 2 as it stands is unsatisfactory for topological purposes, since it contains no reference to continuity of S and T. In topological situations A and B will usually be dense *-subalgebras of Banach *-algebras $A_o$ and $B_o$; and we will only care about those *-representations of A and B which are norm-continuous, i.e., which extend to $A_o$ and $B_o$. Hence the importance of the following result:

*Theorem* 5. *Suppose in addition that* A *is a normed* *-*algebra with norm* $\| \ \|$. *If* S *and* T *correspond under the map (17), the following conditions are equivalent*:

(i) T *is continuous with respect to* $\| \ \|$;

(ii) *for each pair* r, s *of elements of* E, *there is a positive number* k *such that*

$$\| S_{[ar,s]_B} \| \leq k \| a \| \quad \textit{for all} \ a \ \textit{in} \ A.$$

Combining Theorem 5 with its dual version (with A and B interchanged), we find:

*Theorem* 6. *Assume in addition that both* A *and* B *are normed* *-*algebras, and that for each pair of elements* r, s *of* E *the linear maps* $a \to [ar,s]_B$ *of* A *into* B, *and* $b \to [rb,s]_A$ *of* B *into* A *are norm-continuous. Then, if* S *and* T *correspond under* (17), S *is norm-continuous if and only if* T *is*.

Our main example of an imprimitivity bimodule comes from the context of the Imprimitivity Theorem for locally compact groups, and will be treated in the next section. Here is a simpler, quite different example.

Let D be any *-algebra, and I a left ideal of D. Let A and B be the linear spans in D of $II^*$ and $I^*I$ respectively. Thus A and B are *-subalgebras of D (in fact A is a two-sided *-ideal of D). Note that I is stable under left and right multiplication by elements of A and B respectively. If we define $[x,y]_A = xy^*$, $[x,y]_B = y^*x$ $(x,y \in I)$, then I, $[\ ,\ ]_A$, $[\ ,\ ]_B$ is an imprimitivity bimodule for A, B. Theorem 2 sets up a one-to-one correspondence between the non-degenerate *-representations S of B satisfying $S_{x^*x} \geq 0$ for all x in I, and the non-degenerate *-representations T of A satisfying $T_{xx^*} \geq 0$ for all x in I.

If in addition D is a Banach *-algebra, then the continuity conditions of Theorem 6 are certainly satisfied. Hence, by Theorem 6, in that case (17) becomes a correspondence between the completely positive non-degenerate *-representations of the _closures_ (in D) of A and B.

5. _Application to the Imprimitivity Theorem_. We shall now go back to the context and notation of §3, and define a conjugate-bilinear L(G,M)-form on L(G) as follows:

$$\{\phi,\psi\}(x,yH) = \int_H \phi(yh)\psi^*(h^{-1}y^{-1}x)d\nu h \qquad (\phi,\psi \in L(G);\ x,y \in G). \tag{18}$$

(Recall that $M = G/H$.) With this definition we have all the ingredients for a very important imprimitivity bundle.

_Theorem 7. Consider_ L(G) _as a left_ L(G,M)-_module according to_ (15) _and a right_ L(H)-_module according to_ (6). _Let_ A _and_ B _be the linear spans of the ranges of_ { , } _and_ [ , ] _in_ L(G,M) _and_ L(H) _respectively. Then_ $E = (L(G), \{\ ,\ \}, [\ ,\ ])$ _is an imprimitivity bundle for_ A, B.

In this special situation we can prove the following further subsidiary facts:

Theorem 8. (I) $B$ is dense in the $L_1$ group algebra $L_1(H)$. (II) $A$ is dense in the $L_1$ transformation group algebra $L_1(G,M)$. (III) If $S$ and $T$ are completely positive non-degenerate $^*$-representations of $B$ and $A$ respectively which correspond under (17), then $S$ is $L_1(H)$-continuous if and only if $T$ is $L_1(G,M)$-continuous. (IV) Every $L_1(H)$-continuous $^*$-representation of $B$ is completely positive. (V) Every $L_1(G,M)$-continuous $^*$-representation of $A$ is completely positive.

Of the statements in this Theorem all but (III) are quite straightforward to prove. The proof of (III) seems to be more intricate than the direct appeal to Theorem 6 which one would expect.

Combining Theorems 2, 7, and 8, we obtain the classical Imprimitivity Theorem in the form mentioned in the Introduction:

Theorem 9. The inducing map (17), taken in the context of the imprimitivity bimodule of Theorem 7, sets up a one-to-one correspondence between the family of all equivalence classes of non-degenerate $^*$-representations of $L_1(H)$ and the family of all equivalence classes of non-degenerate $^*$-representations of $L_1(G,M)$.

It follows from the observation at the end of §3 that the map (17) in this context is just the classical Mackey-Blattner inducing process.

It should be remarked that Mackey's Imprimitivity Theorem can be generalized to the Banach $^*$-algebraic bundles studied by the author in [3]. (This generalization was carried out in [3] for homogeneous Banach $^*$-algebraic bundles. The Theorem for more or less arbitrary Banach $^*$-algebraic bundles will appear in a forthcoming book by the author.) The bundle generalization can also be proved by combining Theorem 2 with suitable analogues of Theorems 7 and 8.

We should like to conclude with an interesting consequence of Theorem 4. We recall that a unitary representation of a locally compact group $G$ is _compact_ if its integrated form on $L_1(G)$ (or $L(G)$) is compact.

_Theorem 10_. _Let_ $U$ _be a compact unitary representation of the closed subgroup_ $H$ _of_ $G$, _and_ $V$, $P$ _the system of imprimitivity induced by_ $U$. _Then the_ $^*$_-representation of_ $L_1(G,G/H)$ _corresponding to_ $V$, $P$ _is compact. In particular,_ $P_f V_\phi$ _is a compact operator for every_ $f$ _in_ $L(G/H)$ _and every_ $\phi$ _in_ $L(G)$.

If in addition $G/H$ is compact, we can take $f = 1$ in the last statement of Theorem 10, and we obtain:

_Theorem 11_. _Suppose that_ $G/H$ _is compact. If_ $U$ _is a compact unitary representation of_ $H$, _then the unitary representation of_ $G$ _induced by_ $U$ _is compact_.

This result is due to Schochetman [9].

## References

[1] R. J. Blattner, On induced representations, Amer. J. Math. 83 (1961), 79-98.

[2] R. J. Blattner, Positive definite measures, Proc. Amer. Math. Soc. 14 (1963), 423-428.

[3] J. M. G. Fell, An extension of Mackey's method to Banach *-algebraic bundles, Memoir Amer. Math. Soc. No. 90 (1969).

[4] J. Glimm, Families of induced representations, Pac. J. Math. 12 (1962), 885-911.

[5] L. H. Loomis, Positive definite functions and induced representations, Duke Math. J. 27 (1960), 569-580.

[6] G. W. Mackey, Imprimitivity for representations of locally compact groups I, Proc. Nat. Acad. Sci. U.S.A. 35 (1949), 537-545.

[7] G. W. Mackey, Induced representations of locally compact groups, Ann. Math. 55 (1952), 101-139.

[8] K. Morita, Duality for modules and its applications to the theory of rings with minimum conditions, Science Reports of the Tokyo Kyoiku Daigaku Sec. A, vol. 6, No. 150 (1958), 83-142.

[9] Irwin Schochetman, Duals of locally compact groups, Thesis, University of Maryland, 1968.

# FOURIER-STIELTJES TRANSFORMS WITH AN ISOLATED VALUE

by

I. Glicksberg
University of Washington

1. Let $\mu$ be a finite (non-zero) measure on a locally compact abelian group $G$ with dual $\Gamma = G^\wedge$, and suppose $0$ is isolated in the range of the Fourier-Stieltjes transform $\hat{\mu}$, or more precisely,

$$(1.1) \qquad 0 \text{ is isolated in } \{0\} \cup \hat{\mu}(\Gamma).$$

Then what can one say about $\mu$?

Using Cohen's approach to the idempotent problem (as simplified by Amemiya and Ito) we shall show (1.1) implies $\mu$ has a non-zero component of a very nice sort: there is a compact subgroup $H$ of $G$ for which $\mu_H$, the part of $\mu$ carried by the cosets of $H$, is the convolve of a (non-zero) idempotent and an invertible. Thus in particular $\hat{\mu}_H$ (whose range is related to that of $\hat{\mu}$ in general, cf. [8]) is supported by an element of the coset ring of $\Gamma$. As an application, suppose $\mu$ is a measure on $T^1$ whose discrete part is not the convolve of a (non-zero) idempotent and an invertible; if $\{I_n\}$ is a sequence of disjoint intervals in $\mathbb{Z}$ with lengths tending to $\infty$ on each of which $\hat{\mu}$ doesn't vanish identically, then for every $\varepsilon > 0$, $|\hat{\mu}|$ assumes a value in $(0,\varepsilon)$ on all but finitely many $I_n$.

The result arose from another unsuccessful attempt to answer the question raised in [7]; the measures $\mu$ for which $\mu * L_1$ is closed, which satisfy (1.1), may well have the form indicated above for $\mu_H$, but the measures satisfying (1.1) itself need not be of that form. For example, if $E$ is any Sidon set, Drury's construc-

*Work supported in part by the National Science Foundation.

tion [5] yields a measure $\nu$ with $\hat{\nu} = 1$ on $E$ and $|\hat{\nu}| < \frac{1}{2}$ off $E$, so $\delta_o - \nu$ satisfies (1.1), as does $\mu * (\delta_o - \nu)$ for any $\mu$ satisfying (1.1), while $(\mu * (\delta_o - \nu))^\wedge$ vanishes exactly on $E \cup \hat{\mu}^{-1}(0)$. Thus one can alter considerably the set on which the isolated value $0$ is assumed (at least with $G$ compact).

To state our result precisely, recall that for a closed subgroup $H$ of $G$, any $\mu \in M(G)$ can be uniquely decomposed (via exhaustion, as usual) as

$$\mu = \mu_H + \mu' , \tag{1.2}$$

where $\mu_H$ is carried by cosets mod $H$ and $\mu'$ vanishes on all Borel subsets of cosets mod $H$ [10]. Our main result is then

<u>Theorem</u> 1.1. <u>For any non-zero measure</u> $\mu$ <u>on a l.c.a. group</u> $G$ <u>satisfying</u> (1.1) <u>there is a compact subgroup</u> $H$ <u>of</u> $G$ <u>and characters</u> $\gamma_1, \ldots, \gamma_n$ <u>in</u> $\Gamma$ <u>for which</u>

$$\mu_H = \eta * \lambda, \quad \eta = (\sum_1^n \gamma_i) m_H, \quad \lambda \in M(G)^{-1} \tag{1.3}$$

(<u>where</u> $m_H$ <u>is the normalized Haar measure on</u> $H$ <u>and</u> $M(G)^{-1}$ <u>denotes the invertible elements of</u> $M(G)$).

In what follows it will be convenient to use multiplicative notation for the group operation in $\Gamma$, and to omit the usual conjugation in defining the Fourier-Stieltjes transform: $\hat{\mu}(\gamma) = \int \gamma d\mu = \mu(\gamma)$. Finally $G^d$ will denote the discrete version of $G$.

2. We begin with a key lemma which is essentially proved (but not stated) in Cohen's work [2] and in the Amemiya-Ito paper [1], and which was no doubt known to those authors.

<u>Lemma</u> 2.1. <u>Suppose</u> $X$ <u>is a locally compact Hausdorff space</u>,

$\mu \varepsilon M(X)$, and $E$ is a family of unimodular Borel functions on $X$ for which the $w^*$ closure $(E\mu)^-$ consists of measures all of norm $= \|\mu\|$. Then the $w^*$ and norm topologies coincide on $(E\mu)^-$.

One can easily read off a proof from that of [1,ii]), but a much simpler proof (due to Ito) was pointed out to me at this Conference by John Fournier: clearly $(E\mu)^- = E_o\mu$, where $E_o$ consists of functions of unit modulus, and if $e_\delta\mu \to e\mu$ $w^*$, and $\mu = \Theta|\mu|$, $|\Theta| = 1$, $e_\delta\Theta \to e\Theta$ $w$ in $L_2(|\mu|)$. Since $w$ convergence of elements of unit norm in $L_2(|\mu|)$ implies strong convergence, $\|e_\delta\Theta - e\Theta\|_2 \to 0$, whence $\|e_\delta\Theta - e\Theta\|_1 = \|e_\delta\mu - e\mu\| \to 0$.

Besides borrowing from the Cohen-Amemiya-Ito approach we shall at one point have need of a result [4,6.6] which we shall restate here in the form in which it is needed, and which probably lies a bit deeper and is certainly less accessible than the other results used.

Lemma 2.2. Let $G$ be a compact abelian group, $H$ a closed subgroup, and $G_1$ the l.c.a. group formed from $G$ by declaring all cosets mod $H$ to be open, with $\Gamma_1$ its dual (so $\Gamma$ is dense in $\Gamma_1$ since $G_1 \to G$ is 1-1). Then for any $\nu \varepsilon M(G)$ which vanishes on all Borel subsets of $G_1$ (i.e., of cosets mod $H$) there is a net $\{\beta_\delta\}$ in $H^\perp$ with $\beta_\delta\nu \to 0$ $w^*$ in $M(G)$.

(Note that our $G$, $H$ are the subgroups $G^{d\wedge}$, $H^\wedge$ of [4,6.6]. Also note that when $\Gamma$ is taken in the relative topology of $\Gamma_1$, $V = H^\perp$ is a neighborhood of $1$ since $H$ remains compact in $G_1$. Finally it should be pointed out that to understand [4.6.6] it should suffice to read only §2 to 2.5, §5 to Th. 5.2, and p. 179 to Th. 6.6 of [4]; moreover the basic fact that the kernel $K$ of a compact abelian separately continuous semigroup is a compact group used there in §2 is essentially a triviality, that $K$ is algebraically a group as $kK = K = Kk$ for $k \varepsilon K$ shows, combined with Ellis' theorem, that then multiplication on $K$ is jointly continuous

([6], or [3,Appendix]).)

To begin our proof of 1.1 we shall assume our group $G$ is compact, the (deferred) reduction to that case being quite easy. Normalizing, suppose $\mu \varepsilon M(G)$ and

$$(2.1) \qquad |\hat{\mu}(\gamma)| \geq 1 \text{ or } \hat{\mu}(\gamma) = 0 \ \underline{\text{for each}} \ \gamma \varepsilon \Gamma,$$

and let $E_\mu = \{\gamma: |\hat{\mu}(\gamma)| \geq 1\}$, the support of $\hat{\mu}$. Since $\gamma\mu(1) = \mu(\gamma) = \hat{\mu}(\gamma)$ has modulus $\geq 1$ for $\gamma \varepsilon E_\mu$, $0 \notin (E_\mu\mu)^-$, the w* closure of $E_\mu\mu$, so $(E_\mu\mu)^-$ contains a non-zero element $\nu$ of least norm. Evidently $\nu$ also satisfies (2.1), and if $\gamma_\delta\mu \to \nu$ while $\gamma \varepsilon E = E_\nu$, then $|\nu(\gamma)| \geq 1$ so $|\mu(\gamma\gamma_\delta)| \geq 1$ for $\delta \geq \delta_0$, and $\gamma\gamma_\delta \varepsilon E_\mu$ so that $\gamma\nu = \lim \gamma\gamma_\delta\mu \varepsilon (E_\mu\mu)^-$. Hence

$$(2.2) \qquad (E\nu)^- = (E_\nu\nu)^- \subset (E_\mu\mu)^-$$

and all the elements of $(E\nu)^-$, necessarily of norm $\leq \|\nu\|$, must have the same norm since $\|\nu\|$ was minimal. Thus lemma 2.1 applies to $(E\nu)^-$, which is therefore norm compact.

Now since $\nu$ satisfies (2.1), if $|\nu(\alpha\gamma) - \nu(\beta\gamma)| < 1$ then evidently $\alpha\gamma \varepsilon E$ iff $\beta\gamma \varepsilon E$, $E$ being the support of $\hat{\nu}$. Thus $\|\alpha\nu - \beta\nu\| < 1$ implies $(\alpha\nu)^\wedge$ and $(\beta\nu)^\wedge$ have the same support $\alpha^{-1}E = \beta^{-1}E$. Evidently $\alpha \sim \beta$ iff $\alpha^{-1}E = \beta^{-1}E$ defines an equivalence relation on $E$, and the compactness of $(E\nu)^-$ now implies $E$ consists of finitely many equivalence classes $E_1,\ldots,E_n$. Moreover, for any $i \leq n$, $\alpha \sim \beta$ implies

$$(2.3) \qquad \alpha^{-1}E_i = \beta^{-1}E_i \ :$$

for if $\gamma \varepsilon E_i$, $\alpha^{-1}\gamma \varepsilon \alpha^{-1}E = \beta^{-1}E$, so $\alpha^{-1}\gamma = \beta^{-1}\sigma$, $\sigma = \beta\alpha^{-1}\gamma \varepsilon E$. But $\sigma^{-1}E = \gamma^{-1}(\beta^{-1}\alpha E) = \gamma^{-1}E$ so $\sigma \sim \gamma$ and $\sigma \varepsilon E_i$, whence $\alpha^{-1}E_i \subset \beta^{-1}E_i$, and 2.3 follows.

As a consequence each $E_i$ is a coset: for $\alpha \varepsilon E_i$, $\beta \varepsilon \alpha^{-1}E_i$, we have $\alpha\beta \in E_i$ (i.e., $\alpha\beta \sim \alpha$) so that $\alpha^{-1}\beta^{-1}E_i = \alpha^{-1}E_i$ by (2.3), and thus $\alpha^{-1}E_i = \beta(\alpha^{-1}E_i)$ so that $\alpha^{-1}E_i$ is closed under multiplication. Moreover $1 \in \alpha^{-1}E_i$, so $\alpha^{-1}E_i = \beta(\alpha^{-1}E_i)$ implies $\beta^{-1} \in \alpha^{-1}E_i$ as well, and $\alpha^{-1}E_i$ is a subgroup $\Lambda_i$ of $\Gamma$. In fact all the $\Lambda_i$ coincide, for if $\lambda \in \Lambda_i$, $\lambda = \alpha^{-1}\beta$, $\alpha,\beta \in E_i$, so that $\alpha^{-1}E_j = \beta^{-1}E_j$ implies $\lambda E_j = E_j$, whence $\lambda\Lambda_j = \Lambda_j$, and $\Lambda_i \subset \Lambda_i\Lambda_j \subset \Lambda_j$.

Thus

(2.4)
$$E = E_\nu = \bigcup_{i=1}^{n} \gamma_i H^\perp$$

where $H$ is a (compact) subgroup of $G$, and so the support of $\hat{\nu}$ coincides with that of $\hat{\eta}$, where $\eta$ is the idempotent $(\sum_1^n \bar{\gamma}_i)m_H$, $m_H$ the normalized Haar measure of $H$. Hence $\nu = \eta * \nu$, which in particular implies

(2.5) $h \to \delta_h * \nu$ <u>is norm continuous on</u> $H$.

As a consequence we can conclude that any net $\{\gamma_\delta\}$ in $E_\mu$ for which $\gamma_\delta\mu \to \nu$ eventually lies in finitely many cosets mod $H^\perp$, i.e., that for no cofinal subnet will $\gamma_\delta H^\perp \to \infty$ in $\Gamma/H^\perp$. For suppose that weren't the case, so that replacing $\{\gamma_\delta\}$ by that subnet, we would have $\gamma_\delta\mu \to \nu$, $\gamma_\delta H^\perp \to \infty$ in $\Gamma/H^\perp$. We can decompose $\mu$ as $\mu_F + \mu_{F'} = \rho + \sigma$, by exhaustion exactly as in the Lebesgue decomposition, so that $h \to \delta_h * \rho$ is strongly continuous on $H$ while $\sigma$ is singular with respect to all measures with that property, and since $h \to \delta_h * \rho$ is strongly continuous we can approximate $\rho$ arbitrarily well in norm by $\int_H \delta_h * \rho \cdot f(h)m_H(dh) = \rho * fm_H$, $f \in L_1(H)$. Since $\gamma_\delta H^\perp \to \infty$ in $\Gamma/H^\perp = \hat{H}$, $\gamma_\delta fm_H \to 0$ w* in $M(G)$ by Riemann-Lebesgue (w* convergence being just pointwise convergence of Fourier-Stieltjes transforms on bounded sets), so that $\gamma_\delta(\rho * fm_H) = \gamma_\delta\rho * \gamma_\delta fm_H \to 0$ w* as well. Thus

$\gamma_\delta \rho \to 0$ w* also, and $\nu = \lim \gamma_\delta \mu = \lim \gamma_\delta \sigma$.

Now if $g$ is any w* cluster point of $\{\gamma_\delta\}$ in $L_\infty(|\sigma|)$ then evidently $g\sigma$ is a w* cluster point of $\{\gamma_\delta \sigma\}$ in $M(G)$ (since $C_o(G) \subset L_1(|\sigma|)$), so $\nu = g\sigma$, and thus $\nu$, exactly as $\sigma$, must be singular with respect to all measures $\lambda$ for which $h \to \delta_h * \lambda$ is norm continuous on $H$. In particular $\nu(\neq 0)$ is singular with respect to itself by (2.5), a contradiction showing <u>any net</u> $\{\gamma_\delta H^\perp\}$ <u>in</u> $^\Gamma/H^\perp$ <u>with</u> $\gamma_\delta \mu \to \nu$ w* <u>is eventually finite</u>. In particular we can find one net with

(2.6) $$\gamma_\delta \mu \to \nu \quad w^*, \text{ and } \gamma_\delta \in E_\mu \cap \gamma_o H^\perp .$$

Now we recall our decomposition (1.2) of $\mu$ into mutually singular measures $\mu = \mu_H + \mu'$, where $\mu_H$, the "part of $\mu$ carried by cosets of H" is supported by cosets of H and $\mu'$ vanishes on all Borel subsets of cosets of H. Evidently, passing if necessary to a subnet of our $\{\gamma_\delta\}$ in (2.6) we can assume the net $\{\bar{\gamma}_o \gamma_\delta\}$ in $H^\perp = (^G/H)^\wedge$ converges pointwise (and so cosetwise) on G to a character $\chi$ of $(^G/H)^d$, and thus

(2.7) $$\gamma_o \chi \mu_H = \lim \gamma_\delta \mu_H$$

w*, or even in norm. We can also assume $\gamma_\delta \mu' \to \nu'$ w* so $\nu = \gamma_o \chi \mu_H + \nu'$, where necessarily $\nu' << |\mu'|$, as in the paragraph before (2.6), so vanishes on all Borel subsets of cosets mod H. By lemma 2.2 we have a net $\{\beta_\rho\}$ in $H^\perp$ for which $\beta_\rho \nu' \to 0$ w*, and for $\gamma_1$ (as in (2.4)) with $\gamma_1 H^\perp \subset E_\nu$, $\gamma_1 \beta_\rho \nu' \to 0$, whence any w* cluster point $\lambda$ of $\{\gamma_1 \beta_\rho \nu\} \subset E_\nu \nu$ in the w* compact set $(E_\nu \nu)^-$ is actually a w* cluster point of $\{\gamma_1 \beta_\rho \gamma_o \chi \mu_H\}$ since $\nu = \gamma_o \chi \mu_H + \nu'$. Thus $\|\lambda\| \leq \lim \|\gamma_1 \beta_\rho \gamma_o \chi \mu_H\| = \|\gamma_o \chi \mu_H\|$, while, since the elements of $(E_\nu \nu)^-$ all have norm $= \|\nu\|$ as we saw in (2.2), $\|\nu\| = \|\lambda\| \leq \|\gamma_o \chi \mu_H\| \leq \|\gamma_o \chi \mu_H\| + \|\nu'\| = \|\nu\|$ since $\nu'$

and $\gamma_o \chi\mu_H$ are mutually singular. We conclude that $\nu' = 0$, so $\gamma_\delta\mu' \to 0$ w*, and

$$(2.8) \qquad \gamma_o\chi\mu_H = \lim \gamma_\delta\mu_H = \lim \gamma_\delta\mu = \nu \ .$$

By (2.4), $H^\perp E_\nu = E_\nu$, and since we can find a $\{\beta_\delta\}$ in $H^\perp$ with $\beta_\delta \to \bar{\chi}$ pointwise (cf. (2.7)), and $\beta_\delta\gamma_1 \in E_\nu$, $\beta_\delta\gamma_1\nu = \beta_\delta\gamma_1\gamma_o\chi\mu_H \in E_\nu\nu$, so $\gamma_1\gamma_o\mu_H = \lim \beta_\delta\gamma_1\nu \in (E_\nu\nu)^- \subset (E_\mu\mu)^-$, and thus $\gamma_1\gamma_o\mu_H$ is another element of $(E_\mu\mu)^-$ of least norm, $\|\nu\|$, since $\|\gamma_1\gamma_o\mu_H\| = \|\gamma_o\chi\mu_H\| = \|\nu\|$ by (2.8).

Now exactly as in the case of $\nu$ in (2.4) we have the support of $(\gamma_1\gamma_o\mu_H)^\wedge$ that of the transform $\hat{\eta}_1$ of an idempotent, a set $\bigcup_{j=1}^m \beta_j H_1^\perp$; in particular $\gamma_1\gamma_o\mu_H = \eta_1 * \gamma\gamma_o\mu_H$. As we shall see in a moment, both $H$ and $H_1$ can be taken to be the same. Indeed, since $\gamma_1\gamma_o\mu_H \in (E_\nu\nu)^-$ we have a net $\{\beta_\delta\}$ in $E_\nu = \bigcup_{i=1}^n \gamma_i H^\perp$ with $\beta_\delta\nu \to \gamma_1\gamma_o\mu_H$, and we can of course assume the net lies in one coset $\gamma_k H^\perp$. Thus, since $|(\beta_\delta\nu)^\wedge| \geq 1$ precisely on $\bigcup_{i=1}^n \gamma_k\gamma_i H^\perp$ and otherwise vanishes, that set is precisely the support of $(\gamma_1\gamma_o\mu_H)^\wedge$, and

$$\bigcup_{j=1}^m \beta_j H_1^\perp = \bigcup_{i=1}^n \gamma_k\gamma_i H^\perp,$$

which implies each of the groups $H^\perp$ and $H_1^\perp$ lies in a finite union of cosets modulo the other. Now the union of the finitely many cosets mod $H_1^\perp$ which meet $H^\perp$ (or those of $H^\perp$ which meet $H_1^\perp$) forms a subgroup $H_2^\perp = H^\perp + H_1^\perp$ of $\Gamma$ in which both $H^\perp$ and $H_1^\perp$ are of finite index. Thus the subgroup $H_2$ of $G$ is of finite index in the subgroups $H$ and $H_1$ of $G$.

In particular then $\mu_H = \mu_{H_2}$ on the one hand, while

$(\gamma_1\gamma_o\mu_H)^\wedge = (\gamma_1\gamma_o\mu_{H_2})^\wedge$ has its support of the form $\bigcup_{j=1}^{k} \alpha_j' H_2^\perp$, so that we can replace $H$ by $H_2$ and write $\gamma_1\gamma_o\mu_H = \eta_1 * \gamma_1\gamma_o\mu_H$ with $\eta_1 = (\sum_{j=1}^{k} \bar{\alpha}_j')m_H$.

Therefore we can write

$$\mu_H = \eta * \mu_H, \quad \eta = (\sum_{j=1}^{k} \bar{\alpha}_j)m_H,$$

with $\bigcup_1^k \alpha_j H^\perp$ the common support of $\hat{\mu}_H$ and $\hat{\eta}$. On the other hand $\eta$ is a multiple of $\mu_H$. For $(\alpha_j\mu_H) * m_H = (\sum_1^\infty c_j\delta_{x_j}) * m_H$, and with $\sigma = \Sigma c_j\delta_{x_j}$, since $|\hat{\sigma}| = |(\alpha_j\mu_H)^\wedge| \geq 1$ on $H^\perp = (G/H)^\wedge$, the discrete image measure $\rho^*\sigma = \Sigma c_j\delta_{\rho(x_j)}$ of $\sigma$ on $G/H$ (where $\rho: G \to G/H$ is the canonical map) has its transform of modulus $\geq 1$, hence is invertible, with a discrete inverse. Thus, choosing a discrete measure $\tau$ on $G$ which $\rho^*$ maps onto $(\rho^*\sigma)^{-1}$, we have $\rho^*(\sigma * \tau) = \rho^*\sigma * \rho^*\tau$ the identity of $M(G/H)$, so that the discrete measure $\sigma * \tau$, restricted to $H$ is of the form $\Sigma d_j\delta_{y_j}$, where $\Sigma d_j = 1$, while on any other coset of $H$ the restriction is of the same form with $\Sigma d_j = 0$. Consequently $m_H = \tau * (\sigma * m_H) = \tau * (\alpha_j\mu_H) * m_H$ and therefore $\bar{\alpha}_j\tau * \mu_H * \bar{\alpha}_j m_H = \bar{\alpha}_j m_H$, so that $\bar{\alpha}_j m_H$ is a multiple of $\mu_H$, and therefore $\eta$ is.

Finally since each of $\mu_H$ and $\eta$ is a multiple of the other, their Gelfand representatives on the spectrum (maximal ideal space) $\mathfrak{M}$ of $M(G)$ have the same support, and each is bounded away from zero on that support; thus for $\varphi \in \mathfrak{M}$, $\varphi(\mu_H + \delta_o - \eta) = \varphi(\mu_H)$ if $\varphi(\mu_H) \neq 0$, or $= 1$ if $\varphi(\mu_H) = 0 = \varphi(\eta)$, so the Gelfand representative of $\lambda = (\mu_H + \delta_\sigma - \eta)$ is bounded away from $0$ on $\mathfrak{M}$ and $\lambda \in M(G)^{-1}$. Clearly $\mu_H = \eta * \mu_H = \eta * (\mu_H + \delta_o - \eta)$ now yields theorem 1.1 when $G$ is compact.

If $G$ is not compact we view it as a subset of the Bohr compactification $G^a$ of $G$, and $\mu$ as a measure on $G^a$ in the usual fashion. From the compact case we have $\mu_H = \eta * \lambda$, for $\eta = (\sum_1^n \bar{\gamma}_i)m_H$, $\lambda \varepsilon M(G^a)^{-1}$, where H is a subgroup of $G^a$, and $\eta$ is also a multiple of $\mu_H$ in $M(G^a)$. Since $\mu_H$ is carried by G, $\hat{\mu}_H$ is continuous, so its common support with $\hat{\eta}$, $F = \bigcup_1^n \gamma_i H^\perp$, is open in $\Gamma$; it's also closed since $\hat{\mu}_H$ satisfies (2.1) as we saw above (or know from [8]). Thus $\gamma_i(H^\perp)^- = (\gamma_i H^\perp)^- \subset F$, and F is a finite union of cosets mod $(H^\perp)^-$, each of which has interior since F has. Since the smaller subgroup $H_o = H^{\perp - \perp}$ of $G^a$ has finite index in H, $\mu_H = \mu_{H_o}$, so we may as well replace H by $H_o$, and thus assume $H^\perp$ is an open subgroup of $\Gamma$, which means of course that H is a compact subgroup of G itself, whence $\eta \varepsilon M(G)$.

Since $\eta$ and $\mu_H$ are multiples of one another in $M(G^a)$ their Gelfand representatives on the spectrum of this algebra have the same support on which they are bounded away from zero; thus $\mu_H + \delta_o - \eta$ is invertible in $M(G^a)$ with an inverse whose Fourier-Stieltjes transform is $(\hat{\mu}_H + 1 - \hat{\eta})^{-1}$, hence continuous, so also in M(G) by the well known result of Bochner-Schoenberg-Eberlein [10]. Again $\mu_H = \eta * (\mu_H + \delta_o - \eta)$ yields our factorization, and our proof is complete.

We note the following consequence since it extends a key fact in Helson's solution of the idempotent problem for $G = T^1$ [10].

<u>Corollary</u> 2.3. <u>If</u> $\mu$ <u>vanishes on all</u> (<u>Borel subsets of</u>) <u>cosets of compact proper subgroups of</u> G <u>and satisfies</u> (1.1) <u>then</u> G <u>is compact and</u> $\mu = (\sum_{i=1}^n \gamma_i)m_G$.

3. A few applications should be noted. We shall consider only the case of G compact, the general case following from a shift into $G^a$ as before. The following extension of Helson's lemma, which was proved in the two paragraphs before (2.6) above, will be

useful.

Lemma 3.1. Let $\nu = \lim_{w^*} \gamma_\delta\mu$, $\mu \in M(G)$, and suppose $\nu$ has a non-zero continuously H-translating component (i.e., $\nu = \nu_1 + \nu_2$, $\nu_1 \neq 0$, and $h \to \delta_h * \nu_1$ is norm continuous while $\nu_2$ is singular with respect to all such measures). Then $\{\gamma_\delta\}$ eventually lies in finitely many cosets of $H^\perp$. (A slight modification of the proof, using the conditional weak compactness of $\Gamma\mu$, works if G is non-compact, but we have no need of any but the compact case.)

Corollary 3.2. Suppose $S \subset \Gamma$, and i) 0 is isolated in $\{0\} \cup \hat{\mu}(S)$, ii) there is a net $\{\gamma_\delta\}$ in S with $\hat{\mu}(\gamma_\delta) \neq 0$ and the characteristic functions $\chi_{\gamma_\delta^{-1}S} \to 1$ pointwise (i.e., $\gamma \in \gamma_\delta^{-1}S$ for all $\delta \geq \delta_\gamma$). Then the conclusion of theorem 1.1 applies to $\mu$, and a cofinal subnet of $\{\gamma_\delta\}$ lies in one coset of $H^\perp$.

Proof. Suppose $|\hat{\mu}(\gamma)| \geq 1$ or $= 0$ for each $\gamma$ in S. Let $\nu$ be any $w^*$ cluster point of $\{\gamma_\delta\mu\}$; then $|\nu(1)| \geq 1$ and i) and ii) combine to show $\nu$ satisfies the hypothesis of 1.1. Consequently we have our compact subgroup H for which $0 \neq \nu_H = \eta * \lambda$, $\eta = (\Sigma\gamma_i)m_H$, $\lambda \in M(G)^{-1}$.

We can of course assume $\gamma_\delta\mu \to \nu$. Since $\nu_H \neq 0$, $\nu = \lim \gamma_\delta\mu$ satisfies the hypothesis of lemma 3.1 so we can assume $\gamma_\delta \in \gamma H^\perp$, all $\delta$, again passing to a cofinal subnet, and indeed that $\bar{\gamma}\gamma_\delta \to \chi \in (G/H)^d$ , doing so once more. Now writing $\mu = \mu_H + \mu'$, $\nu = \nu_H + \nu'$, as earlier we have $\gamma_\delta\mu_H \to \gamma\chi\mu_H$, while as before $\nu' = \lim \gamma_\delta\mu' \ll |\mu'|$ (cf. (2.7)). Thus $\nu_H = \gamma\chi\mu_H$, and now $\chi\mu_H$ has the form $\eta * \lambda$. Since we have a net $\{\beta_\delta\}$ in $H^\perp$ which tends pointwise (hence cosetwise mod $H^\perp$) to $\bar{\chi}$, we have $\mu_H = \lim \beta_\delta(\eta * \lambda) = \lim \beta_\delta\eta * \beta_\delta\lambda = \eta * \lim(\beta_\delta\lambda)$, again passing to a subnet. But any $w^*$ cluster point, and so our presumed limit

$\lambda_o$ of $\{\beta_\delta\lambda\}$, lies in $M(G)^{-1}$, since convolution is jointly w* continuous: if $\lambda_1$ is a w* cluster point of $\{\beta_\delta\lambda^{-1}\}$ then $\lambda_o * \lambda_1$ is one of $\{\beta_\delta\lambda * \beta_\delta\lambda^{-1}\} = \{\beta_\delta(\lambda * \lambda^{-1})\} = \{\delta_o\}$, so $\lambda_o * \lambda_1 = \delta_o$. Hence $\mu_H$ has the desired form, completing our proof.

Given a measure $\mu$, and $\varepsilon > 0$, one candidate for $S$ in 3.2 is $\hat{\mu}^{-1}(0) \cup E_\varepsilon$, $E_\varepsilon = \{\gamma \in \Gamma\colon |\hat{\mu}(\gamma)| \geq \varepsilon\}$, and we can preclude a variety of possible candidates for pairs $\hat{\mu}^{-1}(0)$, $E_\varepsilon$, by assuming $\mu_H$ is not of the correct form for certain subgroups $H$, while no net in $E_\varepsilon$ is eventually constant mod $H^\perp$ for the remaining $H$. For example, with $G = T^2$ no measure $\mu$ can vanish on the first quadrant $Q$ of $\mathbb{Z}^2$ less $E_\varepsilon$, if $E_\varepsilon$ is confined in $Q$ to say $\{(m,n) \in Q\colon cm^2 < n < dm^2\}$, $0 < c < d$, unless $E_\varepsilon \cap Q$ is finite.

Indeed suppose $E_\varepsilon \cap Q$ is infinite. Computing the mean square of $|\hat{\mu}|$ from averages over squares $[0,m] \times [0,m]$ shows it has no discrete part, while since $E_\varepsilon \cap Q$ is infinite we have a sequence $\{(m_i,n_i)\}$ therein with $(m_i,n_i) \to \infty$, in fact modulo any line of rational slope, and of course $\chi_{-(m_i,n_i)+Q} \to 1$. Thus 3.2 applies to yield our $\mu_H$, which cannot be discrete, so that $H$ cannot be finite.

Now the remaining candidates for $H$ are $T^2$ (which cannot occur since then $\eta = (\sum_1^n \gamma_i) m_{T^2}$, and $\hat{\mu}_H = \hat{\mu}$ has finite support) and proper subgroups of $T^2$ which contain a circle, so that $H^\perp$ lies in a line of rational slope, and thus $(m_i,n_i) \to \infty$ modulo $H^\perp$, contradicting the final part of 3.2 and yielding our assertion.

The same sort of result can be phrased another way. Suppose $(m_i,n_i) \to \infty$ modulo any line of rational slope in $\mathbb{Z}^2$, while $\hat{\mu}(m_i,n_i) \neq 0$ and $\mu_d$, the discrete part of $\mu$, is not of the form $\eta * \lambda$, $\eta = \eta * \eta$, $\lambda \in M(T^2)^{-1}$. Then for any $\varepsilon > 0$, $|\hat{\mu}|$ has a value in $(0,\varepsilon)$ on each of the squares $[m_i-i, m_i+i] \times [n_i-i, n_i+i]$, for $i \geq i_o$. (If $\mu_d = 0$, the conclusion with $[0,\varepsilon)$ in place of $(0,\varepsilon)$ follows easily from the fact that $|\hat{\mu}|$ has mean square zero, so the point is that we obtain small non-zero values.) Indeed if the assertion fails we have a sequence of our squares on each of which $|\hat{\mu}(m,n)| \geq \varepsilon$, or $= 0$, and we take $S$ as the union of these. Clearly 3.2 applies with a subsequence of $(m_i, n_i)$ as our net, and our hypothesis forces $H$ to be infinite, hence to contain a circle. So we find $H^{\perp}$ lies in a line of rational slope, which the final assertion of 3.2 precludes.

Something slightly better can be claimed if we replace the torus by the circle: if $\{I_n\}$ is a sequence of disjoint intervals in $\mathbb{Z}$ with lengths $\to \infty$ on each of which $\mu$ does not vanish identically, then either $\mu_d = \eta * \lambda$, $\eta = \eta * \eta$, $\lambda \in M(T)^{-1}$, or for each $\varepsilon > 0$, $|\hat{\mu}|$ assumes a value in $(0,\varepsilon)$ on all but finitely many $I_n$.

Suppose not, and replace our sequence by a subsequence for which $|\hat{\mu}(m)| \geq \varepsilon$ or $= 0$ for each $m \in I_n$, with $m_n \in I_n$ such that $|\hat{\mu}(m_n)| \geq \varepsilon$. We can pass to a further subsequence for which $\chi_{-m_n + I_n} \to 1$ on a half line (at least), say on $\mathbb{Z}_+$. Then any w* cluster point $\nu$ of $\{e^{im_n} \cdot \mu\}$ is non-zero, singular (by Helson's lemma), and satisfies $|\hat{\nu}(m)| \geq \varepsilon$ or $= 0$ for each $m$ in $\mathbb{Z}_+$. Because of the singularity of $\nu$ and the F. and M. Riesz theorem, $|\hat{\nu}(n)| \geq \varepsilon$ for an infinite sequence $\{n_j\}$ in $\mathbb{Z}_+$, $n_j \to \infty$, and this net, along with $S = \mathbb{Z}_+$ and $\nu$ as our measure, satisfy the hypotheses of 3.2, whence $\nu_H$ has the usual form. Since $H = T^1$ makes $\nu_H$ absolutely continuous, $H$ is a finite subgroup of $T^1$, and $\nu_H = \nu_d$. As in the proof of 3.2, $\nu_d = \chi\mu_d$, where $\chi$ is a character of $(T^1/H)^d$, and as there this implies $\mu_d = \eta * \lambda$.

So far we have not noted the special role 0 plays as an isolated "value": if $a \neq 0$ is isolated in $\{a\} \cup \hat{\mu}(\Gamma)$ our result applies to $a\delta_0 - \mu$, but may yield no information about $\mu$ itself. Even so there are some applications. For example, suppose $\mu \in M(T)$, $\mu_d \neq 0$, and $\hat{\mu}(\mathbb{Z})$ is infinite with zero as its only accumulation point. Then $\hat{\mu}_d$ assumes each of its non-zero values on a finite union of cosets, and on each such union $\hat{\mu}_c$ has finite support (where $\mu_c$ is the continuous component of $\mu$).

Indeed if $a \in \hat{\mu}(Z)$, $a \neq 0$, we have $a\delta_0 - \mu_H = (\sum_1^n \bar{\gamma}_i)m_H * \lambda \neq 0$, $\lambda \in M^{-1}$, for some subgroup $H$ of $T = T^1$. If $H = T$ then $a - \hat{\mu} = (a\delta_0 - \mu_H)^\wedge$ would have finite support and $\hat{\mu}(\mathbb{Z})$ would be finite. Thus $H$ is finite and

$$a\delta_0 - \mu_d = (\sum_1^n \bar{\gamma}_i)m_H * \lambda, \quad \lambda \in M^{-1},$$

so that $\hat{\mu}_d^{-1}(a)$ is a (possibly void) union of cosets $\bigcup_{j=1}^k (\beta_j + H^\perp) = \mathbb{Z} \setminus \bigcup_{i=1}^n (\gamma_i + H^\perp)$. Since $\mu_d \neq 0$ and $\hat{\mu}_d(Z) \subset \hat{\mu}(\mathbb{Z}) \cup \{0\}$ [8] we have some $a \neq 0$ for which $\hat{\mu}_d^{-1}(a) \neq \emptyset$, and for each such $a$, and $\beta_j$ as above, $(m_H * \beta_j\mu_d)^\wedge = a\chi_{H^\perp}$, which says that if $\rho : T \to T/H$ then $\rho^*(\beta_j\mu_d)^\wedge \equiv a$ on $(T/H)^\wedge = H^\perp$. So of course $\rho^*(\beta_j\mu_c)^\wedge$ has its range $(\beta_j\mu_c)^\wedge(H^\perp)$ contained in $(\beta_j\mu)^\wedge(H^\perp) - a \subset \hat{\mu}(\mathbb{Z}) - a$, in which 0 is isolated. Now since $\rho^*(\beta_j\mu_c)$ is a continuous measure on the circle $T/H$, the subgroup theorem 1.1 now yields (if $\rho^*(\beta_j\mu_c) \neq 0$) cannot be finite, so coincides with $T/H$, and

$$\rho^*(\beta_j\mu_c) = (\sum_1^m \gamma_i)m_{T/H} \qquad (\gamma_i \in H^\perp = (H/T)^\wedge).$$

Alternatively $m_H * \beta_j\mu_c = (\sum_1^m \gamma_i)m_T + \nu$, where $\rho^*\nu = 0$, and so $m_H * \nu = 0$, whence $m_H * \beta_j\mu_c = (\sum_1^m \gamma_i)m_T$ since $m_H * (\sum_1^m \gamma_i)m_T = (\sum_1^m \gamma_i)m_T$. Evidently then $\hat{\mu}_c$ has finite support on $\beta_j + H^\perp$, yielding our assertion.

4. A final word should be added concerning the problem of when $\mu * L_1$ is closed. The present result imitates the first step in Cohen's approach to the idempotent problem, and the continuation here requires that one show $\mu' * L_1$ is also closed ($\mu'$ as in (1.2)) which would yield the desired result. In the case of $G = T^1$ our first step only shows that $\mu_d$ has the form of the convolve of a non-zero idempotent and an invertible; but for the "non-zero", this was noted earlier by Wik.

As I learned from Professor Y. Meyer at the Conference, some questions related to those treated here are to be found in his forthcoming book "Algebraic Numbers and Harmonic Analysis", (North-Holland Publishers).

## References


[1] I. Amemiya and T. Ito, A simple proof of a theorem of P. J. Cohen, B.A.M.S. 70(1964), 774-776.

[2] P. J. Cohen, On a conjecture of Littlewood and idempotent measures, Am. J. Math. 82(1960).

[3] K. de Leeuw and I. Glicksberg, Applications of almost periodic compactifications, Acta Math. 105(1961), 63-97.

[4] ________, The decomposition of certain group representations, J. D'Analyse Math; SV(1965), 135-192.

[5] S. W. Drury, Sur les ensembles de Sidon, C. R. 271 Ser. A (1970), 162-163.

[6] R. Ellis, Locally compact transformation groups, Duke Math. J. 24(1957), 119-126.

[7] I. Glicksberg, When is $\mu * L_1$ closed, T.A.M.S. (to appear)

[8] I. Glicksberg and I. Wik, The range of Fourier-Stieltjes transforms of parts of measures, These Proceedings.

[9] H. Helson, Note on harmonic functions, PAMS 4(1953), 686-691.

[10] W. Rudin, Fourier Analysis on Groups, Interscience, New York, 1962.

# THE RANGE OF FOURIER-STIELTJES TRANSFORMS OF PARTS OF MEASURES

by

I. Glicksberg and I. Wik

University of Washington and University of Umeå

In recent articles [2], [3], Dunkl and Ramirez have proved that

(1) $$\|\hat{\mu}_d\|_\infty \leq \|\hat{\mu}\|_\infty ,$$

where $\mu_d$ is the discrete part of a measure $\mu$ in the measure algebra $M(G)$ of a locally compact abelian group $G$. They also proved that $\hat{\mu}_d$ can be replaced by a more general component $\hat{\mu}_K$, [4]. In this note we prove the stronger result

(2) $$\hat{\mu}_d(\Gamma) \subset \hat{\mu}(\Gamma)^-$$

(where the bar denotes closure). We also show that $\hat{\mu}_d$ in (2) can be replaced by $\hat{\mu}_K$.

These facts had apparently not been noticed by Dunkl and Ramirez, although as they pointed out to us, they follow from their result by noting that $\hat{\mu} \to \hat{\mu}_K(\gamma)$ is a multiplicative linear functional on $M(G)^\wedge$, which extends to its uniform closure by virtue of their extended version of (1). Moreover they do not need $K$ closed as we do. However our approach is considerably more elementary and yields some further information for the discrete part $\mu_d$.

The proof of Theorem 1 is based on the well-known fact that a continuous measure has a transform which averages to 0 [5, p. 118].

Theorem 1. (2) holds for every $\mu \in M(G)$.

Proof. Let $\varepsilon > 0$ and $\gamma_0 \in \Gamma$ be arbitrary. A sequence $\{\gamma_n\}_1^\infty$ in $\Gamma$ is constructed inductively, such that

(3) $$|\hat{\mu}_c(\gamma_0 + \gamma_n - \gamma_j)| < \varepsilon \quad \text{for} \quad j < n \quad (\mu_c = \mu - \mu_d).$$

This is possible because

$$(4) \qquad h(\gamma) = \sum_{j=1}^{n-1} |\hat{\mu}_c(\gamma_0 + \gamma - \gamma_j)|^2$$

is the Fourier-Stieltjes transform of a continuous measure and thus $\inf_{\gamma\varepsilon\Gamma} h(\gamma) = 0$ .

Let $\hat{\mu}_d(\gamma) = \sum_{p=1}^{\infty} a_p(x_p, \gamma)$, $\sum_1^{\infty} |a_p| < \infty$ .

Choose $N$ so large that $\sum_{N+1}^{\infty} |a_p| < \varepsilon$ and pass, if necessary, to a subsequence of $\{\gamma_n\}_1^{\infty}$ such that

$$(x_p, \gamma_n) \longrightarrow c_p \quad \text{as} \quad n \longrightarrow \infty, \quad 1 \leq p \leq N .$$

For these $p$, $|1 - (x_p, \gamma_n - \gamma_m)|$ tends to $0$ as $n$ and $m$ tend to infinity. Hence for large $m$ and $n > m$ we get

$$|\hat{\mu}_d(\gamma) - \hat{\mu}_d(\gamma + \gamma_n - \gamma_m)| < 2\varepsilon + \sum_{p=1}^{N} |a_p| \; |1 - (x_p, \gamma_n - \gamma_m)| < 3\varepsilon .$$

For such $n$ and $m$ we have

$$(5) \qquad |\hat{\mu}_d(\gamma_0) - \hat{\mu}(\gamma_0 + \gamma_n - \gamma_m)| < 4\varepsilon.$$

Since $\varepsilon > 0$ and $\gamma_0$ are arbitrary, (2) follows.

Our more general version of (2) is a simple corollary. Suppose $K$ is a closed subgroup of $G$ and $\mu = \mu_K + \nu$ is our (unique) decomposition in which $\mu_K$ is carried by a countable union of cosets of $K$ while $\nu$ vanishes on all Borel subsets of cosets mod $K$. Then

<u>Corollary</u> 2. <u>For any</u> $\mu$ <u>in</u> $M(G)$, $\hat{\mu}_K(\Gamma) \subset \hat{\mu}(\Gamma)^-$.

Let $\pi$ denote the map of measures on $G$ into measures on $G/K$ induced by $G \longrightarrow G/K$. We observe that $(\pi\mu)_d = \pi\mu_K$, and since we may identify the dual of $G/K$ with the subgroup $K^{\perp}$ of $\Gamma$, while $(\pi\mu)^{\wedge} = \hat{\mu}|K^{\perp}$, by (2) we know

$$(\pi\mu)^{\wedge}_d = (\pi\mu_K)^{\wedge} = \hat{\mu}_K|K^{\perp}$$

has range contained in the closure of the range of $(\pi\mu)^{\wedge} = \hat{\mu}|K^{\perp}$, i.e., that $\hat{\mu}_K(K^\perp) \subset \hat{\mu}(K^\perp)^-$. Evidently $\bar{\gamma}\mu_K$ is the part of $\bar{\gamma}\mu$ carried by cosets of $K$ (i.e., $(\bar{\gamma}\mu)_K = \bar{\gamma}\mu_K$), so that

$$\hat{\mu}_K(\gamma + K^\perp) = (\bar{\gamma}\mu_K)^\wedge (K^\perp) \subset (\bar{\gamma}\mu)^\wedge (K^\perp)^- = \hat{\mu}(\gamma + K^\perp)^- ,$$

which of course implies our corollary.

At the Conference, R. Kaufman raised the question of whether (2) could be obtained with $\Gamma$ replaced by a subset, for example a halfplane or the points inside a parabola in $\Gamma = R^2$. In fact the argument of Theorem 1 applies and gives an answer in the affirmative, in part because such a set is large enough to determine mean values of Fourier-Stieltjes transforms.

Theorem 3. *Suppose $E$ is a subset and $S$ a subsemigroup of $\Gamma$ for which $E + S \subset E$ and $\Gamma = E - S$. Then for every $\mu$ in $M(G)$ we have*

$$\hat{\mu}_d(E) \subset \hat{\mu}(E)^- . \tag{6}$$

Proof. We have only to show that for a given $\gamma_0$ in $E$ we can construct a sequence $\{\gamma_n\}_1^\infty$, satisfying (3) with $\gamma_0 + \gamma_n - \gamma_j \in E$ for $j < n$, since then our original argument leads directly to (5). In order to construct such a sequence inductively we have to know that for the non-negative Fourier-Stieltjes transform $h$, given in (4), there exists a $\gamma$ with $h(\gamma) < \varepsilon^2$ and $\gamma_0 - \gamma - \gamma_j \in E$ for $j < n$.

Now $h$ is the transform of a continuous measure on $G$ and so of mean zero. Thus zero appears as the (unique) constant in the closed convex hull $K$ of the set of translates of $h$. (The existence of such a constant is an immediate and well-known consequence of the Markov-Kakutani theorem [1, p. 456] and the weak compactness of $K$.) So we have $\alpha_i$ in $\Gamma$ and $c_i \geq 0$, $i = 1, 2, \ldots, m$, with $\sum_1^m c_i = 1$, and

(7) $\sum_1^m c_i\ h(\alpha_i + \gamma) < \varepsilon^2$, all $\gamma \in \Gamma$ .

Since $\Gamma = E - S$, we can write $\alpha_i + \gamma_0 - \gamma_j = e_{ij} - s_{ij}$, with $e_{ij} \in E$ and $s_{ij} \in S$ for each $i \leq m$ and $j < n$ . The element $\gamma_n^* = \sum_{i,j} s_{ij} \in S$ has the property that $\gamma_n^* + \alpha_i + \gamma_0 - \gamma_j \in E + S \subset E$, $i \leq m$, $j < n$, while (7) implies $h(\alpha_i + \gamma_n^*) < \varepsilon^2$ for some i. For that i we put $\gamma_n = \gamma_n^* + \alpha_i$ to obtain (3) and $\gamma_0 + \gamma_n - \gamma_j \in E$, $j < n$, which thus completes our proof.

In the same way one obtains

$$\hat{\mu}_K(E) \subset \hat{\mu}(E)^-,$$

if, in addition to the conditions in Theorem 3, we assume for example that $K^\perp \cap (E - \gamma) - K^\perp \cap S = K^\perp$ for all $\gamma$ in E. For as in Corollary 2 we have

$$\hat{\mu}_K(K^\perp \cap (E - \gamma)) \subset \hat{\mu}(K^\perp \cap (E - \gamma))^-$$

as a consequence of Theorem 3, and replacing $\mu$ by $\bar{\gamma}\mu$ this yields

$$\hat{\mu}_K(\gamma + K^\perp \cap (E - \gamma)) \subset \hat{\mu}(\gamma + K^\perp \cap (E - \gamma))^-.$$

Now we only have to note that $E = \bigcup_{\gamma \in E} (\gamma + K^\perp \cap (E - \gamma))$.

We also note that for smaller sets E, like R as subset of $R^2$, (6) does not hold. A continuous measure on $R^2$ may have a discrete projection on R and we may have $\hat{\mu}(E) = 1$ but $\hat{\mu}_d(E) = 0$ on a one-dimensional subgroup E of $R^2$.

## References


[1] Dunford, N. and Schwartz, J., Linear Operators, Part I, Interscience, N.Y. 1958.

[2] Dunkl, C. and Ramirez, D., $C^*$-algebras generated by measures, Bull. Amer. Math. Soc. 77 (1971), 411-412.

[3] Dunkl, C. and Ramirez, D., $C^*$-algebras generated by Fourier-

Stieltjes transforms, (to appear).

[4] Dunkl, C. and Ramirez, D., Bounded projections on Fourier-Stieltjes transforms, Proc. Amer. Math. Soc., Jan. 1972.

[5] Rudin, W., Fourier Analysis on Groups, Interscience, N.Y. 1962.

# FOURIER BESSEL TRANSFORMS AND HOLOMORPHIC DISCRETE SERIES

by

Kenneth I. Gross and Ray A. Kunze *

Dartmouth College and University of California, Irvine

## §1. Introduction

This is the first in a series of papers concerned with generalizations of classical Bessel functions and the role they play in describing discrete series representations and limits of discrete series representations of certain classical groups. In the present paper, we are concerned with the case in which the Bessel functions are defined on $\mathbb{C}^{n \times n}$, the space of all complex $n \times n$ matrices, and $U(n, n)$ is the appropriate group. We also limit our considerations to those and only those Bessel functions and discrete series which appear in the decomposition of the Weil representation for $U(n, n)$. Roughly speaking, this accounts for about half of the holomorphic discrete series.

The projective Weil representation [1] for groups of 'symplectic type' is actually a unitary representation $R$ of $U(n, n)$ in the usual sense, the cocycle being trivial. An outline of the construction of $R$ for $U(n, n)$ and its explicit description is given in §2.

In §3 we show that $R$ has a discrete direct sum decomposition

$$R = \sum_{\lambda \in \tilde{U}} \oplus \, d_\lambda \, R(\cdot \, , \lambda)$$

* This research was partially supported by the National Science Foundation under Grant No. GP-22795 and Grant No. GP-30061.

in which $R(\cdot,\lambda)$ is an irreducible unitary representation of $U(n,n)$ , $U = U(n)$, $\tilde{U}$ is the dual of $U$ , and $d_\lambda$ is the degree of $\lambda$ .

The generalized Bessel functions are introduced in §4. If $\lambda \in \tilde{U}$, the Bessel function $J_\lambda$ of order $\lambda$ is defined on $\mathbb{C}^{n\times n}$ by

$$J_\lambda(z) = \lambda(i) \int_U \exp\left(-i \operatorname{Re} \operatorname{tr}(zu)\right) \lambda(u^{-1})\, du \,,$$

where $du$ denotes normalized Haar measure on $U$ . The connection between Bessel functions and representations is due to the following fact: If $\delta(u) = (\det u)^n$ $(u \in U)$ and $p$ is the element of $U(n,n)$ which is defined by (2.1), then up to a constant $R(p,\lambda)$ is an integral operator with kernel $J_{\delta^{-1}\lambda}$ .

In §5 we show that some of the representations $R(\cdot,\lambda)$ are equivalent to representations $T(\cdot,\lambda)$ in spaces of holomorphic vector-valued functions on the upper-half plane in $\mathbb{C}^{n\times n}$ . We then show that a subset of these are square integrable.

Many analogous results are valid more generally for the groups associated with Hermitian symmetric spaces of tube type, and these extensions of our work will be pursued in future papers. Similar results for $Sp(n,\mathbb{R})$ have been obtained by rather different methods in recent, but as yet unpublished, work by S. Gelbart.

## §2. Weil Representations of $U(n, n)$

In this section we use the classical euclidian Fourier transform on the space $\mathbb{C}^{k\times n}$, equipped with the standard inner product

$$(z|w) = \operatorname{Re}\operatorname{tr}(zw^*),$$

to construct unitary representations of $U(n, n)$, analogous to those first constructed by Shale [2] and Weil [1]. Since the construction is in principle well known, and since a similar construction is carried out in detail for $Sp(n, \mathbb{C})$, in [3], we shall limit ourselves to a sketch of the main arguments.

Let $M = \mathbb{C}^{n\times n}$, $M^{2\times 2}$ the space of $2\times 2$ matrices over $M$, and

$$p = \begin{bmatrix} 0 & 1 \\ -1 & 0 \end{bmatrix} \in M^{2\times 2} . \tag{2.1}$$

In addition, let $Y = \mathbb{C}^{k\times n} \times \mathbb{C}^{k\times n}$. Then the set

$$Z = Y \times \mathbb{R},$$

with the multiplication

$$(y_1, t_1)(y_2, t_2) = (y_1+y_2, t_1+t_2-\operatorname{Re}\operatorname{tr} y_1 p y_2^*), \tag{2.2}$$

is a nilpotent (Heisenberg) group with center $\{0\} \times \mathbb{R}$.

The restriction of an infinite dimensional irreducible unitary representation of $Z$ to the center of $Z$ is defined by a non-trivial character of $\mathbb{R}$. Two infinite dimensional irreducible unitary representations of $Z$ are equivalent if and only if they agree on the center of $Z$. Thus the Plancherel dual $\hat{Z}$ of $Z$ may be indexed by the set $\mathbb{R}^\times$ of non-zero real numbers. In fact, $\hat{Z}$ may be realized concretely by representations $V_\alpha$ $(\alpha \in \mathbb{R}^\times)$ on $L^2(\mathbb{C}^{k\times n})$ which reduce to the character $t \to \exp i\alpha t$ $(t \in \mathbb{R})$ on the center of $Z$. These representations are defined for $(z_1, z_2, t)$ in $Z$ and $f$ in $L^2(\mathbb{C}^{k\times n})$ by

(2.3) $$V_\alpha(z_1, z_2, t)\, f(w) = \exp i\alpha[t + (z_1|2w + z_2)] \cdot f(w + z_2)$$

Let $\mathrm{Aut}_o(Z)$ denote the group of automorphisms of $Z$ which leave the center of $Z$ fixed and

$$G = U(n, n) = \left\{g \in M^{2\times 2} : g p g^* = p\right\} .$$

Then each of the maps

$$\tilde{g} : (w, t) \to (w g^{-1}, t) \quad , \; g \in G$$

is an automorphism of $Z$ fixing the center. Hence, $g \to \tilde{g}$ is an isomorphism of $G$ into $\mathrm{Aut}_o(Z)$. Therefore, the following is true.

PROPOSITION 1. _For each_ $g$ _in_ $G$ _and_ $\alpha$ _in_ $\mathbb{R}^\times$, _there is a unitary operator_ $R_\alpha(g)$ _on_ $L^2(\mathbb{C}^{k\times n})$ _such that_

(2.4) $$R_\alpha(g)\, V_\alpha(z)\, R_\alpha(g)^{-1} = V_\alpha(\tilde{g}(z))$$

_for all_ $z$ _in_ $Z$; _moreover,_ $R_\alpha(g)$ _is unique up to a scalar factor of absolute value_ 1.

From the fact that $R_\alpha(g)$ is essentially unique, it is easy to derive the following result.

COROLLARY. _Given particular choices for_ $R_\alpha(g_1 g_2)$, $R_\alpha(g_1)$, _and_ $R_\alpha(g_2)$, _there exists a scalar_ $c$ _such that_

$$R_\alpha(g_1 g_2) = c\, R_\alpha(g_1)\, R_\alpha(g_2) \quad .$$

We shall indicate how it is possible to specify the operators $R_\alpha(g)$ once and for all in such a way that the resulting map, $g \to R_\alpha(g)$ $(g \in G)$, is actually a representation.

For this purpose, we need some simple facts about $G$. A matrix

$$g = \begin{bmatrix} a & b \\ c & d \end{bmatrix}$$

in $M^{2\times 2}$ lies in $G$ if and only if

$$\begin{bmatrix} d^* & -b^* \\ -c^* & a^* \end{bmatrix} = g^{-1} . \tag{2.5}$$

If $s = s^*$ it follows that the matrix

$$v(s) = \begin{bmatrix} 1 & 0 \\ s & 1 \end{bmatrix} \tag{2.6}$$

belongs to $G$, and it is clear that

$$V = \{v(s) : s = s^*\}$$

is a closed subgroup of $G$ isomorphic to the vector group $S$ of self-adjoint elements of $M$. Let $A = GL(n, \mathbb{C})$. If $a \in A$, (2.3) implies that the diagonal matrix

$$c(a) = \begin{bmatrix} \check{a} & 0 \\ 0 & a \end{bmatrix}, \quad \check{a} = (a^*)^{-1}, \tag{2.7}$$

is also in $G$. The diagonal elements of $G$ are all of the form (2.5), and they form a closed subgroup

$$C = \{c(a) : a \in A\}$$

which normalizes $V$ and is isomorphic to $A$. It follows that $L = VC$ is also a subgroup of $G$. In fact, it is easy to see that $L$ is the entire lower triangular subgroup of $G$.

The element $p$ defined by (2.1) also lies in $G$ and plays a key role in what follows. The apparent coincidence that both p and the usual euclidian Fourier transform

(2.8) $$\hat{f}(w) = (2\pi)^{-kn} \int_{\mathbb{C}^{k\times n}} e^{-i(w|v)} f(v)\,dv$$

on $\mathbb{C}^{k\times n}$ have order 4 is of crucial importance. In fact, $R_\alpha(p)$ is defined in terms of the Fourier transform.

PROPOSITION 2. _Let_ $s \in S$ _and_ $a \in A$. _Then the equations_

(2.6) $$R_\alpha(v(s))\,f(w) = \exp(i\alpha(w|ws)) \cdot f(w)$$

(2.7) $$R_\alpha(c(a))\,f(w) = (\det a)^k f(wa)$$

(2.8) $$R_\alpha(p)\,f(w) = (-2i\alpha)^{kn}\hat{f}(2\alpha w)$$

_define operators satisfying_ (2.4).

This can be verified by straightforward but somewhat lengthy computations.

On the other hand, it is easy to check that the equation

(2.9) $$R_\alpha(v(s)\,c(a))\,f(w) = (\det a)^k \exp(i\alpha(w|ws))\,f(wa)$$

defines a continuous unitary representation of L on $L^2(\mathbb{C}^{k\times n})$, and (2.9) extends (2.6) and (2.7) in an obvious fashion.

The next step is to extend $R_\alpha$ to LpV. For this one shows that

$$LpV = \left\{\begin{bmatrix} a & b \\ c & d\end{bmatrix} \in G : b \in A\right\}$$

an open set in G, and that the map

$$(\ell, v) \to \ell p v$$

is a homeomorphism of $L \times V$ onto LpV. It follows, in particular, that

(2.10) $$R_\alpha(\ell p v) = R_\alpha(\ell) R_\alpha(p) R_\alpha(v)$$

extends $R_\alpha$ to LpV .

THEOREM 1 . $R_\alpha$ *is multiplicative on* LpV *and extends uniquely to a continuous unitary representation of* G *on* $L^2(\mathbb{C}^{k\times n})$ .

As announced earlier, we shall only sketch the main ideas of the proof, because an analogous result for $Sp(n, \mathbb{C})$ is proved in [3] and simple modifications of that argument may be used to prove the present theorem.

The first step is to prove the following result.

LEMMA 1. $R_\alpha$ *is multiplicative on* LpV *if and only if*

(2.11) $$R_\alpha(p) R_\alpha(c(a)) R_\alpha(p)^{-1} = R(c(\check{a}))$$

*for all* a *in* A , *and*

(2.12) $$R_\alpha(p) R_\alpha(v(e)) R_\alpha(p) = R_\alpha(c(e)) R_\alpha(v(-e)) R_\alpha(p) R_\alpha(v(-e))$$

*for all* n×n *diagonal matrices* e *such that* $e^2 = 1$ .

The next step is to establish the validity of (2.11) and (2.12) . Now (2.11) is easy, but (2.12) is more difficult and involves a computation of the Fourier transform of

$$w \to \exp i\alpha(w|w) \ , \ w \in \mathbb{C}^{k\times n}$$

in the sense of distribution theory and some technical arguments.

After this one shows that G - LpV is a set of Haar measure 0 . The fact that $R_\alpha$ extends uniquely to a representation of G then follows immediately from Lemma 6 of [1] .

Some additional comments are in order. First, it is easy to see that

$R_{\alpha_1}$ and $R_{\alpha_2}$ are unitarily equivalent if and only if $\alpha_1$ and $\alpha_2$ have the same sign. Thus, it suffices, for most purposes, to consider $R_1$ or $R_{-1}$. Second, in what follows, we shall only consider $R_1$ and the case $k = n$. For this case it is technically convenient to work with an equivalent representation $R$ on $L^2(A)$.

For $z$ in $M$ set $\delta(z) = (\det z)^n$ and let $dz$ denote Lebesque measure on $M$. Then $|\delta(z)|^{-2} dz$ is a Haar measure on $A$. Hence, the equation

$$(\Phi f)(a) = \delta(a) f(a)$$

defines a unitary map of $L^2(M)$ onto $L^2(A)$.

Next, set $R(g) = \Phi R_1(g) \Phi^{-1}$ $(g \in G)$, and $X(z) = \operatorname{Re} \operatorname{tr}(z)$ $(z \in M)$. Then $X(zw) = X(wz)$, $X(z^*) = X(z)$ and $(z|w) = X(zw^*)$. With this notation

$$(2.13) \qquad R(p) f(a) = (i\pi)^{-n^2} \int_A \exp(-2i\, X(ab^*))\, \delta(ab^*)\, f(b)\, db$$

for all sufficiently nice $f$, for example, for $f$ in $C_c(A)$. In (2.13), $db$ denotes Haar measure on $A$, and in general we make the convention that integrals over groups are always taken relative to the appropriate Haar measure. Since $\operatorname{tr} bsb^*$ is real when $s = s^*$, it follows that

$$(2.14) \qquad R(v(x)\, c(a))\, f(b) = \exp(i \operatorname{tr} b s b^*)\, f(ba)$$

for all $(s, a)$ in $S \times A$.

## §3. The Decomposition of R

In this section we use the results of [4] to show that R decomposes discretely into a direct sum of irreducible representations.

Let U denote the group U(n) of unitary matrices in A and $L_r$ the left regular representation of U on $L^2(A)$:

$$L_r(u) f(a) = f(u^{-1} a) \text{ , } a \in A$$

for $u \in U$ and $f \in L^2(A)$. Since $X(aza^{-1}) = X(z)$ for all $(a, z)$ in $A \times M$, it follows from (2. 13) and the invariance properties of the Haar measure on A that R(p) commutes with each of the operators $L_r(u)$ $(u \in U)$. The same is evidently true of the operators given by (2. 14). Since LpV is dense in G, it follows by continuity that $R(g) L_r(u) = L_r(u) R(g)$ for all g in G. This also follows algebraically from the fact that $G = (LpV)^2$.

Now let $\lambda$ denote an arbitrary irreducible unitary matrix representation of U and $d_\lambda$ its degree. If T is any unitary representation of U, the orthogonal projection of the representation space for T on the subspace that transforms according to $\lambda$ is given by the equation

$$P_\lambda = d_\lambda \int_U \operatorname{tr} \overline{\lambda}(u) \, T(u) \, du \text{ .}$$

If one applies this to $L_r$, it is then immediate that

$$P_\lambda R(g) = R(g) P_\lambda$$

for all g in G. Therefore, the subspaces $P_\lambda L^2(A)$ are invariant under R, and R is the direct sum of the corresponding sub-representations.

These subspaces and representations may be realized in a form which is particularly convenient for our purpose.

Departing slightly from the terminology of [4], we say that a function

$$f : A \to \mathbb{C}^{d_\lambda \times 1},$$

or one with values in $\mathbb{C}^{d_\lambda \times d_\lambda}$, is $\lambda$-covariant if

$$f(u\,a) = \lambda(u)\, f(a)$$

for all $(u, a)$ in $U \times A$. Let $C(A, \lambda)$ denote the vector space of continuous $\lambda$-covariant functions from $A$ to $\mathbb{C}^{d_\lambda \times 1}$, and $d_\lambda \cdot C(A, \lambda)$ the space of continuous $\lambda$-covariant functions from $A$ to $\mathbb{C}^{d_\lambda \times d_\lambda}$. The spaces $L^2(A, \lambda)$ and $d_\lambda \cdot L^2(A, \lambda)$ are defined in a similar fashion. Thus $d_\lambda \cdot L^2(A, \lambda)$ is the space of $\lambda$-covariant Baire functions $f : A \to \mathbb{C}^{d_\lambda \times d_\lambda}$ such that

$$\int_A \mathrm{tr}(f(a^*)\, f(a))\, da < \infty,$$

equipped with the inner product

$$(h|k)_\lambda = \int_A \mathrm{tr}(k(a^*)\, h(a))\, da .$$

Every $\lambda$-covariant function $f$ with values in $\mathbb{C}^{d_\lambda \times d_\lambda}$ is specified by a sequence $f_1, f_2, \ldots, f_{d_\lambda}$ of $\lambda$-covariant functions

$$f_j : A \to \mathbb{C}^{d_\lambda \times 1}, \quad 1 \le j \le d_\lambda ;$$

moreover, $(f^*(a)\, f(a))_{ij} = f_i^*(a)\, f_j(a)$. Thus $d_\lambda L^2(A, \lambda)$ is a direct sum of $d_\lambda$ copies of $L^2(A, \lambda)$.

We shall show that $d_\lambda \cdot L^2(A, \lambda)$ is canonically isomorphic to $P_\lambda L^2(A)$. Therefore, the subrepresentation $R^\lambda$ of $R$ which is defined by $P_\lambda L^2(A)$ may be realized in $d_\lambda \cdot L^2(A, \lambda)$ . To complete the decomposition of $R$ , we show that $R^\lambda$ is a multiple of an irreducible representation $R(\cdot, \lambda)$ on $L^2(A, \lambda)$ .

Although $R(\cdot, \lambda)$ may be defined in terms of the final decomposition, it is easy to describe directly.

For this purpose note that the integral in (2. 13) is absolutely convergent for any continuous function of compact support in $C(A, \lambda)$ ; moreover, by the invariance properties of the measure and the Plancherel theorem, it defines a $\lambda$-covariant function of the same norm. Since the set $C_c(A, \lambda)$ of all such functions is dense in $L^2(A, \lambda)$ , there is a unique operator $R(p, \lambda)$ on $L^2(A, \lambda)$ such that

$$R(p, \lambda) f(a) = (i\pi)^{-n^2} \int_A \exp(-2i\, X(ab^*))\, \delta(a b^*) f(b)\, db \tag{3.1}$$

for $f$ in $C_c(A, \lambda)$ . For $(s, a)$ in $S \times A$ set

$$R(v(s)\, c(a), \lambda) f(b) = \exp(i \operatorname{tr} b\, s\, b^*) f(b\, a) , \tag{3.2}$$

as in (2. 14) . Then $R(v(s)\, c(a), \lambda)$ is also a unitary operator on $L^2(A, \lambda)$ . Proceeding as in §2 , we now define $R(\cdot, \lambda)$ on $LpV$ by

$$R(\ell p v, \lambda) = R(\ell, \lambda) R(p, \lambda) R(v, \lambda) . \tag{3.3}$$

THEOREM 2. $R(\cdot, \lambda)$ <u>is irreducible on</u> $L$ <u>and extends uniquely from</u> $LpV$ <u>to a continuous unitary representation of</u> $G$ .

Proof. The irreducibility on $L$ is a direct consequence of Mackey's results on semi-direct products. For $V$ is a self dual abelian normal subgroup of $L$ , and

$$c(a)\, v(s)\, c(a)^{-1} = v(a^* s\, a) \tag{3.4}$$

for all $(s, a)$ in $S \times A$ . It follows that $u \to c(u)$ $(u \in U)$ is an identification of of $U$ with the stability group in $C$ of the identity in $V$. From this it is easy to see that $\ell \to R(\ell, \lambda)$ $(\ell \in L)$ is equivalent to the irreducible unitary representation of $L$ induced by $\lambda$, [5]. The proof that $R(\cdot, \lambda)$ extends to a representation of $G$ is completely analogous to the proof of Theorem 1.

To show that $R$ decomposes into representations of the form $R(\cdot, \lambda)$ let $U^{\sim}$ be a set of representatives, fixed once and for all, for the equivalence classes of continuous irreducible unitary matrix representations of $U$. Let

$$L^2(A)^{\sim} = \sum_{\lambda \in U^{\sim}} \oplus (d_\lambda \cdot L^2(A, \lambda))\, d_\lambda .$$

Thus, $L^2(A)^{\sim}$ is the set of all functions $F$ on $U^{\sim}$ such that $F(\lambda) \in d_\lambda L^2(A, \lambda)$ for every $\lambda$ and

$$\Sigma_\lambda \|F(\lambda)\|_\lambda^2 d_\lambda < \infty .$$

The Hilbert space structure of $L^2(A)^{\sim}$ is given by the formula

$$(H|K) = \Sigma_\lambda (H(\lambda)|K(\lambda))_\lambda\, d_\lambda$$

for the inner product of the elements $H$ and $K$.

Next we define a kind of Fourier transform which we shall refer to as the Peter-Weyl transform. If $f \in C(A)$ its <u>Peter-Weyl transform</u> is the function on $U^{\sim}$ whose value at $\lambda$ is the element of $C(A, \lambda)$ which is given by

$$\tilde{f}(\lambda)(a) = \int_U f(u^{-1}a)\, \bar{\lambda}(u)\, du , \quad a \in A . \tag{3.5}$$

THEOREM 3. *The Peter-Weyl transform $f \to \tilde{f}$ maps $C(A) \cap L^2(A)$ into $L^2(A)^{\sim}$, and its restriction to $C(A) \cap L^2(A)$ extends uniquely to a unitary transformation $\mathcal{T}$ of $L^2(A)$ onto $L^2(A)^{\sim}$ which maps $P_\lambda L^2(A)$ onto $L^2(A, \bar{\lambda})$.*

This is proved in [4].

THEOREM 4. Let $R^{\lambda}(g)$ denote the restriction of $\mathcal{T}R(g)\mathcal{T}^{-1}$ to $d_{\lambda}L^2(A, \lambda)$ and $f$ a continuous function of compact support in $d_{\lambda}L^2(A, \lambda)$. Then

$$R^{\lambda}(p)f(a) = (i\pi)^{-n^2} \int_A \exp(-2iX(ab^*))\,\delta(ab^*)\,f(b)\,db \tag{3.6}$$

and

$$R^{\lambda}(v(s)c(a))f(b) = \exp(i\,\mathrm{tr}\,b s b^*)\,f(ba) \tag{3.7}$$

for all $(s, a)$ in $S \times A$.

Proof. By [4, Lemma 6] there is a continuous function $h$ of compact support on $A$ such that $\tilde{h}(\lambda) = f$. Now apart from constants

$$\begin{aligned}
(R(p)h)^{\sim}(\lambda)(a) &= \int_U (R(p)h)(u^{-1}a)\,\overline{\lambda}(u)\,du \\
&= \int_U \left(\int_A \exp(-2iX(u^{-1}ab^*))\,\delta(u^{-1}ab^*)\,h(b)\,db\right) \overline{\lambda}(u)\,du \\
&= \int_U \left(\int_A \exp(-2iX(ab^*))\,\delta(ab^*)\,h(u^{-1}b)\,db\right) \overline{\lambda}(u)\,du \\
&= \int_A \exp(-2iX(ab^*)\,\delta(ab^*) \left(\int_U h(u^{-1}b)\,\overline{\lambda}(u)\,du\right) db \\
&= \int_A \exp(-2iX(ab^*))\,\delta(ab^*)\,f(b)\,db \;.
\end{aligned}$$

Therefore, $R^{\lambda}(p)f$ is given by (3.6). By a similar, but easier, computation, $R^{\lambda}(v(s)c(a))f$ is defined by (3.7).

COROLLARY. $R^{\lambda} = d_{\lambda}R(\cdot, \lambda)$

This follows immediately from the formulas defining the representations on $L \cup \{p\}$.

## §4. The Bessel Functions $J_\lambda$

Since the operator $R(p)$ on $L^2(A)$ defined by (2.13) is unitarily equivalent to the operator $R_1(p)$ on $L^2(M)$ given by (2.8), and since $R_1(p)$ is just a slight variant of the usual Fourier transform, the results of §3 may be regarded as giving in particular a decomposition of the Fourier transform.

When $n = 1$, $A = \mathbb{C}^\times$ and $U$ is the unit circle in $\mathbb{R}^2$. Hence, $\tilde{U} = \mathbb{Z}$ in the sense that an irreducible representation is of the form $u \to u^k$ $(k \in \mathbb{Z})$. Each such representation is of degree 1, so that $R^k = R(\cdot, k)$ and

$$L^2(\mathbb{C}^\times) = \sum_{k=-\infty}^{\infty} \oplus L^2(\mathbb{C}^\times, k) .$$

A function $f$ in $L^2(\mathbb{C}^\times)$ lies in $L^2(\mathbb{C}^\times, k)$ if and only if $f(e^{i\theta} z) = e^{ik\theta} f(z)$ for all $\theta$ and $z$. Thus a function in $L^2(\mathbb{C}^\times, k)$ is determined by its values on $(0, \infty)$; moreover, it is easy to see that in $L^2(\mathbb{C}^\times, k)$, the Fourier transform reduces to a Hankel transform defined by the Bessel function $J_{k-1}$. Recall that

$$J_k(z) = \frac{1}{2\pi} \int_{-\pi}^{\pi} e^{iz\sin\theta} e^{-ik\theta} \, d\theta$$

for any complex $z$. When $z$ is real, we can also write

$$J_k(z) = (i)^k \int_U e^{-i \operatorname{Re} z u} u^{-k} \, du .$$

Now suppose, as in §3, that $\lambda$ is a continuous irreducible unitary matrix representation of $U$ and let $d_\lambda$ be its degree. We define the Bessel function $J_\lambda : M \to \mathbb{C}^{d_\lambda \times d_\lambda}$ by

(4. 1) $$J_\lambda(z) = \lambda(i)\int_U \exp(-iX(z\,u)\,\lambda(u^{-1})\,du \quad .$$

Then $J_\lambda$ is not complex-analytic, but in the case $n=1$, its restriction to $\mathbb{R}$ is a classical Bessel function. In the present case, the restriction of $J_\lambda$ to $S$ has a holomorphic extension to $M$, but (4. 1) is more convenient for our purpose.

PROPOSITION 3. $J_\lambda$ _has the following two properties:_

(4. 2) $$J_\lambda(z)^* = J_\lambda(z^*), \quad z \in M$$

(4. 3) $$J_\lambda(u_1 z u_2) = \lambda(u_1)\, J_\lambda(z)\, \lambda(u_2)$$

_for arbitrary_ $u_1, u_2$ _in_ $U$ _and_ $z$ _in_ $M$.

Proof. Taking adjoints on both sides of (4. 1), we find that

$$\begin{aligned} J_\lambda(z)^* &= \lambda(i)^* \int_U \exp(iX(zu)\,\lambda(u)\,du \\ &= \lambda(-i)\int_U \exp(iX(xu^*))\,\lambda(u^{-1})\,du \\ &= \lambda(i)\int_U \exp(-iX(z^* u))\,\lambda(u^{-1})\,du \\ &= J_\lambda(z^*) \quad . \end{aligned}$$

Now suppose $u_1, u_2$ belong to $U$. Then

$$J_\lambda(u_1 z u_2) = \lambda(i)\int_U \exp(-iX(z\,u_2 u\,u_1))\,\lambda(u^{-1})\,du$$

and making the transformations $u \to u u_1^{-1}$ and $u \to u_2^{-1} u$, in the integral, we obtain (4. 3).

A continuous unitary representation of $U$ extends uniquely to a holomorphic representation of A . Conversely, every irreducible finite dimensional holomorphic matrix representation of A is similar to the extension of an irreducible unitary representation of U , [6] . Thus, we may regard the Bessel functions in (4.1) as being indexed by holomorphic representations $\lambda$ of A which are irreducible and unitary on U .

The following result indicates the connection between Fourier transforms and Bessel functions.

THEOREM 5. *For any integrable function* f *in* $L_2(A, \lambda)$,

$$R(p, \lambda)f(a) = \pi^{-n^2} \lambda(-i) \int_A J_{\delta^{-1}\lambda}(2ab^*)\, \delta(ab^*)\, f(b)\, db \quad . \tag{4.4}$$

Proof. By (3.1),

$$R(p, \lambda)f(a) = (i\pi)^{-n^2} \int_A \exp(-2iX(ab^*)\, \delta(ab^*)\, f(b)\, db$$

for all f in $C_c(A, \lambda)$ . By standard measure theoretic arguments, this also holds for all f in $L^2(A, \lambda)$ such that

$$\int_A \|f(a)\|\, da < \infty \, .$$

For any such f ,

$$\int_A \exp(-2iX(ab^*)\, \delta(ab^*)\, f(b)\, db$$

$$= \int_A \exp(-2iX(ab^* u)\, \delta(ab^*)\, \delta(u)\, f(u^{-1}b)\, db \, , \quad u \in U$$

$$= \int_A \delta(ab^*) \left( \int_U \exp(-2iX(ab^* u)\, \delta(u)\, \lambda(u^{-1})\, du \right) f(b)\, db$$

$$= \delta(i)\, \lambda(-i) \int_A J_{\delta^{-1}\lambda}(2ab^*)\, \delta(ab^*)\, f(b)\, db \quad .$$

Now let B denote the upper triangular subgroup of A whose

elements have positive diagonal entries. Then $U \cap B = \{1\}$ and $A = UB$ (the Iwasawa decomposition). Moreover, the right Haar measure $db$ on $B$ may be normalized in such a way that

$$\int_A f(a)\, da = \int_U \int_B f(ub)\, du\, db \tag{4.5}$$

for every integrable function $f$, [7].

Since each $a$ in $A$ has a unique decomposition of the form $a = ub$ with $u \in U$ and $b \in B$, it follows that a $\lambda$-covariant function $f$ is completely determined by its restriction $f_B$ to $G$; moreover, $f_B$ is arbitrary. Therefore, (4.5) implies that the map $f \to f_B$ is an isometry of $L^2(A, \lambda)$ onto $L^2(B, \mathbb{C}^{d_\lambda \times 1})$. In addition, it follows from (4.3) and (4.5) that

$$R(p, \lambda) f(a) = \pi^{-n^2} \lambda(-i) \int_B J_{\delta^{-1}\lambda}(2ab^*)\, \delta(ab^*)\, f(b)\, db \tag{4.6}$$

for any integrable $f$ in $L^2(A, \lambda)$. Thus the unitary map

$$f_B \to (R(p, \lambda) f)_B$$

is given explicitely by (4.6) for integrable functions in $L^2(A, \lambda)$.

Next, let $P$ denote the cone of positive definite elements of $A$. Since $A = UP$, it is immediate that a $\lambda$-covariant function is also completely determined by its restriction to $P$. In addition, there is a unitary map of $L^2(A, \lambda)$ onto an $L^2$ space of functions on $P$, and it is interesting to realize the operator $R(p, \lambda)$ on that space.

For this purpose we need a number of integration formulas. Since $M = S + iS$, the standard Lebesque measure on $M$ may be expressed as the product measure obtained from an appropriately normalized Lebesgue measure dx on S. Because P is an open set in S, dx restricts to a non-trivial measure dr on P.

LEMMA 1. *There is a constant* $\beta > 0$ *such that for any integrable or non-negative Baire function* f *on* A,

$$\int_A f(a)\,da = \beta \int_U \int_P f(u r^{\frac{1}{2}})\,du \frac{dr}{\delta(r)} . \tag{4.7}$$

*If* f *is an integrable or non-negative Baire function on* M, *then with the same* $\beta$,

$$\int_M f(z)\,dz = \beta \int_U \int_P f(u r^{\frac{1}{2}})\,du\,dr . \tag{4.8}$$

Proof. The transformations

$$r \to a^* r a , \quad a \in A ,$$

map P onto P, and it is easy to see that $\delta^{-1}(r)\,dr$ is a measure invariant under the action of A. In particular, it is invariant under B. By standard linear algebra, the mapping $b \to b^* b$ $(b \in B)$ is a homeomorphism of B onto P; in terms of this mapping, right multiplication on B corresponds to the action of B on P. Thus there is a constant $\beta > 0$ such that

$$\int_B \varphi(b^* b)\,db = \beta \int_P \varphi(r) \frac{dr}{\delta(r)} \tag{4.9}$$

for all $\varphi$ in $C_c(P)$.

For an arbitrary element a of A let u(a) and b(a) denote its components in the Iwasawa decomposition. Then

$$a = u(a)\,b(a) , \quad u(a) \in U , \quad b(a) \in B . \tag{4.10}$$

Each r in P has a unique square root $r^{\frac{1}{2}}$ in P and $r^{\frac{1}{2}} = u(r^{\frac{1}{2}})\,b(r^{\frac{1}{2}})$. Thus $r = b(r^{\frac{1}{2}})^* b(r^{\frac{1}{2}})$. Therefore, $r \to b(r^{\frac{1}{2}})$ $(r \in P)$ is the inverse of the map

$b \to b^* b$ $(b \in B)$. From this and (4.9) it is easy to conclude that for $f$ in $C_c(B)$,

(4.11)
$$\int_B f(b)\, db = \beta \int_P f(b(r^{\frac{1}{2}}))\, \frac{dr}{\delta(r)} .$$

Now suppose $f \in C_c(A)$. Then

$$\int_A f(a)\, da = \int_U \int_B f(ub)\, du\, db$$

by (4.5). From this and (4.11),

$$\int_A f(a)\, da = \beta \int_U \int_B f(u\, b(r^{\frac{1}{2}}))\, du\, \frac{dr}{\delta(r)}$$

$$= \beta \int_U \int_B f(u\, u(r^{\frac{1}{2}})^{-1}\, r^{\frac{1}{2}})\, du\, \frac{dr}{\delta(r)} ,$$

and making the transformation $u \to u(r^{\frac{1}{2}})$, we see that (4.7) is valid for $f$ in $C_c(A)$. On the other hand, the general case follows from this by standard measure theory.

To prove (4.8) suppose $f \in C_c(M)$. Then

$$\int_M f(z)\, dz = \int_A f(a)\, |\delta(a)|^2\, da ,$$

and applying (4.7), we find that

$$\int_M f(z)\, dz = \beta \int_U \int_P f(u\, r^{\frac{1}{2}})\, \delta(r^{\frac{1}{2}})^2\, du\, \frac{dr}{\delta(r)}$$

$$= \beta \int_U \int_P f(u\, r^{\frac{1}{2}})\, du\, dr .$$

Now suppose that $f$ is an integrable function in $L^2(A, \lambda)$. Then by (4.3), (4.4) and (4.7), we have

(4.12)
$$R(p, \lambda) f(r^{\frac{1}{2}}) = \beta \pi^{-n^2} \lambda(-i) \int_P J_{\delta^{-1}\lambda}(2 r^{\frac{1}{2}} s^{\frac{1}{2}})\, \delta(r^{\frac{1}{2}})\, \delta(s^{-\frac{1}{2}})\, f(s^{\frac{1}{2}})\, ds .$$

This suggests the following result which may be regarded as a generalization of the classical Hankel transform theory.

THEOREM 6. <u>For</u> $\varphi$ <u>in</u> $L^1(P, \mathbb{C}^{d_\lambda \times 1})$ <u>define</u> $\varphi^\#$ <u>on</u> $P$ <u>by</u>

$$\varphi^\#(r) = \beta(-i\pi)^{-n^2}\lambda(-i)\int_P J_\lambda(2r^{\frac{1}{2}}s^{\frac{1}{2}})\varphi(s)\,ds \tag{4.13}$$

and set $D = L^1(P, \mathbb{C}^{d_\lambda \times 1}) \cap L^2(P, \mathbb{C}^{d_\lambda \times 1})$.

<u>Then the mapping</u>

$$\varphi \to \varphi^\# \ , \ \varphi \in D$$

<u>extends (uniquely) to a unitary transformation</u> $W_\lambda$ <u>on</u> $L^2(P, \mathbb{C}^{d_\lambda \times 1})$.

Proof. It follows from (4.1) that $J_\lambda$ is a bounded function. Hence, if $\varphi \in L^1(P, \mathbb{C}^{d_\lambda \times 1})$, the integral defining $\varphi^\#$ is absolutely convergent.

Next note that any function $\varphi : P \to \mathbb{C}^{d_\lambda \times 1}$ extends uniquely to a $\delta\lambda$ - covariant function on $A$. If $f$ is a $\delta\lambda$ - covariant Baire function, it follows from (4.7) that

$$\|f\|^2_{\delta\lambda} = \beta\int_P \|f(r^{\frac{1}{2}})\|^2 \frac{dr}{\delta(r)} \ . \tag{4.14}$$

Thus, there is a unitary map $E$ of $L^2(P, \mathbb{C}^{d_\lambda \times 1})$ onto $L^2(A, \delta\lambda)$ such that

$$(E\varphi)(r^{\frac{1}{2}}) = \beta^{-\frac{1}{2}}\delta(r^{\frac{1}{2}})\varphi(r) \ , \ r \in P \ . \tag{4.15}$$

Now suppose $\varphi \in C_c(P, \mathbb{C}^{d_\lambda \times 1})$ and set $f = E\varphi$. Then $f \in C_c(A, \delta\lambda)$, and

$$\beta^{-\frac{1}{2}}\delta(r^{\frac{1}{2}})\varphi^\#(r) = \beta(-i\pi)^{-n^2}\lambda(-i)\int_P J_\lambda(2r^{\frac{1}{2}}s^{\frac{1}{2}})\,\delta(r^{\frac{1}{2}})\,\delta(s^{-\frac{1}{2}})\,f(s^{\frac{1}{2}})\,ds \ .$$

Comparing this with (4.12), we see that

$$\beta^{-\frac{1}{2}}\delta(r^{\frac{1}{2}})\varphi^\#(r) = R(p, \delta\lambda)\,f(r^{\frac{1}{2}}) \ , \ r \in P \ ,$$

and by (4.14), that

$$\|R(p, \delta\lambda)f\|_{\delta\lambda} = \|\varphi^{\#}\|_2 .$$

Since $f \in C_c(A, \delta\lambda)$ and $R(p, \delta\lambda)$ is a unitary operator on $L^2(A, \delta\lambda)$, it follows that $\varphi^{\#} \in L^2(P, \mathbb{C}^{d_\lambda \times 1})$. Therefore, $E\varphi^{\#} = R(p, \delta\lambda) f$. Hence

$$\varphi^{\#} = E^{-1} R(p, \delta\lambda) E \varphi .$$

and by standard measure theory this also holds for $\varphi$ in $D$. Thus $\varphi \to \varphi^{\#}$ is isometric on $D$ and extends uniquely to the unitary operator $W_\lambda = E^{-1} R(p, \delta\lambda) E$.

At this stage further progress is dependent on additional properties of the representations $\lambda$.

Suppose $\lambda$ is an irreducible holomorphic matrix representation of $A$ which is unitary on $U$. Then by analytic continuation one sees that

$$\lambda(a)^* = \lambda(a^*) \tag{4.16}$$

for all $a$ in $A$, so $\lambda$ maps Hermitian matrices to Hermilian matrices and positive definite matrices to positive definite matrices. Moreover, there is an integer $\sigma$ and a polynomial representation $\lambda_o$ such that

$$\lambda(a) = (\det a)^{\sigma} \lambda_o(a) , \quad a \in A , \tag{4.17}$$

and $\lambda_o$ is not divisible by the determinant representation, $a \to \det a$, [6]. Thus $\lambda$ extends to a holomorphic function on $M$ if and only if $\sigma \geq 0$, in which case the extended function (still denoted $\lambda$) is itself polynomial and has the property that

$$\lambda(z\,w) = \lambda(z)\,\lambda(w) \tag{4.18}$$

for all $z, w$ in $M$.

The following lemma is analogous to results of Bochner [8] and Herz [9]. It is used in the proof of Theorem 7 which plays a fundamental role in the development of the material in §5 .

LEMMA 2. If the representation $\lambda$ extends to a holomorphic function on $M$, then

(4.19) $$\lambda(-i)\exp(-\mathrm{tr}\, z z^*)\lambda(z) = \pi^{-n^2}\int_M \exp(-2iX(z w^*))\exp(-\mathrm{tr}\, w w^*)\lambda(w)\, dw,$$

for every $z$ in $M$.

Proof. Suppose $\lambda$ extends to a polynomial function on $M$. Then for each $z$ in $M$, $w \to \exp(-\mathrm{tr}\, w w^*)\lambda(w+z)$ $(w \in M)$ is obviously an integrable function on $M$, and by (4.8),

$$\int_M \exp(-\mathrm{tr}\, w w^*)\lambda(w+z)\, dw = \beta\int_P \exp(-\mathrm{tr}\, r)\left(\int_U \lambda(u r^{\frac{1}{2}} + z)\, du\right) dr .$$

By the mean value theorem for holomorphic functions [10, Th 4.7.3.],

$$\lambda(z) = \int_U \lambda(u r^{\frac{1}{2}} + z)\, du .$$

Since (4.8) implies

$$\beta\int_P \exp(-\mathrm{tr}\, r)\, dr = \int_M \exp(-\mathrm{tr}\, w w^*)\, dw = \pi^{n^2},$$

it follows that

$$\lambda(z) = \pi^{-n^2}\int_M \exp(-\mathrm{tr}\, w w^*)\lambda(w+z)\, dw$$

$$= \pi^{-n^2}\int_M \exp(-\mathrm{tr}(w-z)(w-z)^*)\lambda(w)\, dw .$$

Recalling that $X(a) = \mathrm{Re\,tr}\, a$, we see that

$$\mathrm{tr}(w-z)(w-z)^* = \mathrm{tr}\, w w^* - 2X(z w^*) + \mathrm{tr}\, z z^* .$$

Therefore

$$\exp(\mathrm{tr}\, z z^*)\lambda(z) = \pi^{-n^2}\int_M \exp(2X(z w^*))\exp(\mathrm{tr}\, w w^*)\lambda(w)\, dw .$$

In particular, we have

$$\exp(\operatorname{tr} x^2)\,\lambda(x) = \pi^{-n^2}\int_M \exp(2X(xw^*))\exp(\operatorname{tr} ww^*)\,\lambda(w)\,dw$$

when $x = x^*$ . The left side of this equation, regarded as a function of $S$ , has a unique analytic continuation to $M$ , namely $\exp(\operatorname{tr} z^2)\lambda(z)$ . To continue the right side, set

$$[x + iy, w] = X(xw^*) + i\,X(yw^*)$$

for $x, y$ in $S$ and $w$ in $M$ . Then $[\cdot\,,\,\cdot]$ is a form on $M \times M$ which is complex-linear in its first argument. From this it follows that

$$z \to \pi^{-n^2}\int_M \exp(2[z, w])\exp(\operatorname{tr}(ww^*))\,\lambda(w)\,dw$$

is holomorphic on $M$ . Therefore

$$\exp(\operatorname{tr} z^2)\,\lambda(z) = \pi^{-n^2}\int_M \exp(2[z, w])\exp(\operatorname{tr} ww^*)\,\lambda(w)\,dw$$

for all $z$ in $M$ . In particular, for $z = -iy$ $(y \in S)$

$$\exp(-\operatorname{tr} y^2)\,\lambda(-iy) = \pi^{-n^2}\int_M \exp(-2iX(yw^*)\exp(\operatorname{tr} ww^*)\,\lambda(w)\,dw\ .$$

Now multiply both sides of this equation by $\lambda(u)$ where $u \in U$ and make the transformation $w \to u^{-1}w$ in the integral. The resulting equation can be written in the form

$$\exp(-\operatorname{tr}(uy(uy)^*))\,\lambda(-i\,uy)$$
$$= \pi^{-n^2}\int_M \exp(-2iX(uyw^*)\exp(\operatorname{tr} ww^*)\,\lambda(w)\,dw\ .$$

Since each $z$ in $M$ is a product $z = uy$ with $u \in U$ and $y \geq 0$ , this completes the proof of (4.19) .

Now let $H$ denote the <u>upper half plane</u> in $M$ , i.e., the set of all elements $z$ of the form $x + iy$ with $x, y$ in $S$ and $y > 0$ .

Since $x + iy = y^{\frac{1}{2}}(y^{-\frac{1}{2}}xy^{-\frac{1}{2}} + i)y^{\frac{1}{2}}$ , it follows that the elements of $H$

are all invertible.

THEOREM 7. *If the representation* $\lambda$ *extends to a holomorphic function on* $M$, *then for a given* $z$ *in* $H$

$$\exp(-i\,\mathrm{tr}\, z^{-1} v^* v)\, \lambda(v)\, \lambda(z^{-1})\, \delta(z^{-1}) = (i\pi)^{-n^2} \int_M \exp(-2iX(v\,w\,x))\, \exp(i\,\mathrm{tr}\, z\, w^* w)\, \lambda(w)\, dw \tag{4.20}$$

*for every* $v$ *in* $M$.

Proof. Let $v \in M$. Then by (4.19),

$$\exp(-\mathrm{tr}\, v\, v^*)\, \lambda(-iv) = \pi^{-n^2} \int_M \exp(-2iX(v\, w^*))\, \exp(-\mathrm{tr}\, w\, w^*)\, \lambda(w)\, dw\,.$$

Now replace $v$ by $vy^{-\frac{1}{2}}$ where $y > 0$, and in the integral, make the transformation $w \to w\,y^{\frac{1}{2}}$. Since $|\delta(y^{\frac{1}{2}})|^2 = \delta(y)$, this results in the equation

$$\exp(-\mathrm{tr}\, v\, y^{-1} v^*)\, \lambda(-iv\,y^{-\frac{1}{2}}) = \pi^{-n^2} \int_M \exp(-2iX(v\, w^*))\, \exp(-\mathrm{tr}\, w\, y\, w^*)\, \lambda(w\, y^{\frac{1}{2}})\, \delta(y)\, dw\,.$$

We can rewrite this in the form

$$\exp(-i\,\mathrm{tr}(iy)^{-1} v^* v)\, \lambda(v)\, \lambda((iy)^{-1})\, \delta((iy)^{-1})\, \delta(i) = \pi^{-n^2} \int_M \exp(-2iX(v\, w^*)\, \exp(i\,\mathrm{tr}\, iy\, w^* w)\, \lambda(w)\, dw\,,$$

and (4.20) follows by analytic continuation.

COROLLARY. *Suppose* $\lambda$ *extends to a holomorphic function on* $M$. *Then for a given* $v$ *in* $M$,

$$\exp(-i\,\mathrm{tr}\, z^{-1} v^* v)\, \lambda(z^{-1})\, \delta(z^{-1}) = \beta(i\pi)^{-n^2} \lambda(-i) \int_P \exp(i\,\mathrm{tr}\, z\, r)\, \lambda(r^{\frac{1}{2}})\, J_\lambda(2r^{\frac{1}{2}} v^*)\, \lambda(v^{*-1})\, dr \tag{4.21}$$

for all $z$ in $H$.

Proof. By (4.20) and (4.8)

$$\exp(-i\,\mathrm{tr}\, z^{-1} v^* v)\, \lambda(z^{-1})\, \delta(z^{-1})$$

$$= (i\pi)^{-n^2} \int_M \exp(-2iX(v w^*))\, \exp(i\,\mathrm{tr}\, z w^* w)\, \lambda(v^{-1})\, \lambda(w)\, dw$$

$$= \beta (i\pi)^{-n^2} \int_P \exp(i\,\mathrm{tr}\, z\, r) \Big( \int_U \exp(-2iX(v r^{\frac{1}{2}} u^{-1}))\lambda(v^{-1})\lambda(u) du \Big) \lambda(r^{\frac{1}{2}}) dr$$

$$= \beta (i\pi)^{-n^2} \lambda(-i) \int_P \exp(i\,\mathrm{tr}\, z\, r)\, \lambda(v^{-1})\, J_\lambda(2 v r^{\frac{1}{2}})\, \lambda(r^{\frac{1}{2}})\, dr .$$

Now taking adjoints and using (4.2), we see that

$$\exp(i\,\mathrm{tr}\, z^{*-1} v^* v)\, \lambda(z^{*-1})\, \delta(z^{*-1})$$

$$= \beta(-i\pi)^{-n^2} \lambda(i) \int_P \exp(i\,\mathrm{tr}\, z^* r)\, \lambda(r^{\frac{1}{2}})\, J_\lambda(2 r^{\frac{1}{2}} v^*)\, \lambda(v^{*-1})\, dr \quad .$$

Since the transformation $z \to -z^*$ maps $H$ onto $H$, we can replace $z$ by $-z^*$ in the equation above. Then we have

$$\exp(-i\,\mathrm{tr}\, z^{-1} v^* v)\, \lambda(-1)\, \lambda(z^{-1})\, \delta(-1)\, \delta(z^{-1})$$

$$= \beta(i\pi)^{-n^2} \lambda(i) \int_P \exp(i\,\mathrm{tr}\, z\, r)\, \lambda(r^{\frac{1}{2}})\, J_\lambda(2 r^{\frac{1}{2}} v^*)\, \lambda(v^{*-1})\, dr$$

and this implies (4.21).

## §5. Holomorphic Discrete Series

In this section we show that certain of the representations $R(\cdot,\lambda)$ defined in §3 are equivalent to representations $T(\cdot,\lambda)$ on spaces of holomorphic functions on $H$, and show that some of these are part of the discrete series for $G$.

We proceed by constructing a 'Laplace transform' $\mathcal{L}_\lambda$ (for suitable $\lambda$) which maps $L^2(A,\lambda)$ onto a space $\mathcal{H}_\lambda$ of holomorphic functions $F:H \to \mathbb{C}^{d_\lambda \times 1}$ and intertwines $R(\cdot,\lambda)$ and $T(\cdot,\lambda)$. For the existence of $\mathcal{L}_\lambda$ we need to assume that $\lambda$ extends to a holomorphic function on $M$ which vanishes sufficiently rapidly at $0$.

The precise condition is related to the absolute convergence of the integral defining the generalized $\Gamma$ function. For a given $\lambda$ we set

(5.1) $$\Gamma(\lambda) = \int_P \exp(-\operatorname{tr} r)\lambda(r)\frac{dr}{\delta(r)},$$

provided the integral is absolutely convergent. By the remarks following (4.16), $\Gamma(\lambda) > 0$ whenever it exists.

It is convenient to write $\lambda = (\sigma,\lambda_o)$ when $\sigma$ is an integer, $\lambda_o$ is an irreducible polynomial representation not divisible by the determinant, and $\lambda$, $\sigma$, and $\lambda_o$ are related by (4.17). We then write $\Gamma(\lambda) = \Gamma(\sigma,\lambda_o)$, and more generally set

(5.2) $$\Gamma(\sigma,\lambda_o) = \int_P \exp(-\operatorname{tr} r)\lambda_o(r)(\det r)^{\sigma-n}dr$$

whenever $\sigma$ is a real number for which the integral is absolutely convergent.

THEOREM 8. $\Gamma(\sigma,\lambda_o)$ *is well defined for all real* $\sigma > n-1$. *If* $\lambda = (\sigma,\lambda_o)$ *with* $\sigma$ *an integer* $\geq n$, *then* $\Gamma(\lambda)$ *is a scalar multiple of the identity, and*

$$\Gamma(\lambda)\,\lambda\big((aa^*)^{-1}\big) = \int_P \exp(-\mathrm{tr}\, aa^* r)\,\lambda(r)\frac{dr}{\delta(r)} \tag{5.3}$$

*for all* $a \in A$ .

Proof. According to [10, Theorem 2.2.1],

$$\int_{0<r<1} (\det r)^q \, dr < \infty \quad \text{for } q > -1 \ . \tag{5.4}$$

From this and the fact that $\lambda_o$ is polynomial, it is easy to see that $\Gamma(\sigma, \lambda_o)$ exists for $\sigma > n-1$ .

Now suppose $\lambda = (\sigma, \lambda_o)$ with $\sigma > n-1$ . Then because the measure $\delta^{-1}(r)\,dr$ is A-invariant, we have

$$\int_P \exp(-\mathrm{tr}\, a^* r a)\lambda(r)\frac{dr}{\delta(r)} = \lambda(a^{*-1})\Gamma(\lambda)\lambda(a^{-1}) \tag{5.5}$$

for all $a$ in $A$ . In particular,

$$\Gamma(\lambda) = \lambda(u)\Gamma(\lambda)\lambda(u^{-1})$$

for all $u$ in $U$ . Since $\lambda$ is irreducible on $U$ , it follows that $\Gamma(\lambda)$ is a scalar multiple of the identity; hence, (5.5) reduces to (5.3).

For a $\lambda$-covariant function $f$ on $A$ , we define $\mathcal{L}_\lambda f$ in $H$ by

$$(\mathcal{L}_\lambda f)(z) = \int_A \exp(i\,\mathrm{tr}\, z\, a^* a)\lambda(a^*) f(a)\, da \tag{5.6}$$

for all $z$ in $H$ for which the integral is absolutely convergent.

To state the next result we need the fact that the transformation $z \to -z^{-1}$ maps $H$ into $H$ . But this follows at once from the formula

$$(-z^{-1}) - (-z^{-1})^* = z^{-1}(z - z^*)(z^{-1})^* \ .$$

THEOREM 9. Suppose $\lambda = (\sigma, \lambda_o)$ with $\sigma$ an integer $\geq n$, and let $f \in L^2(A, \lambda)$. Then $\mathcal{L}_\lambda f$ is a holomorphic function on $H$, and $\mathcal{L}_\lambda f = 0$ if and only if $f = 0$ a.e. If $F = \mathcal{L}_\lambda f$ then

$$(5.7) \qquad \mathcal{L}_\lambda R\left(v(s)c(a), \lambda\right)f(z) = \lambda(\check{a})F\left(a^{-1}(z+s)\check{a}\right), \quad z \in H,$$

for all $(s, a)$ in $S \times A$, and

$$(5.8) \qquad \mathcal{L}_\lambda R(p, \lambda)f(z) = \lambda(z^{-1})F(-z^{-1}), \quad z \in H.$$

Proof. For the Hilbert-Schmidt norm of the integrand in (5.6), we have

$$\left\|\exp\left(i \operatorname{tr} a(x+iy)a^*\right)\lambda(a^*)f(a)\right\| \leq \exp(-\operatorname{tr} aya^*)\left\|\lambda(a^*)f(a)\right\|$$
$$\leq \exp(-\operatorname{tr} aya^*)\left\|\lambda(a^*)\right\| \left\|f(a)\right\| .$$

Now for a given $\varepsilon$ in $P$, let

$$H_\varepsilon = \{x+iy \in H : y > \varepsilon\} .$$

Then

$$\left\|\exp(i \operatorname{tr} z\, a^*a)\lambda(a^*)f(a)\right\| \leq \exp(-\operatorname{tr} \varepsilon\, a^*a)\left\|\lambda(a^*)\right\| \left\|f(a)\right\|$$

for all $z$ in $H_\varepsilon$. By the Schwarz inequality,

$$\int_A \exp(-\operatorname{tr} \varepsilon\, a^*a)\left\|\lambda(a^*)\right\| \left\|f(a)\right\| da$$
$$\leq \left(\int_A \exp(-2\operatorname{tr} \varepsilon\, a^*a)\left\|\lambda(a^*)\right\|^2 da\right)^{\frac{1}{2}} \|f\|_\lambda .$$

On the other hand,

$$\int_A \exp(-2\operatorname{tr} \varepsilon\, a^*a)\left\|\lambda(a^*)\right\|^2 da$$

$$= \int_A \exp(-2\mathrm{tr}\,\epsilon\, a^*a)\mathrm{tr}\,\lambda(a^*a)\,da$$

$$= \beta \int_P \exp(-2\mathrm{tr}\,\epsilon\, r)\mathrm{tr}\,\lambda(r)\frac{dr}{\delta(r)}$$

$$= \beta\, \mathrm{tr}\Big(\Gamma(\lambda)\lambda(2\epsilon)\Big) \qquad \text{(by 5.3)}$$

$$< \infty \qquad \text{(by Theorem 8)} .$$

It follows that

$$\int_A \exp(-\mathrm{tr}\,\epsilon\, a^*a)\|\lambda(a^*)\|\,\|f(a)\|\,da \le \Big(\beta\, \mathrm{tr}\,\Gamma(\lambda)\lambda(2\epsilon)\Big)^{\frac{1}{2}}\|f\|_\lambda < \infty .$$

Therefore, $\mathcal{L}_\lambda f(z)$ exists, and

$$\|\mathcal{L}_\lambda f(z)\| \le \Big(\beta\, \mathrm{tr}\,\Gamma(\lambda)\lambda(2\epsilon)\Big)^{\frac{1}{2}}\|f\|_\lambda \tag{5.9}$$

for all $z$ in $H_\epsilon$. Since every point of $H$ lies in some $H_\epsilon$, it follows that $\mathcal{L}_\lambda f$ is a function $F$ on $H$.

To prove that $F$ is holomorphic, we first consider the case in which $f$ has compact support, and for a given $w$ in $M$, we set

$$D_w F(z) = \int_A \exp(i\,\mathrm{tr}\, z\, a^*a)(i\,\mathrm{tr}\, w\, a^*a)\lambda(a^*)f(a)\,da .$$

Then for $z$ in $H_\epsilon$ and $\zeta$ in $\mathbb{C}$ with $|\zeta| \le 1$, it is easy to check that

$$\|F(z+\zeta w) - F(z) - \zeta D_w F(z)\|$$
$$\le |\zeta|^2 \int_A \exp(-\mathrm{tr}\,\epsilon\, a^*a)\exp(|\mathrm{tr}\, w\, a^*a|)\|\lambda(a^*)\|\,\|f(a)\|\,da .$$

Because $f$ has compact support, it follows that

$$D_w F(z) = \lim_{\zeta\to 0} \frac{F(z+\zeta w) - F(z)}{\zeta}$$

for every $z$ in $H$. Applying Hartog's theorem to the obvious coordinate system on $H$, we conclude that $F$ is holomorphic.

In the general case, there is a sequence of compactly supported functions $f_m$ in $L^2(A, \lambda)$ such that $\|f - f_m\|_\lambda \to 0$. Setting $F_m = \mathcal{L}_\lambda f_m$, we then have

$$F(z) - F_m(z) = \int_A \exp(i \operatorname{tr} z\, a^*a)\lambda(a^*)\big(f(a) - f_m(a)\big)\, da \ .$$

It now follows from (5.9) that $F_m \to F$ uniformly on $H_\varepsilon$ for every $\varepsilon$. Therefore, $F$ is holomorphic on $H$.

Next observe that we can also write

$$(5.10) \qquad F(z) = \beta \int_P \exp(i \operatorname{tr} z\, r)\lambda(r^{\frac{1}{2}})f(r^{\frac{1}{2}})\frac{dr}{\delta(r)} \ .$$

For $y$ in $P$, define $f_y$ on $P$ by

$$f_y(r) = \exp(-\operatorname{tr} y\, r)\delta(r^{-1})\lambda(r^{\frac{1}{2}})f(r^{\frac{1}{2}})$$

and set $f_y = 0$ on $S - P$. Then $f_y \in L^1\big(S, \mathbb{C}^{d_\lambda \times 1}\big)$, and by (5.10),

$$x \to F(x+iy) \quad , \quad x \in S ,$$

is a constant times the inverse Fourier transform of $f_y$. Therefore, $F(x+iy) = 0$ for all $x$ in $S$ if and only if $f_y = 0$ a.e.. On the other hand, since

$$\|f\|_\lambda^2 = \beta \int_P \|f(r^{\frac{1}{2}})\|^2 \frac{dr}{\delta(r)} \ ,$$

it follows that $f_y = 0$ a.e. in $S$ if and only if $f = 0$ a.e. in $A$.

The proof of (5.7) is a simple computation:

$$\mathcal{L}_\lambda R\big(v(s)c(a), \lambda\big)f(z) = \int_A \exp(i \operatorname{tr} b\, zb^*) \exp(i \operatorname{tr} b\, sb^*)\, \lambda(b^*)\, f(ba)\, db$$

$$= \lambda(\check{a})\int_A \exp(i\,\mathrm{tr}\, ba^{-1} z \check{a} b^*)\exp(i\,\mathrm{tr}\, ba^{-1} s \check{a} b^*)\lambda(b^*)f(b)\, db$$

$$= \lambda(\check{a})\int_A \exp\Big(i\,\mathrm{tr}\, a^{-1}(z+s)\check{a} b^* b\Big)\, \lambda(b^*)\, f(b)\, db \quad .$$

In proving (5.8), we first consider the case in which $f$ has compact support. Then by (4.4), (5.10), and the Fubini theorem,

$$\mathcal{L}_\lambda R(p,\lambda)f(z) = \beta\int_P \exp(i\,\mathrm{tr}\, z\, r)\lambda(r^{\frac{1}{2}})\delta(r^{-1})\, R(p,\lambda)f(r^{\frac{1}{2}})\, dr$$

$$= \beta\pi^{-n^2}\lambda(-i)\int_P \exp(i\,\mathrm{tr}\, z\, r)\delta^{-1}(r^{\frac{1}{2}})\lambda(r^{\frac{1}{2}})\Big(\int_A J_{\delta^{-1}\lambda}(2r^{\frac{1}{2}}b^*)\delta(b^*)f(b)\, db\Big)\, dr$$

$$= \beta\pi^{-n^2}\lambda(-i)\int_A \Big(\int_P \exp(i\,\mathrm{tr}\, z\, r)\delta^{-1}(r^{\frac{1}{2}})\lambda(r^{\frac{1}{2}})J_{\delta^{-1}\lambda}(2r^{\frac{1}{2}}b^*)\, dr\Big)\, \delta(b^*)f(b)\, db$$

Since $\delta^{-1}\lambda = (\sigma-n, \lambda_o)$ and $\sigma-n \geq 0$, i.e., since $\delta^{-1}\lambda$ extends to a holomorphic function on $M$, it follows from (4.21) that

(5.11)
$$\exp(-i\,\mathrm{tr}\, z^{-1}\, b^* b)\, \lambda(z^{-1})\, \lambda(b^*)\, \delta^{-1}(b^*)$$

$$= \beta\pi^{-n^2}\lambda(-i)\int_P \exp(i\,\mathrm{tr}\, z\, r)\delta^{-1}(r^{\frac{1}{2}})\lambda(r^{\frac{1}{2}})J_{\delta^{-1}\lambda}(2r^{\frac{1}{2}}b^*)\, dr \quad .$$

Therefore

$$\mathcal{L}_\lambda R(p,\lambda)f(z) = \int_A \exp(-i\,\mathrm{tr}\, z^{-1} b^* b)\, \lambda(z^{-1})\, \lambda(b^*)\, f(b)\, db$$

$$= \lambda(z^{-1})\, F(-z^{-1}) \quad .$$

For the general case, first note that (5.9) has the following implication: If $h_m \to h$ in $L^2(A,\lambda)$, then $\mathcal{L}_\lambda h_m \to \mathcal{L}_\lambda h$ pointwise on $H$ and uniformly on $H_\varepsilon$.

Now let $f$ be an arbitrary element of $L^2(A,\lambda)$, and choose $f_m$ of compact support so that $f_m \to f$ in $L^2(A,\lambda)$. Then $R(p,\lambda)f_m \to R(p,\lambda)f$ in $L^2(A,\lambda)$. Let $F_m = \mathcal{L}_\lambda f_m$. Then

$$\mathcal{L}_\lambda R(p,\lambda) f_m(z) = \lambda(z^{-1}) F_n(-z^{-1}) \quad , \quad z \in H ,$$

and by the remark above ,

$$\mathcal{L}_\lambda R(p,\lambda) f_m(z) \to \mathcal{L}_\lambda R(p,\lambda) f(z) \ .$$

Since $\lambda(z^{-1}) F_m(-z^{-1}) \to \lambda(z^{-1}) F(-z^{-1})$ , it follows that $\mathcal{L}_\lambda R(p,\lambda) f(z) = \lambda(z^{-1}) F(-z^{-1})$ , and this completes the proof.

When $\lambda = (\sigma, \lambda_o)$ with $\sigma$ an integer $\geq n$ , we set

$$\mathcal{K}_\lambda = \mathcal{L}_\lambda\left(L^2(A,\lambda)\right) \ .$$

Since the restriction of $\mathcal{L}_\lambda$ to $L^2(A,\lambda)$ is a 1 - 1 linear map of $L^2(A,\lambda)$ onto $\mathcal{K}_\lambda$ , there is a well defined and unique inner product on $\mathcal{K}_\lambda$ such that

$$(5.12) \qquad (\mathcal{L}_\lambda h \mid \mathcal{L}_\lambda k)_\lambda = (h \mid k)_\lambda$$

for all $h$ , $k$ in $L^2(A,\lambda)$ .

THEOREM 10. _With respect to the inner product given by_ (5.12), $\mathcal{K}_\lambda$ _is a Hilbert space with the property that for each_ $z$ _in_ $H$ , _the linear map_

$$E_z : F \to F(z) \quad , \quad F \in \mathcal{K}_\lambda ,$$

_is continuous from_ $\mathcal{K}_\lambda$ _to_ $\mathbb{C}^{d_\lambda \times 1}$ ; _moreover_

$$E_z(\mathcal{K}_\lambda) = \mathbb{C}^{d_\lambda \times 1} \ .$$

Proof. It is obvious that $\mathcal{K}_\lambda$ is complete. By (5.9), if $F$ is any element of $L^2(A,\lambda)$ , then for any $z$ in $H_\varepsilon$ ,

$$(5.13) \qquad \|F(z)\| \leq \left(\beta \operatorname{tr} \Gamma(\lambda)\, \lambda(2\varepsilon)\right)^{\frac{1}{2}} \|F\|_\lambda \ .$$

Since $H = \bigcup_{\epsilon>0} H_\epsilon$ , this implies the continuity of $E_z$ for every z in H . To show that $E_z(\mathcal{H}_\lambda) = \mathbb{C}^{d_\lambda \times 1}$ , suppose that $\varphi \in \mathbb{C}^{d_\lambda \times 1}$ and $\varphi^* F(z) = 0$ for all F in $\mathcal{H}_\lambda$ . Then

$$\int_A \exp(i \operatorname{tr} z\, a^*a)\Big(\varphi^* \lambda(a^*)\, f(a)\Big)\, da = 0$$

for all f in $\mathcal{H}_\lambda$ . Let $z = x+iy$ , $x \in S$ , $y \in P$ , and set

$$f(a) = \exp(-i \operatorname{tr} z^* a^*a)\, \lambda(a)\varphi \quad , \quad a \in A \ .$$

Then $f \in L^2(A, \lambda)$ because $\sigma \geq n$ , and

$$\int_A \exp(i \operatorname{tr} z\, a^*a)\Big(\varphi^* \lambda(a^*)\, f(a)\Big) da = \int_A \exp(-2 \operatorname{tr} y\, a^*a) \|\lambda(a)\varphi\|^2\, da = 0 \ .$$

Therefore, $\lambda(a)\varphi = 0$ for every a ; hence $\varphi = 0$ , and $E_z(\mathcal{H}_\lambda) = \mathbb{C}^{d_\lambda \times 1}$ .

It follows that $E_z$ has a continuous non-singular adjoint $E_z^* : \mathbb{C}^{d_\lambda \times 1} \to \mathcal{H}_\lambda$ such that

$$\varphi^* F(z) = (F \,|\, E_z^* \varphi)_\lambda$$

for all F in $\mathcal{H}_\lambda$ and $\varphi$ in $\mathbb{C}^{d_\lambda \times 1}$ . In fact, if $F = \mathcal{L}_\lambda f$ , then

$$\varphi^* F(z) = \int_A \Big(\exp(-i \operatorname{tr} z^* a^*a)\, \lambda(a)\varphi\Big)^* f(a)\, da \ ,$$

so that $E_z^* \varphi = \mathcal{L}_\lambda(\tilde{E}_z \varphi)$, where

(5.14) $$(\tilde{E}_z \varphi)(a) = \exp(-i \operatorname{tr} z^* a^*a)\, \lambda(a)\varphi \quad , \quad a \in A \ .$$

A function F in $\mathcal{H}_\lambda$ is orthogonal to $E_z^*\left(\mathbb{C}^{d_\lambda \times 1}\right)$ if and only if $F(z) = 0$ .

Therefore, the functions which may be expressed as finite sums

$$F = \Sigma_j E^*_{w_j}(\varphi_j)$$

with $w_j$ in H and $\varphi_j$ in $\mathbb{C}^{d_\lambda \times 1}$ form a dense linear subspace $\mathscr{H}^\circ_\lambda$ of $\mathscr{H}_\lambda$. For any such F,

$$\|F\|^2_\lambda = \Sigma_{ij}\left(E_{w_i} E^*_{w_j} \varphi_j \,|\, \varphi_i\right)_\lambda .$$

Since $\|F\|_\lambda \geq 0$, the operator-valued kernel

$$Q_\lambda(z, w) = E_z E^*_w \tag{5.15}$$

is <u>positive definite</u> in the sense that $Q_\lambda(w, w) > 0$ for all w in H and

$$\Sigma_{ij}\left(Q_\lambda(w_i, w_j)\varphi_j \,|\, \varphi_i\right)_\lambda \geq 0$$

for all finite sequences $w_1, \cdots, w_n$ in H and $\varphi_1, \cdots, \varphi_n$ in $\mathbb{C}^{d_\lambda \times 1}$.

THEOREM 11. <u>For any</u> z <u>and</u> w <u>in</u> H,

$$Q_\lambda(z, w) = \beta\Gamma(\lambda)\,\lambda(i)\,\lambda^{-1}(z - w^*) . \tag{5.16}$$

Proof. Let $\varphi \in \mathbb{C}^{d_\lambda \times 1}$. Then by (5.15) and (5.14),

$$\begin{aligned} Q_\lambda(z, w)\varphi &= E_z E^*_w \varphi = E_z \mathscr{L}_\lambda \tilde{E}_w \varphi \\ &= \int_A \exp(i\,\mathrm{tr}\, z a^*a) \exp(-i\,\mathrm{tr}\, w^* a^*a)\, \lambda(a^*a) \cdot \varphi \\ &= \int_A \exp\left(i\,\mathrm{tr}\,(z - w^*)\, a^*a\right) \lambda(a^*a) \cdot \varphi . \end{aligned}$$

For z in H, set

$$Q^\circ_\lambda(z) = \int_A \exp(i\,\mathrm{tr}\, z a^*a)\, \lambda(a^*a)\, da .$$

Then $Q_\lambda(z, w) = Q^\circ_\lambda(z - w^*)$, $Q^\circ_\lambda$ is holomorphic on H, and

$$Q^{\circ}_{\lambda}(z) = \beta \int_P \exp(i\,\mathrm{tr}\, z\, r)\, \lambda(r) \frac{dr}{\delta(r)} \quad .$$

For y in P ,

$$Q^{\circ}_{\lambda}(iy) = \beta \int_P \exp(-\mathrm{tr}\, y\, r)\, \lambda(r) \frac{dr}{\delta(r)}$$

$$= \beta\, \Gamma(\lambda)\, \lambda(y^{-1}) \qquad \text{(by 5.3)}$$

$$= \beta\, \Gamma(\lambda)\, \lambda(i)\, \lambda\big((iy)^{-1}\big) \quad .$$

It now follows by analytic continuation that

$$(5.17) \qquad Q^{\circ}_{\lambda}(z) = \beta\, \Gamma(\lambda)\, \lambda(i)\, \lambda(z^{-1})$$

$$= \beta \int_P \exp(i\,\mathrm{tr}\, z\, r)\, \lambda(r) \frac{dr}{\delta(r)}$$

$$= \int_A \exp(i\,\mathrm{tr}\, z\, a^*a)\, \lambda(a^*a)\, da$$

for all z in H . Thus $Q_\lambda$ is given by (5.16).

To construct the representations $T(\cdot, \lambda)$ , we need some additional properties of G .

Recall that $g \in G$ if and only if $gpg^* = p$ where p is defined by (2.1). If $g \in G$ it follows that

$$(z_1, z_2)\, p(w_1, w_2)^* = \big((z_1, z_2)g\big)\, p\big((w_1, w_2)g\big)^*$$

for all $z_1$ , $z_2$ , $w_1$ , $w_2$ in M . In other words, for

$$g = \begin{bmatrix} a & b \\ c & d \end{bmatrix}$$

in G , and arbitrary $z_1$ , $z_2$ , $w_1$ , $w_2$ in M

$$(5.19) \qquad z_1 w_2^* - z_2 w_1^* = (z_1 a + z_2 c)(w_1 b + w_2 d)^* - (z_1 b + z_2 d)(w_1 a + w_2 c)^* \quad .$$

In the special case, $z_1 = w_1$ and $z_2 = w_2 = 1$, we have

$$z - z^* = (za+c)(zb+d)^* - (zb+d)(za+c)^* \quad . \tag{5.20}$$

If $z \in H$, then $-i(z - z^*) > 0$. From this and (5.20), it is easy to see that $zb+d$ and $z^*b+d$ are invertible whenever $z \in H$. If $z \in M$ and $zb+d$ is invertible, we set

$$z \circ g = (zb+d)^{-1}(za+c) \quad . \tag{5.21}$$

Whenever $zb+d$ is invertible, (5.20) can be written in the form

$$z - z^* = (zb+d)\Big(z \circ g - (w \circ g)^*\Big)(zb+d)^* \quad . \tag{5.22}$$

It follows that the 'linear fractional' transformation $z \to z \circ g$ maps $H$ into $H$ and $H^*$ into $H^*$.

If $z_1 = z$, $w_1 = z^*$, and $z_2 = w_2 = 1$, then (5.19) reduces to the equation

$$(za+c)(z^*b+d)^* = (zb+d)(z^*a+c)^* \quad .$$

This has the implication that for any $z$ in $H$,

$$(z \circ g)^* = z^* \circ g \, , \tag{5.23}$$

which may be interpreted as the statement that inverse points relative to the 'circle' $S$ are mapped to inverse points by $g$.

Next, we take $z_1 = z \in H$, $w_1 = w \in H$, and $z_2 = w_2 = 1$ in (5.19). It then reduces to the equation

$$z - w^* = (zb+d)\Big(z \circ g - (w \circ g)^*\Big)(wb+d)^* \quad . \tag{5.24}$$

This also holds if we replace $z$ by $z^*$ and $w$ by $w^*$. Then taking adjoints and using (5.23), we see that

(5.25) $$z - w^* = (w^*b+d)\left(z \circ g - (w \circ g)^*\right)(z^*b+d)^*$$

for all $z$ and $w$ in $H$ .

Some of the properties of $G$ in regards to its action on $H$ , that were obtained above, may be expressed more group theoretically in the following way.

PROPOSITION 3. _For any element_ $g \in G$ _of the form_ (5.18) _and any_ $z \in H$ ,

$$\begin{bmatrix} 1 & 0 \\ z & 1 \end{bmatrix}\begin{bmatrix} a & b \\ c & d \end{bmatrix} = \begin{bmatrix} (z^*b+d)^{\vee} & b \\ 0 & zb+d \end{bmatrix}\begin{bmatrix} 1 & 0 \\ z \circ g & 1 \end{bmatrix}$$

_and_ $z \circ g$ _is the unique element of_ $w$ _of_ $H$ _such that_

$$\begin{bmatrix} 1 & 0 \\ z & 1 \end{bmatrix}\begin{bmatrix} a & b \\ c & d \end{bmatrix}\begin{bmatrix} 1 & 0 \\ -w & 1 \end{bmatrix}$$

_is upper triangular._

Proof. To prove (5.26) we must show that

$$a = (z^*b+d)^{\vee} + b(z \circ g) .$$

By (5.23), $z \circ g = (z^* \circ g)^*$ . Thus it is enough to show that

$$a(z^*b+d)^* = 1 + b(z^*a+c)^*$$

or that

$$(ab^* - ba^*)z + ad^* - bc^* = 1 .$$

But this is a direct consequence of (2.5). The uniqueness of the decomposition follows by a simple computation.

COROLLARY. _The equation_

(5.26) $$m(z, g) = \begin{bmatrix} (z^*b+d)^{\vee} & 0 \\ 0 & zb+d \end{bmatrix}$$

*defines a map* $m: H \times G \to C$ *such that*

$$m(z, g_1 g_2) = m(z, g_1) m(z \circ g_1, g_2) \tag{5.27}$$

*for all* $z$ *in* $H$ *and all* $g_1$, $g_2$ *in* $G$.

This follows easily from Proposition 3 and implies the following result.

PROPOSITION 4. *If* $\lambda$ *is a holomorphic matrix representation of* $A$ *of degree* $d_\lambda$, *then the equation*

$$m_\lambda(z, g) = \lambda\left((z^*b+d)^\vee\right) \tag{5.28}$$

*defines a continuous map* $m_\lambda : H \times G \to GL(d_\lambda, \mathbb{C})$ *with the following two properties: For each* $g$ *in* $G$, *the map* $z \to m_\lambda(z, g)$ $(z \in H)$ *is holomorphic, and*

$$m_\lambda(z, g_1 g_2) = m_\lambda(z, g_1) m_\lambda(z \circ g_1, g_2) \tag{5.29}$$

*for all* $z$ *in* $H$ *and all* $g_1$, $g_2$ *in* $G$.

The desired representations $T(\cdot, \lambda)$ are now characterized by the following result.

THEOREM 12. *If* $\lambda = (\sigma, \lambda_o)$ *with* $\sigma$ *an integer* $\geq n$, *then the equation*

$$T(g, \lambda) F(z) = m_\lambda(z, g) F(z \circ g) \quad , \quad z \in H , \tag{5.30}$$

*defines a continuous unitary representation* $T(\cdot, \lambda)$ *of* $G$ *on* $\mathcal{H}_\lambda$.

Proof. Let $Q_\lambda$ be the positive definite kernel defined by (5.16), and suppose $z$ and $w$ are elements of $H$. Then for an arbitrary $g \in G$ of the form (5.18),

$$m_\lambda(z, g) Q_\lambda(z \circ g, w \circ g) m_\lambda(w, g)^*$$
$$= \beta \Gamma(\lambda) \lambda(i) \lambda\left((z^*b+d)^\vee \left(z \circ g - (w \circ g)^*\right)^{-1} (w^*b+d)^{-1}\right)$$

$$= \beta\,\Gamma(\lambda)\,\lambda(i)\,\lambda^{-1}\Big((w^*b+d)\big(z\circ g - (w\circ g)^*\big)(z^*b+d)^*\Big)$$

and applying (5.25), we see that

$$Q_\lambda(z,w) = m_\lambda(z,g)\,Q_\lambda(z\circ g, w\circ g)\,m_\lambda(w,g)^* \quad . \tag{5.31}$$

It now follows from [11, Theorem (2)] that (5.30) defines a continuous unitary representation of $G$ on $H$ .

The connection between the representations $R(\cdot,\lambda)$ and $T(\cdot,\lambda)$ is given by the following result.

THEOREM 13. *If* $\lambda = (\sigma, \lambda_o)$ *with* $\sigma$ *an integer* $\geq n$ , *then*

$$\mathcal{L}_\lambda R(g,\lambda) = T(g,\lambda)\mathcal{L}_\lambda \tag{5.32}$$

*for all* $g$ *in* $G$ .

Proof. Suppose $(s,a) \in S \times A$ . Then it follows from (5.21), (5.28), and (5.30) that for any $z$ in $H$,

$$T\big(v(s)\,c(a),\lambda\big)\,F(z) = \lambda(\check{a})\,F\big(a^{-1}(z+s)\check{a}\big) \quad . \tag{5.33}$$

By the same equations,

$$T(p,\lambda)\,F(z) = \lambda(z^{-1})\,F(-z^{-1}) \quad . \tag{5.34}$$

Comparing these equations with (5.7) and (5.9), we find that (5.32) is true for $g$ in $L \cup \{p\}$ . If (5.32) is valid for $g_1$ and $g_2$ , it also holds for their product $g_1 g_2$ . Thus, (5.32) is true for all $g$ in $LpV$ , and since $G = (LpV)^2$ , it holds for all $g$ in $G$ .

Since $R(\cdot,\lambda)$ is irreducible, in fact, irreducible on $L$ , we have the

COROLLARY. $T(\cdot,\lambda)$ *is irreducible on* $L$ .

Some, but not all, of the representations $T(\cdot,\lambda)$ are members of the holomorphic discrete series for $G$. The following theorem gives sufficient conditions for square-integrability.

THEOREM 14. *If* $\lambda = (\sigma, \lambda_o)$ *with* $\sigma$ *an integer* $\geq 2n$, *then*

$$\int_G \left|\left(T(g,\lambda)F \mid F\right)_\lambda\right|^2 dg < \infty \tag{5.34}$$

*for all* $F$ *in* $\mathscr{H}_\lambda$.

LEMMA 3. *Suppose* $\lambda = (\sigma, \lambda_o)$ *with* $\sigma$ *an integer* $\geq 2n$, *and let* $z = x+iy$ *with* $x \in S$ *and* $y \in P$. *Then for any* $F$ *in* $\mathscr{H}_\lambda$,

$$\int_L \|T(\ell,\lambda)F(z)\|^2 d\ell \leq (2\pi)^{n^2} \beta^2 \operatorname{tr}\left(\Gamma(\delta^{-1}\lambda)\lambda^{-1}(2y)\right)\|F\|_\lambda^2 \tag{5.35}$$

*where* $d\ell$ *denotes right Haar measure on* $L$.

Proof. Recall that the lower triangular subgroup $L$ is the semi-direct product $VC$ where $V$ and $C$ are the subgroups defined by (2.6) and (2.7). Let $dv$ and $dc$ denote Haar measures on $V$ and $C$ respectively. Then it is easy to check that $dvdc$ is a right Haar measure on $L$. Since the mapping $(s,a) \to v(s)c(a)$ is a homeomorphism of $S \times A$ onto $L$, we may identify functions on $L$ with functions on $S \times A$ and set $d\ell = dsda$.

Now suppose $F = \mathscr{L}_\lambda f$ with $f \in L^2(A,\lambda)$. Then by (5.33) and (5.7),

$$T(s,a,\lambda)F(z) = \int_A \exp(i \operatorname{tr} z\, b^*b) \exp(i\, s\, b^*b)\, \lambda(b^*) f(ba)\, db$$

for an arbitrary $z$ in $H$. By (5.10) we can also write

$$T(s,a,\lambda)F(z) = \beta \int_P \exp(i \operatorname{tr} s\, r) \exp(i \operatorname{tr} z\, r)\, \lambda(r^{\frac{1}{2}}) f(r^{\frac{1}{2}}a) \frac{dr}{\delta(r)}$$

If $z = x+iy$ with $x \in S$ and $y \in P$, then

$$\int_A \int_P \|\exp(i\,\mathrm{tr}\, z\, r)\, \lambda(r^{\frac{1}{2}})\, f(r^{\frac{1}{2}} a)\|^2\, \delta^{-2}(r)\, dr\, da$$

$$\leq \int_P \int_A \exp(-2\,\mathrm{tr}\, y\, r)\, \delta^{-2}(r)\, \|\lambda(r^{\frac{1}{2}})\|^2\, \|f(r^{\frac{1}{2}} a)\|^2\, da\, dr$$

$$= \|f\|_\lambda^2 \int_P \exp(-2\,\mathrm{tr}\, y\, r)\, \mathrm{tr}\Big(\delta^{-1}(r)\, \lambda(r)\Big) \frac{dr}{\delta(r)} \quad .$$

Since $\delta^{-1}\lambda = (\sigma - n, \lambda_o)$ and $\sigma - n \geq n$, it follows from Theorem 8 that

$$\int_P \exp(-2\,\mathrm{tr}\, y\, r)\, \delta^{-1}(r)\, \lambda(r) \frac{dr}{\delta(r)} = \Gamma(\delta^{-1}\lambda)\, \lambda^{-1}(2y) \quad .$$

Therefore

$$\int_P \|\exp(i\,\mathrm{tr}\, z\, r)\, \lambda(r^{\frac{1}{2}})\, f(r^{\frac{1}{2}} a)\|^2\, \delta^{-2}(r)\, dr < \infty$$

for almost every $a$ in $A$. When the integral is finite, it follows from the Plancherel theorem that

$$\int_S \|T(s, a, \lambda)\, F(z)\|^2\, ds = (2\pi)^{n^2} \beta^2 \int_P \|\exp(i\,\mathrm{tr}\, z\, r)\lambda(r^{\frac{1}{2}}) f(r^{\frac{1}{2}} a)\|^2 \delta^{-2}(r)\, dr \; .$$

Hence, integrating over $A$, we see that

$$\int_L \|T(\ell, \lambda)\, F(z)\|^2\, d\ell \leqq (2\pi)^{n^2} \beta^2\, \mathrm{tr}\Big(\Gamma(\delta^{-1}\lambda)\, \lambda^{-1}(2y)\Big) \|F\|_\lambda^2 \quad .$$

LEMMA 4. <u>Suppose</u> $\lambda = (\sigma, \lambda_o)$ <u>with</u> $\sigma$ <u>an integer</u> $\geq n$, <u>and let</u> $\varphi \in \mathbb{C}^{d_\lambda \times 1}$. <u>Then for any</u> $g$ <u>in</u> $G$ <u>and any</u> $w$ <u>in</u> $H$,

(5.36) $$T(g, \lambda)\, Q_\lambda(\cdot\,, w)\, \varphi = Q_\lambda(\cdot\,, w \circ g^{-1})\, m_\lambda(w, g^{-1})^* \varphi \quad .$$

This follows from (5.31) and is proved in [11]. The next lemma is well known, but we give a proof.

LEMMA 5. <u>Let</u> $K = G \cap U(2n)$. <u>Then</u> $K$ <u>is the stability group of the</u>

point i in H and G = KL .

Proof. An element g of the form (5.18) lies in K if and only if

$$\begin{bmatrix} d^* & -b^* \\ -c^* & a^* \end{bmatrix} = \begin{bmatrix} a^* & c^* \\ b^* & d^* \end{bmatrix} = g^{-1} .$$

Thus $d = a$ , $c = -b$ , and $ia - b = i(ib+a)$ ; hence $(ib+a)^{-1}(ia - b) = i$ . Conversely, if g has the form (5.18) and $(ib+d)^{-1}(ia+c) = i$ , then $i(a - d) = -b - c$ . Since $g^{-1}$ also fixes i , we have $i(d^* - a^*) = b^*+c^*$ , as well. But this implies $i(a - d) = b+c$ , so that $a = d$ and $c = -b$ .

Now suppose $g \in G$ . Then, since L evidently acts transitively on H , there exists an element $\ell$ in L such that $i \circ g = i \circ \ell$ . Then $i \circ g\ell^{-1} = i$ and $g\ell^{-1}$ is an element k of K .

Proof of Theorem 14. Since $T(\cdot, \lambda)$ is irreducible it suffices to prove (5.34) for just one non-zero F in $\mathcal{H}_\lambda$ , [12]. We take $F = Q_\lambda(\cdot, i)\varphi$ where $\varphi$ is a non-zero element of $\mathbb{C}^{d_\lambda \times 1}$ . Since G = KL , it follows from [ 7 , Lemma 12] that

$$\int_G |(T(g,\lambda)F \mid F)_\lambda|^2 \, dg = \int_L \left( \int_K |(T(k\ell,\lambda)F \mid F)_\lambda|^2 \, dk \right) d\ell$$

$$= \int_L \left( \int_K |(T(\ell,\lambda)F \mid T(k^{-1},\lambda)F)_\lambda|^2 \, dk \right) d\ell .$$

By (5.36) and Lemma 5 ,

$$T(k^{-1},\lambda)F = Q_\lambda(\cdot, i)\, m_\lambda(i,k)^* \varphi$$

and

$$T(\ell,\lambda)F = Q_\lambda(\cdot, i\circ \ell^{-1})\, m_\lambda(i,\ell^{-1})^* \varphi .$$

Thus

$$\left(T(\ell,\lambda)F \,|\, T(k^{-1},\lambda)\right)_\lambda = \left(m_\lambda(i,k)^*\varphi\right)^* Q_\lambda(i, i\circ \ell^{-1})\, m_\lambda(i,\ell^{-1})^*\varphi$$

$$= \left(m_\lambda(i,k)^*\varphi\right)^*\left(T(\ell,\lambda)F(i)\right) \quad .$$

Since the elements of K fix i , it follows from (5.29) and (5.28) that

$$k \to m_\lambda(i,k) \quad , \quad k \in K \ ,$$

is an irreducible unitary representation of K . Therefore

$$\int_K \left|\left(T(\ell,\lambda)F \,|\, T(k^{-1},\lambda)F\right)_\lambda\right|^2 dk = \int_K \left|\left(m_\lambda(i,k)\varphi\right)^*\left(T(\ell,\lambda)F(i)\right)\right|^2 dk$$

$$= d_\lambda^{-1}\|\varphi\|^2 \, \|T(\ell,\lambda)F(i)\|^2$$

by the Schur orthogonality relations. Thus

$$\int_G \left|\left(T(g,\lambda)F \,|\, F\right)_\lambda\right|^2 dg = d_\lambda^{-1}\|\varphi\|^2 \int_L \|T(\ell,\lambda)F(i)\|^2 d\ell$$

$$\leq d_\lambda^{-1}(2\pi)^{n^2}\beta^2\, \mathrm{tr}\left(\Gamma(\delta^{-1}\lambda)\lambda^{-1}(2)\right)\|F\|_\lambda^2$$

by Lemma 3, and this completes the proof.

REMARKS. There is another proof of Theorem 14 which shows that $T(\cdot,\lambda)$ is square-integrable, provided

$$\int_A \|J_{\delta^{-1}\lambda}(a)\|^2 da < \infty \ ;$$

and in general it can be shown that

$$\int_A \|J_\lambda(a)\|^2 da < \infty \ ,$$

whenever $\lambda = (\sigma, \lambda_o)$ with $\sigma \geq n$ . Thus, square-integrability is a property that only depends on the nature of the single operator $T(p, \lambda)$ . We shall present such a proof and a more complete discussion of the holomorphic discrete series in another paper.

## REFERENCES


[1] A. Weil, "Sur certains groupes d'opérateurs unitaires," Acta Math. 111 (1965), 143-211.

[2] D. Shale, "Linear symmetries of free boson fields," Trans. Amer. Math. Soc. 103 (1962), 149-167.

[3] K.I. Gross, "The dual of a parabolic subgroup and a degenerate principal series of $Sp(n, \mathbb{C})$," Amer. J. Math. 93 (1971), 398-428.

[4] K.I. Gross and R.A. Kunze, "Fourier decompositions of certain representations," Analysis and Geometry of Symmetric Spaces, Marcel Dekker, 1972.

[5] G.W. Mackey, "Unitary representations of group extensions I," Acta Math. 99 (1958), 265-311.

[6] H. Boerner, Representations of Groups, North Holland Publishing Company, Amsterdam, 1962.

[7] R.A. Kunze and E.M. Stein, "Uniformly bounded representations III," Amer. J. Math. 89 (1967), 385-442.

[8] S. Bochner, Harmonic Analysis and the Theory of Probability, University of California Press, 1955.

[9] C. Herz, "Bessel functions of a matrix argument," Ann. Math. 61 (1955), 474-523.

[10] L.K. Hua, Harmonic Analysis of Functions of Several Complex Variables in the Classical Domains, American Mathematical Society, 1963.

[11] R.A. Kunze, "Positive definite operator-valued kernels and unitary representations," Proceedings of the Conference on Functional Analysis at Irvine, California, Thompson Book Company, 1966.

[12] G.W. Mackey, "Infinite-dimensional group representations," Bull. Amer. Math. Soc. 69 (1963), 628-686.

# ON THE THEORY OF THE EISENSTEIN INTEGRAL

by

Harish-Chandra

The Institute for Advanced Study
Princeton, N. J.

## §1. Introduction

Let $F$ be a field and $\underset{\sim}{G}$ a connected, reductive algebraic group defined over $F$. Let $G_F$ denote the subgroup of all $F$-rational points of $G$. If $F$ is a local field (i.e. $F = \underset{\sim}{R}$, $\underset{\sim}{C}$ or a $\mathfrak{p}$-adic field), $G_F$ has a natural locally compact topology. (We write $G_F = G_{\mathfrak{p}}$ when $F$ is a $\mathfrak{p}$-adic field.) On the other hand if $F$ is a finite field, $G_F$ is a finite group. Finally if $F$ is a global field (i.e. a number-field or a function-field in one variable over a finite field), we have the adèle group $G_{\underset{\sim}{A}}$ associated to $\underset{\sim}{G}$. From the point of view of harmonic analysis and the theory of automorphic forms, the study of the following five cases is important.

1) $G_{\underset{\sim}{R}}/\Gamma$ where $\Gamma$ is an arithmetic subgroup of $G$,

2) $G_{\underset{\sim}{R}}$,

3) $G_{\mathfrak{p}}$,

4) $G_F$ where $F$ is a finite field,

5) $G_{\underset{\sim}{A}}/G_F$ where $F$ is a global field.

Actually 5) is the most difficult case and the other four may, in fact, be regarded merely as its several facets. Nevertheless it is useful to pursue their study individually since a knowledge of one case enables us to guess, often quite accurately, analogous results in another case. This similarity from the standpoint of harmonic analysis, is most striking between $G_{\underset{\sim}{R}}$ and $G_{\mathfrak{p}}$ (see [4(e)]). But in this lecture I propose to bring out the resemblance between $G_{\underset{\sim}{R}}/\Gamma$ and $G_{\underset{\sim}{R}}$.

The theory of Eisenstein Series over $G_{\underset{\sim}{R}}/\Gamma$, which is largely due to Selberg, Gelfand-Piatetsky-Shapiro and Langlands (see [4(a)]), has an exact counterpart in the theory of Eisenstein Integrals over $G_{\underset{\sim}{R}}$. These integrals have functional equations (see §6) which are governed by the coefficients

$c_{P_2|P_1}(s:\nu)$ appearing in their asymptotic expansion. Following Gelfand [2] (see also [4(b), §5]), one is tempted to call these coefficients the local zeta-functions at infinity. Since a similar theory holds for $G_{\mathfrak{p}}$, one gets in this way local zeta-functions at each prime. It seems likely that the global zeta-functions (i.e. the corresponding coefficients appearing in the Eisenstein Series for $G_{\mathbb{A}}/G_F$) are actually products built out of the local factors. This leads one to expect that the local factors are "elementary" or "Eulerian." Therefore, in particular, $c_{P_2|P_1}(s:\nu)$ should be expressible in terms of gamma factors. Although this conjecture remains unproved, we do obtain a rather simple formula for the "absolute value" of this operator and therefore for the Plancherel measure $\mu_\omega(\nu)d\nu$ (see §13). Moreover there is some evidence (see §14) that the zeros of the function $\mu_\omega$ are related to the occurrence of the exceptional series (see also [6(a), (b)] and [7]).

In the first few sections of this paper, we restate in a precise form those results of [4(c)] and [4(d)], which are necessary for the understanding of our main theorems.

## §2. The assumptions on G

For any Lie group $G$, we denote by $G^0$ the connected component of 1 in $G$ and by $X(G)$ the group of all continuous homomorphisms of $G$ into $\mathbb{R}^\times$. Put

$$^0G = \bigcap_{\chi \in X(G)} \ker|\chi| \quad .$$

Then $G/{}^0G$ is an abelian Lie group. By the parabolic rank of $G$ we mean $\dim G/{}^0G$ and denote it by prk $G$. A split component of $G$ is a closed subgroup $A$ of $G$ such that

$$G = {}^0G.A\,, \quad {}^0G \cap A = \{1\} \quad .$$

Note that $A$, if it exists, is abelian. In fact $A \simeq G/{}^0G$ and therefore dim A = prk G. By a vector subgroup V of G, we mean a closed subgroup which is topologically isomorphic to the additive group of $\mathbb{R}^n$ for some $n \geq 0$.

Let G be a Lie group with Lie algebra $\mathfrak{g}$. Let $G_c$ denote the connected complex adjoint group[1)] of $\mathfrak{g}_c$. We make the following assumptions on G.

1) $\mathfrak{g}$ is reductive and $\mathrm{Ad}(G) \subset G_c$.

2) Let $G_1$ denote the analytic subgroup of G corresponding to $\mathfrak{g}_1 = [\mathfrak{g}, \mathfrak{g}]$. Then the center of $G_1$ is finite.

3) $[G : G^o] < \infty$.

Fix a maximal compact subgroup K of G. (Such a subgroup exists and is unique up to conjugacy by $G^o$.) Let C be the center of $G^o$. Fix a maximal vector subgroup $C_2$ of C. Then $C = C_1 . C_2$ where $C_1 = C \cap K$. Let $\mathfrak{k}$, $\mathfrak{c}_1$, $\mathfrak{c}_2$ be the Lie algebras of K, $C_1$, $C_2$ respectively and $\theta$ the Cartan involution of $\mathfrak{g}_1$ with respect to $\mathfrak{k}_1 = \mathfrak{k} \cap \mathfrak{g}_1$. Extend $\theta$ to an involution of $\mathfrak{g}$ by setting

$$\theta(X_1 + X_2) = X_1 - X_2 \qquad (X_i \in \mathfrak{c}_i,\ i = 1, 2) .$$

Let $\mathfrak{p}$ be the set of all points $X \in \mathfrak{g}$ such that $\theta(X) = -X$.

Lemma 1. <u>The mapping</u> $(k, X) \mapsto k \exp X$ $(k \in K, X \in \mathfrak{p})$ <u>is an analytic diffeomorphism of</u> $K \times \mathfrak{p}$ <u>onto</u> G <u>and</u> $\theta$ <u>extends to an automorphism of</u> G <u>such that</u>

$$\theta(k \exp X) = k \exp(-X) \qquad (k \in K, X \in \mathfrak{p}) .$$

We denote by log the inverse of the mapping $\exp : \mathfrak{p} \longrightarrow \exp \mathfrak{p}$.

Lemma 2. <u>There exists a real symmetric bilinear form</u> B <u>on</u> $\mathfrak{g}$ <u>such that:</u>

1) $B(\theta X, \theta Y) = B(X, Y)$ <u>and</u>

$$B([X, Y], Z) + B(Y, [X, Z]) = 0 \qquad (X, Y, Z \in \mathfrak{g}) .$$

2) <u>The quadratic form</u>

$$\|X\|^2 = -B(X, \theta X) \qquad (X \in \mathfrak{g})$$

<u>is positive-definite on</u> $\mathfrak{g}$.

We fix K, θ and B as above, once for all and define $\sigma(x) = \|X\|$ for $x = k \exp X$ $(k \in K,\ X \in \mathfrak{p})$.

## §3. Parabolic subgroups[2)]

A subalgebra $\mathfrak{q}$ of $\mathfrak{g}$ is called parabolic, if $\mathfrak{q}_c$ contains a Borel subalgebra (i.e. a maximal solvable subalgebra) of $\mathfrak{g}_c$. A subgroup Q of G is called parabolic, if it is the normalizer in G of some parabolic subalgebra $\mathfrak{q}$ of $\mathfrak{g}$. Then Q is closed and its Lie algebra is $\mathfrak{q}$.

Lemma 3. <u>Let</u> Q <u>be a parabolic subgroup</u> (psgp) <u>of</u> G. <u>Then</u> G = KQ.

Let $\mathfrak{q}$ be a parabolic subalgebra of $\mathfrak{g}$. By the radical of $\mathfrak{q}$, we mean the maximal ideal $\mathfrak{n}$ of $\mathfrak{q}_1 = \mathfrak{q} \cap [\mathfrak{g}, \mathfrak{g}]$ such that ad X is nilpotent for every $X \in \mathfrak{n}$. If Q is the psgp corresponding to $\mathfrak{q}$ and N the analytic subgroup corresponding to $\mathfrak{n}$, then N is called the radical of Q. Put $M_1 = Q \cap \theta(Q)$ and $M = {}^oM_1$ (see §2). Let A be the maximal θ-stable vector subgroup lying in the center of $M_1$. Then A is a split component both for Q and $M_1$. We call A <u>the</u> split component of Q. Then $M_1 = MA$ is the centralizer of A in G and

$$Q = MAN ,$$

the mapping $(m, a, n) \mapsto man$ being an analytic diffeomorphism of $M \times A \times N$ onto Q. We call this the Langlands decomposition of Q.

Let $\mathfrak{m}$, $\mathfrak{a}$ denote the Lie algebra of M, A respectively and put $\mathfrak{m}_1 = \mathfrak{m} + \mathfrak{a}$, $K_M = K \cap M$. Let $\theta_M$ and $B_M$ be the restrictions of θ and B respectively on $\mathfrak{m}$. Then if we replace (G, K, θ, B) by

$(M, K_M, \theta_M, B_M)$ all the conditions of §2 are fulfilled. The same holds for $(M_1, K_M, \theta_1, B_1)$ where $\theta_1$ and $B_1$ are the restrictions of $\theta$ and $B$ respectively on $\mathfrak{m}_1$.

By a p-pair $(Q, A)$, we mean a psgp $Q$ and its split component $A$. Then $Q = MAN$. By a root of $Q$ (or $(Q, A)$), we mean an element $\alpha$ in $\mathfrak{a}^*$ with the following property. Let $\mathfrak{n}_\alpha$ denote the set of all $X \in \mathfrak{n}$ such that $[H, X] = \alpha(H)X$ for all $H \in \mathfrak{a}$. Then $\mathfrak{n}_\alpha \neq \{0\}$. It is clear that $\operatorname{prk} Q = \dim A \geq \operatorname{prk} G$.

Let $\ell = \operatorname{prk} Q - \operatorname{prk} G$ and let $\Sigma(Q)$ denote the set of all roots of $Q$. A root $\alpha \in \Sigma(Q)$ is called simple if it cannot be written in the form $\alpha = \beta + \gamma$ with $\beta, \gamma \in \Sigma(Q)$. Let $\Sigma^o(Q)$ be the set of all simple roots. Then[5] $\ell = [\Sigma^o(Q)]$. Let $\alpha_1, \ldots, \alpha_\ell$ be all the simple roots. Then they are linearly independent over $\mathbb{R}$ and every $\alpha \in \Sigma(Q)$ can be written in the form

$$\alpha = m_1\alpha_1 + \ldots + m_\ell\alpha_\ell$$

where $m_i \in \mathbb{Z}$ and $m_i \geq 0$ $(1 \leq i \leq \ell)$.

Fix a subset $F$ of $\Sigma^o(Q)$ and let $\mathfrak{a}_F$ denote the set of all $H \in \mathfrak{a}$ such that $\alpha(H) = 0$ for all $\alpha \in F$. Let $\Sigma_F$ be the set of all $\alpha \in \Sigma = \Sigma(Q)$ which vanish identically on $\mathfrak{a}_F$. Put

$$\mathfrak{n}_F = \sum_{\alpha \in \Sigma'_F} \mathfrak{n}_\alpha$$

where $\Sigma'_F$ is the complement of $\Sigma_F$ in $\Sigma$. Then $\mathfrak{n}_F$ is an ideal in $\mathfrak{n}$. Put $N_F = \exp \mathfrak{n}_F$, $A_F = \exp \mathfrak{a}_F$ and let $Q_F$ denote the normalizer of $N_F$ in $G$. Then $(Q_F, A_F)$ is a p-pair in $G$ and

$$Q_F = M_F A_F N_F$$

where $M_F = {}^oZ(A_F)$ and $Z(A_F) = M_F A_F$ is the centralizer of $A_F$ in $G$. We write $(Q, A)_F = (Q_F, A_F)$.

Let $(Q', A')$ be any p-pair in $G$. We write $(Q', A') \succ (Q, A)$ if

$Q' \supset Q$. This implies that $A' \subset A$. Every p-pair $(Q', A') \succ (Q, A)$ is of the form $(Q', A') = (Q, A)_F$ for a unique $F \subset \Sigma^o(Q)$.

Lemma 4. There is a one-one correspondence between psgps $P$ of $G$ which are contained in $Q$ and psgps ${}^*P$ of $M$. This correspondence is given by the relation ${}^*P = P \cap M$. If

$$P = M'A'N' \ , \quad {}^*P = {}^*M\,{}^*A\,{}^*N$$

are the corresponding Langlands decompositions, then

$$M' = {}^*M \ , \quad A' = {}^*A.A \ , \quad N' = {}^*N.N \ ,$$

$${}^*A = M \cap A' \ , \quad {}^*N = M \cap N' \ .$$

Lemma 5. Any two minimal psgps of $G$ are conjugate under $K^o$. Let $P_1$, $P_2$ and $P$ be three psgps of $G$. Suppose $P_1 \cap P_2 \supset P$ and $P_2$ is conjugate to $P_1$ under $G$. Then $P_1 = P_2$.

Let $(P_i, A_i)$ $(i = 1, 2)$ be two p-pairs. We denote by $\mathfrak{w}(A_2|A_1) = \mathfrak{w}(\mathfrak{a}_2|\mathfrak{a}_1)$ the set of all linear mappings

$$s : \mathfrak{a}_1 \longrightarrow \mathfrak{a}_2$$

satisfying the following condition. There should exist an element $y \in G$ such that $sH = \mathrm{Ad}(y)H$ for all $H \in \mathfrak{a}_1$. $y$ is then called a representative of $s$ in $G$. (One can always choose a representative $y \in K$.) We also write $a^s = yay^{-1}$ $(a \in$

$P_1$, $P_2$ are said to be associated if $\mathfrak{w}(A_2|A_1)$ and $\mathfrak{w}(A_1|A_2)$ are both nonempty. This is equivalent to saying that $A_1$ and $A_2$ are conjugate under $G$.

Let $(P, A)$ be a p-pair. We write $\mathfrak{w}(A) = \mathfrak{w}(A|A)$. Then $\mathfrak{w}(A)$ is a finite group and in fact

$$\mathfrak{w}(A) = (\text{Normalizer of } A \text{ in } K)/(\text{Centralizer of } A \text{ in } K) \ .$$

We call $\mathfrak{W}(A)$ the Weyl group of A in G and sometimes denote it by $\mathfrak{W}(G/A)$.

Let P = MAN be the Langlands decomposition of P. Define $\rho_P \in \mathfrak{a}^*$ by

$$\rho_P(H) = \frac{1}{2}\mathrm{tr}(\mathrm{ad}\, H)_{\mathfrak{n}} \qquad (H \in \mathfrak{a}) ,$$

where $(\mathrm{ad}\, H)_{\mathfrak{n}}$ denotes the restriction of ad H on $\mathfrak{n}$. Since G = KP, every $x \in G$ can be written in the form $x = kman$ ($k \in K$, $m \in M$, $a \in A$, $n \in N$). Here a is uniquely determined and we put $H_P(x) = \log a$. Then $H_P : G \longrightarrow \mathfrak{a}$ is an analytic mapping.

Let $d_\ell p$ and $d_r p$ denote the left- and right-invariant Haar measures on P so that $d_r p = d_\ell p^{-1}$. Then

$$d_r p = \delta_p(p) d_\ell p$$

where $\delta_P$ is a continuous homomorphism of P into $\mathbf{R}_+^\times$ . In fact

$$\delta_P(man) = e^{2\rho(\log a)} \qquad (m \in M,\ a \in A,\ n \in N)$$

where $\rho = \rho_P$. Note that $\delta_P = 1$ on $K \cap P = K \cap M$.

Now suppose P is a minimal psgp of G. We extend $\delta_P$ to a function on G by setting $\delta_P(kp) = \delta_P(p)$ ($k \in K$, $p \in P$). Now put

$$\Xi_G(x) = \Xi(x) = \int_K \delta_P(xk)^{-1/2} dk$$

where dk is the normalized Haar measure on the compact group K. It follows from Lemma 5 that $\Xi$ is actually independent of the choice of P.

## §4. Cusp forms and the space $\mathcal{A}(G, \tau)$

Let $\mathfrak{G}$ be the universal enveloping algebra of $\mathfrak{g}_c$. We regard elements of $\mathfrak{G}$ as left-invariant differential operators on G. There is an obvious anti-isomorphism $g \mapsto g'$ of $\mathfrak{G}$ onto the algebra $\mathfrak{G}'$ of right-invariant differential operators on G. If $g_1, g_2 \in \mathfrak{G}$, we write

$$f(g_1; x; g_2) = (g_1' g_2 f)(x) \qquad (x \in G)$$

for any $f \in C^\infty(G)$. Put

$$\nu_{g_1, g_2, r}(f) = \sup_G |f(g_1; x; g_2)| \Xi(x)^{-1}(1+\sigma(x))^r$$

for $r \geq 0$. Then the Schwartz space $\mathcal{C}(G)$ consists of all functions $f \in C^\infty(G)$ such that

$$\nu_{g_1, g_2, r}(f) < \infty$$

for all $g_1, g_2 \in \mathfrak{G}$ and $r \geq 0$. The set of all seminorms $\nu_{g_1, g_2, r}$ defines the structure of a locally convex Hausdorff space on $\mathcal{C}(G)$ which is complete. Moreover $\mathcal{C}(G)$ is contained in $L_2(G)$.

If $P = MAN$ is a psgp of $G$ and $f \in \mathcal{C}(G)$, we put

$$f^P(x) = \int_N f(xn)dn \qquad (x \in G) \ ,$$

where $dn$ is the Haar measure on $N$. (This integral is always convergent.) $f$ is said to be a cusp form if $f^P = 0$ whenever $\operatorname{prk} P > 0$. Let ${}^\circ\mathcal{C}(G)$ denote the space of all cusp forms. Then ${}^\circ\mathcal{C}(G)$ is a closed subspace of $\mathcal{C}(G)$.

Let $\tau$ be a unitary double representation of $K$ on a finite-dimensional Hilbert space $V$. Let $C^\infty(G, \tau)$ denote the subspace of all $f \in C^\infty(G) \otimes V$ such that

$$f(k_1xk_2) = \tau(k_1)f(x)\tau(k_2) \qquad (k_1, k_2 \in K, \ x \in G) \ .$$

Put ${}^\circ\mathcal{C}(G, \tau) = C^\infty(G, \tau) \cap ({}^\circ\mathcal{C}(G) \otimes V)$.

Theorem 1. $\dim {}^\circ\mathcal{C}(G, \tau) < \infty$.

Let $\mathfrak{Z}$ be the algebra of all differential operators on $G$ which commute with both left and right translations of $G$. Then $\mathfrak{Z}$ is the center of $\mathfrak{G}$ and therefore abelian.

Corollary. Every element in $^\circ\mathcal{C}(G, \tau)$ is $\mathfrak{Z}$-finite.

A continuous function $f$ on $G$ is said to satisfy the weak inequality, if there exist numbers $c, r \geq 0$ such that

$$|f(x)| \leq c\,\Xi(x)\,(1+\sigma(x))^r$$

for all $x \in G$.

Let $\mathcal{A}(G, \tau)$ denote the space of all $f \in C^\infty(G, \tau)$ such that:

1) $f$ is $\mathfrak{Z}$-finite;

2) $|f|$ satisfies the weak inequality.

Let $P = MAN$ be a psgp of $G$ and put

$$\gamma(a) = \inf_{\alpha} \alpha(\log a) \qquad (a \in A)\ ,$$

where $\alpha$ runs over all roots of $(P, A)$. $a$ being a variable element of $A$, we write $a \underset{P}{\longrightarrow} \infty$ if 1) $\sigma(a) \longrightarrow \infty$ and 2) we can choose $\varepsilon > 0$ such that $\gamma(a) \geq \varepsilon\sigma(a)$. Let $\tau_M$ denote the restriction of $\tau$ on $K_M = K \cap M$.

Theorem 2. Given $f \in \mathcal{A}(G, \tau)$, there exists a unique element $f_P \in \mathcal{A}(MA, \tau_M)$ such that

$$\lim_{a \underset{P}{\to} \infty} \{\delta_P(ma)^{1/2} f(ma) - f_P(ma)\} = 0$$

for $m \in MA$ and $a \in A$.

We call $f_P$ the constant term of $f$ along $P$.

Let $^*P = {}^*M\,{}^*A\,{}^*N$ be a psgp of $M$ and $P' = M'A'N'$ the psgp of $G$ contained in $P$, which corresponds to $^*P$ according to Lemma 4.

Lemma 6. Fix $f \in \mathcal{A}(G, \tau)$, $a \in A$ and put

$$g(m) = f_P(ma) \qquad (m \in M)\ .$$

Then $g \in \mathcal{A}(M, \tau_M)$ and

$$g_{{}^*P}({}^*m\, {}^*a) = f_{P'}({}^*m . {}^*a . a)$$

for ${}^*m \in {}^*M = M'$ and ${}^*a \in {}^*A$.

We write $S^x = xSx^{-1}$ $(x \in G)$ for any subset $S$ of $G$. If $k \in K$, it is clear that $P^k = M^k A^k N^k$ is the Langlands decomposition of the psgp $P^k$.

Lemma 7. Fix $f \in \mathcal{A}(G, \tau)$ and $k \in K$. Then

$$f_{P^k}(m^k) = \tau(k) f_P(m) \tau(k^{-1})$$

for $m \in MA$.

Let $f \in \mathcal{A}(G, \tau)$. We write $f_P \sim 0$ if

$$\int_M (\phi(m), f_P(ma))_V dm = 0$$

for all $\phi \in {}^\circ\mathcal{C}(M, \tau_M)$ and $a \in A$. Here the scalar product is in $V$, $dm$ is the Haar measure on $M$ and the integral is always convergent.

Theorem 3. Suppose $f \in \mathcal{A}(G, \tau)$ and $f_P \sim 0$ for all psgps $P$ of $G$ (including $P = G$). Then $f = 0$.

Theorem 4. Fix $f \neq 0$ in $\mathcal{A}(G, \tau)$ and choose a psgp $P = MAN$ of $G$ such that:

1) $f_P \not\sim 0$,

2) $P$ is minimal with respect to condition 1).

Then for any $a \in A$, the function $m \longmapsto f_P(ma)$ lies in ${}^\circ\mathcal{C}(M, \tau_M)$. Let $Q$ be a psgp of $G$ such that $Q \subset P$ and $Q \neq P$. Then $f_Q = 0$.

## §5. The Eisenstein Integral and its constant term

Let $P = MAN$ be a psgp of $G$. Fix $\psi \in {}^{\circ}\mathcal{C}(M, \tau_M)$ and extend it to a function on $G = KP$ as follows:

$$\psi(kman) = \tau(k)\psi(m) \qquad (k \in K,\ m \in M,\ a \in A,\ n \in N)\ .$$

Put

$$E(P : \psi : \nu : x) = \int_K \psi(xk)\tau(k^{-1})\exp\{((-1)^{1/2}\nu - \rho_P)(H_P(xk))\}dk$$

for $\nu \in \mathfrak{a}_c^*$ and $x \in G$. Then $E$ is an analytic function on $\mathfrak{a}_c^* \times G$ which, for a fixed $\nu$, is $\mathfrak{Z}$-finite and for a fixed $x$, an entire function of $\nu$.

Lemma 8. Fix $\psi \in {}^{\circ}\mathcal{C}(M, \tau_M)$ and $\nu \in \mathfrak{a}^*$. Then $E(P : \psi : \nu) \in \mathcal{A}(G, \tau)$. Let $P' = M'A'N'$ be another psgp of $G$. Then

$$E_{P'}(P : \psi : \nu) \sim 0$$

unless $P$ and $P'$ are associated.

Let $\mathcal{P}(A)$ be the set of all psgps $P'$ of $G$ such that $A$ is the split component of $P'$. Then $\mathcal{P}(A)$ is a finite set and if $N'$ is the radical of $P'$, it is clear that $P' = MAN'$ is the Langlands decomposition of $P'$. Put

$$\varpi_P = \prod_{\alpha>0} H_\alpha$$

where $\alpha$ runs over all roots of $(P, A)$ and $H_\alpha$ is the element of $\mathfrak{a}$ given by

$$B(H, H_\alpha) = \alpha(H) \qquad (H \in \mathfrak{a})\ .$$

Then $\varpi_P$ may be regarded as a polynomial function on $\mathfrak{a}_c^*$. Put $\mathcal{W} = \mathcal{W}(A)$ and $L = {}^{\circ}\mathcal{C}(M, \tau_M)$. Then by Theorem 1, $\dim L < \infty$.

Theorem 5. We can choose an open connected neighborhood $U$ of zero in $\mathfrak{a}^*$ and an integer $r \geq 0$ with the following properties. Fix $P_1, P_2 \in \mathcal{P}(A)$

and a point $\nu \in \mathfrak{a}^*$ such that $\varpi_P(\nu) \neq 0$. Then there exist unique elements $c_{P_2|P_1}(s:\nu) \in \mathrm{End}\, L$ $(s \in \mathfrak{W})$ such that

$$E_{P_2}(P_1 : \psi : \nu : ma) = \sum_{s \in \mathfrak{W}} (c_{P_2|P_1}(s:\nu)\psi)(m)e^{(-1)^{1/2} s\nu(\log a)}$$

for $\psi \in L$, $m \in M$ and $a \in A$. Moreover for any $s \in \mathfrak{W}$, the function

$$\nu \longmapsto \varpi_P(\nu)^r c_{P_2|P_1}(s:\nu)$$

extends to a holomorphic function of $\nu$ on $\mathfrak{a}^* + (-1)^{1/2}U$.

As usual $s\nu(H) = \nu(s^{-1}H)$ for $\nu \in \mathfrak{a}_c^*$ and $H \in \mathfrak{a}_c$.

For $\nu \in \mathfrak{a}_c^*$, define $\nu_R$ and $\nu_I$ in $\mathfrak{a}^*$ by $\nu = \nu_R + (-1)^{1/2}\nu_I$. Let $\mathfrak{F}_c(P)$ denote the set of all $\nu \in \mathfrak{a}_c^*$ such that $\nu_I(H_a) > 0$ for every root $a$ of $(P, A)$. (The definition of $\mathfrak{F}_c(P)$ given in [4(d), p. 546] is incorrect.) Put $\overline{P} = \theta(P)$, $\overline{N} = \theta(N)$, $\rho = \rho_P$ and $H(x) = H_P(x)$ $(x \in G)$. Any $x \in G$ can be written uniquely in the form $x = kman$ where $k \in K$, $m \in M \cap \exp\mathfrak{p}$, $a \in A$, $n \in N$ (in the notation of Lemma 1). Put $\kappa(x) = k$, $\mu(x) = m$.

Lemma 9. $c_{\overline{P}|P}(1:\nu)$ and $c_{P|P}(1:-\nu)$ extend to holomorphic functions of $\nu$ on $\mathfrak{F}_c(P)$ and they are given there by the following convergent integrals.

$$(c_{\overline{P}|P}(1:\nu)\psi)(m) = \int_{\overline{N}} \tau(\kappa(\overline{n}))\psi(\mu(\overline{n})m)e^{((-1)^{1/2}\nu-\rho)(H(\overline{n}))}d\overline{n} \ ,$$

$$(c_{P|P}(1:-\nu)\psi)(m) = \int_{\overline{N}} \psi(m\mu(\overline{n})^{-1})\tau(\kappa(\overline{n}))^{-1}e^{((-1)^{1/2}\nu-\rho)(H(\overline{n}))}d\overline{n} \ .$$

Here $\psi \in {}^\circ\mathcal{C}(M, \tau_M)$, $\nu \in \mathfrak{F}_c(P)$, $m \in M$ and the Haar measure $d\overline{n}$ on $\overline{N}$ is so normalized that

$$\int_{\overline{N}} e^{-2\rho(H(\overline{n}))}d\overline{n} = 1 \ .$$

The following consequence of the above integral representation was pointed out to me by R. P. Langlands.

Corollary. det $c_{P|P}(1:-\nu)$ is not identically zero in $\nu$.

## §6. The functional equations

Put

$$\|\psi\|_M^2 = \int_M |\psi(m)|^2 dm$$

for $\psi \in L = {}^{\circ}\mathcal{C}(M, \tau_M)$. This defines the structure of a Hilbert space on L.

Theorem 6. Fix $P_1, P_2 \in \mathcal{P}(A)$. Then for any $s \in \mathfrak{W}$, $c_{P_2|P_1}(s:\nu)$ extends to a meromorphic function of $\nu$ on $\mathfrak{a}_c^*$. Put

$$\overset{\circ}{c}_{P_2|P_1}(s:\nu) = c_{P_2|P_1}(s:\nu)c_{P_1|P_1}(1:\nu)^{-1}$$

and

$$E^{\circ}(P:\psi:\nu) = E(P:c_{P|P}(1:\nu)^{-1}\psi:\nu) \qquad (\psi \in L,\ P \in \mathcal{P}(A)) .$$

Then

1) $\overset{\circ}{c}_{P_2|P_1}(s:\nu)$ is holomorphic and unitary on $\mathfrak{a}^*$,

2) $E^{\circ}(P:\psi:\nu)$ is holomorphic for $\nu \in \mathfrak{a}^*$,

3) $c_{P|P}(1:\nu) = (c_{\overline{P}|\overline{P}}(1:\bar{\nu}))^*$, $c_{\overline{P}|P}(1:\nu) = (c_{P|\overline{P}}(1:\bar{\nu}))^*$

where $\bar{\nu} = \nu_R - (-1)^{1/2}\nu_I$ and the star denotes the adjoint of a linear transformation. Moreover we have the following functional equations.

a) $\overset{\circ}{c}_{P_3|P_1}(ts:\nu) = \overset{\circ}{c}_{P_3|P_2}(t:s\nu)\overset{\circ}{c}_{P_2|P_1}(s:\nu)$,

b) $E^{\circ}(P_1:\psi:\nu) = E^{\circ}(P_2:\overset{\circ}{c}_{P_2|P_1}(s:\nu)\psi:s\nu)$

for $P_1, P_2, P_3 \in \mathcal{P}(A)$ and $s, t \in \mathfrak{W}$.

Theorem 7. Fix $P_1, P_2 \in \mathcal{P}(A)$ and suppose $P' = M'A'N'$ is a psgp of G such that $P' \supset P_1 \cup P_2$ and prk $P' \geq 1$. Put ${}^*P_i = M' \cap P_i$ $(i = 1, 2)$, ${}^*A = M' \cap A$ and let ${}^*\mathfrak{W}$ denote the subgroup of all $s \in \mathfrak{W}$ which leave A' pointwise fixed. Then $({}^*P_i, {}^*A)$ $(i = 1, 2)$ are parabolic pairs in $M'$ and ${}^*\mathcal{W}$ may be identified with the Weyl group of ${}^*A$ in $M'$. For any $\nu \in \mathfrak{a}^*$, let ${}^*\nu$ denote the restriction of $\nu$ on ${}^*\mathfrak{a}$. Fix $\psi \in L$ and $\nu \in \mathfrak{a}^*$ and put $f(\nu) = E^0(P_1 : \psi : \nu)$. Then

$$f_{P'}(\nu : m'a') = \sum_{s \in {}^*\mathcal{W}\backslash\mathcal{W}} E^{0,*}({}^*P_2 : c^0_{P_2|P_1}(s : \nu)\psi : {}^*(s\nu) : m')e^{(-1)^{1/2} s\nu(\log a')}$$

for $m' \in M'$, $a' \in A'$. Moreover

$$c^0_{P_2|P_1}(t : \nu) = c^0_{{}^*P_2|{}^*P_1}(t : {}^*\nu)$$

for $t \in {}^*\mathcal{W}$.

Here $s$ runs over a complete system of representatives in $\mathcal{W}$ for ${}^*\mathcal{W}\backslash\mathcal{W}$. Theorems 6 and 7 reduce the determination of $c^0_{P_2|P_1}(s : \nu)$ to that of $c^0_{\bar{P}|P}(1 : \nu)$ in case $\operatorname{prk} P = 1$.

## §7. The Maass-Selberg relations and their consequences

Let $\mathcal{A}_P(G, \tau)$ denote the space of all $f \in \mathcal{A}(G, \tau)$ with the following property. If $Q$ is a psgp of $G$, then $f_Q \sim 0$ unless $Q$ is associated to $P$. The following theorem plays a decisive role in the proof of Theorems 6 and 7. The case when $\operatorname{prk} P = 1$ is especially important.

Theorem 8. Fix $\nu \in \mathfrak{a}^*$ such that $\overline{\varpi_P}(\nu) \neq 0$ and suppose $f \in \mathcal{A}_P(G, \tau)$ and $\phi_{Q,s} \in L$ $(Q \in \mathcal{P}(A),\ s \in \mathcal{W})$ are given functions satisfying the relation

$$f_Q(ma) = \sum_{s \in \mathcal{W}} \phi_{Q,s}(m)e^{(-1)^{1/2} s\nu(\log a)} \qquad (m \in M,\ a \in A)$$

for every $Q \in \mathcal{P}(A)$. Then

$$\|\phi_{P_1, s_1}\|_M = \|\phi_{P_2, s_2}\|_M$$

*for* $P_1, P_2 \in \mathcal{P}(A)$ *and* $s_1, s_2 \in \mathfrak{W}$.

Corollary. *Suppose* $\phi_{Q,s} = 0$ *for some pair* $(Q, s)$. *Then* $f = 0$.

This theorem is a consequence of, what I call, the Maass-Selberg relations when $\operatorname{prk} P = 1$. (These relations are similar to those discussed in [4(a), Chap. IV, §2].) The rest follows by induction on $\operatorname{prk} P$. In fact Theorems 6, 7 and 8 are proved together in this induction (cf. [4(a), Chap. V]).

## §8. The evaluation of $(\phi_a)_\nu^{(P)}$

Put $\mathfrak{F} = \mathfrak{a}^*$ and let $\mathfrak{F}'$ be the set of all $\nu \in \mathfrak{F}$ such that $\varpi_P(\nu) \neq 0$. Let $\mathcal{C}(\mathfrak{F})$ denote the Schwartz space on the finite-dimensional vector space $\mathfrak{F}$.

Lemma 10. *For any* $a \in \mathcal{C}(\mathfrak{F}) \otimes L$, *put*

$$\phi_a(x) = \int E(P : a(\nu) : \nu : x)\,d\nu \qquad (x \in G),$$

*where* $d\nu$ *denotes the Euclidean measure on* $\mathfrak{F}$. *Suppose* $a$ *satisfies the following condition. For any* $s \in \mathfrak{W}$ *and* $P_1, P_2 \in \mathcal{P}(A)$, *the function*

$$\nu \longmapsto \| c_{P_2|P_1}(s : \nu) a(\nu) \|_M \qquad (\nu \in \mathfrak{F}')$$

*remains locally bounded on* $\mathfrak{F}$. *Then* $\phi_a \in \mathcal{C}(G, \tau)$.

In particular the condition of the lemma is fulfilled if $a \in C_c^\infty(\mathfrak{F}') \otimes L$. For any $f \in \mathcal{C}(G, \tau)$, define a function $f^{(P)}$ on $MA$ by

$$f^{(P)}(m) = \delta_P(m)^{1/2} \int_N f(mn)\,dn \qquad (m \in MA).$$

Then $f^{(P)} \in \mathcal{C}(MA, \tau_M)$ and $f \longmapsto f^{(P)}$ is a continuous mapping of $\mathcal{C}(G, \tau)$ into $\mathcal{C}(MA, \tau_M)$.

Theorem 9. *Fix* $P_1, P_2 \in \mathcal{P}(A)$ *and suppose* $a \in \mathcal{C}(\mathfrak{F}) \otimes L$ *satisfies the condition of Lemma* 10. *Put*

$$\phi_a(x) = \int_{\mathfrak{F}} E(P_1 : a(\nu) : \nu : x) d\nu \qquad (x \in G)$$

and

$$\phi_{P_2, a}(m) = \int_{\mathfrak{F}} E_{P_2}(P_1 : a(\nu) : \nu : m) d\nu \qquad (m \in MA) .$$

Extend $\phi_{P_2, a}$ to a function on $G$ by the rule

$$\phi_{P_2, a}(kmn) = \tau(k)\phi_{P_2, a}(m) \qquad (k \in K,\ m \in MA,\ n \in N_2)$$

Then

$$\phi_a^{(\overline{P}_2)}(m) = \int_{\overline{N}_2} e^{-\rho_2(H_2(\overline{n}))} \phi_{P_2, a}(\overline{n}m) d\overline{n} \qquad (m \in MA) .$$

Here $P_i = MAN_i$ $(i = 1, 2)$, $\rho_2 = \rho_{P_2}$, $H_2(x) = H_{P_2}(x)$ $(x \in G)$, and all the integrals are convergent.

Put

$$f_\nu^{(P)}(m) = \int_A f^{(P)}(ma) e^{-(-1)^{1/2}\nu(\log a)} da \qquad (m \in M)$$

for $f \in \mathscr{C}(G, \tau)$ and $\nu \in \mathfrak{F}$.

Theorem 10. Fix $a \in C_c^\infty(\mathfrak{F}')$, $\psi \in L$ and put

$$\phi_a(x) = \int a(\nu) E(P : \psi : \nu : x) d\nu \qquad (x \in G) .$$

Then $\phi_a \in \mathscr{C}(G, \tau)$ and

$$(\phi_a)_\nu^{(P)} = \gamma(P) \sum_{s \in W} a(s^{-1}\nu)(c_{P|\overline{P}}(1 : \nu) c_{\overline{P}|P}(s : s^{-1}\nu)\psi)$$

for $\nu \in \mathfrak{F}$. Here

$$\gamma(P) = \int_N e^{-2\rho(H(\theta(n)))} dn$$

and the measures $da$ and $d\nu$ are assumed to be dual to each other.

## §9. Normalization of the Haar measures

$\mathfrak{q}$ being any linear subspace of $\mathfrak{g}$, we denote by $d\mathfrak{q}$ the Euclidean measure on $\mathfrak{q}$ corresponding to the Euclidean norm on $\mathfrak{g}$ defined in Lemma 2. Let $P_o = M_oA_oN_o$ be a minimal psgp of G. Then $G = KA_oN_o$. Put $\rho_o = \rho_{P_o}$ and normalize the Haar measure dx on G in such a way that

$$dx = e^{2\rho_o(\log a_o)} dk\, da_o\, dn_o \; .$$

Here $x = ka_on_o$ ($k \in K$, $a_o \in A_o$, $n_o \in N_o$) and $da_o$ and $dn_o$ are the Haar measures on $A_o$ and $N_o$ respectively which correspond to the Euclidean measures on their Lie algebras under the exponential mapping. dk is the normalized Haar measure on K (so that the total measure of K is 1). This normalization of dx is independent of the choice of $P_o$. We call dx the standard Haar measure on G.

Now let P = MAN as in §8. We can apply the above procedure to M instead of G and thus obtain the standard Haar measure dm on M.

Lemma 11. Let P = MAN be any psgp of G and dx, dm the standard Haar measures on G and M respectively. Let da and dn denote the Haar measures on A and N which correspond to $d\mathfrak{a}$ and $d\mathfrak{n}$ respectively under the exponential mapping. Then

$$\int_G f(x)dx = \int_{K\times M\times A\times N} f(kman)e^{2\rho(\log a)} dk\, dm\, da\, dn$$

for $f \in C_c(G)$. Here dk is the normalized Haar measure on K and $\rho = \rho_P$.

From now on we shall always assume that the various Haar measures are normalized as in the above lemma.

## §10. The space $\mathcal{C}_\omega(G, \tau)$

Let $\mathcal{E}(G)$ be the set of all equivalence classes of irreducible unitary representations of G and $\mathcal{E}_2(G)$ the subset of those classes $\omega$ which are square-integrable. For any $\omega \in \mathcal{E}_2(G)$, let $d(\omega)$ denote the formal degree of $\omega$.

Let $G_1 = ZG^o$ where $Z$ is the centralizer of $G^o$ in $G$. Then $G/G_1$ is a finite group. Fix $\omega_1 \in \mathcal{E}_2(G_1)$ and let $\operatorname{Ind}\omega_1$ denote the class of the representation of $G$ obtained from $\omega_1$ by inducing from $G_1$ to $G$. Then $\operatorname{Ind}\omega_1$ is irreducible and $\omega_1 \longrightarrow \operatorname{Ind}\omega_1$ is a surjective mapping of $\mathcal{E}_2(G_1)$ on $\mathcal{E}_2(G)$. Taking into account the results of [4(f), §41], one can give an explicit formula for $d(\omega)$ $(\omega \in \mathcal{E}_2(G))$.

Now fix $\omega \in \mathcal{E}_2(G)$ and let $\mathcal{H}_\omega$ denote the smallest closed subspace of $L_2(G)$ containing all K-finite matrix coefficients of the class $\omega$. Put $\mathcal{C}_\omega(G) = \mathcal{C}(G) \cap \mathcal{H}_\omega$ and

$$\mathcal{C}_\omega(G, \tau) = \mathcal{C}(G, \tau) \cap (\mathcal{C}_\omega(G) \otimes V) .$$

Then $\mathcal{C}_\omega(G, \tau) \subset {}^o\mathcal{C}(G, \tau)$.

## §11. The characters $\Theta_{\omega, \nu}$

Let $P = MAN$ be a psgp of $G$. Fix $\omega \in \mathcal{E}_2(M)$, $\nu \in \mathfrak{a}^*$ and define a tempered distribution $\Theta_{\omega, \nu}$ on $G$ as follows:

$$\Theta_{\omega, \nu}(f) = \theta_\omega(g_{f, \nu}) \qquad (f \in \mathcal{C}(G)) .$$

Here $\theta_\omega$ is the character of $\omega$,

$$g_{f, \nu}(m) = \int_{A \times N} \bar{f}(man)\exp\{((-1)^{1/2}\nu + \rho)(\log a)\}da\, dn \qquad (m \in M)$$

and

$$\bar{f}(x) = \int_K f(kxk^{-1})dk .$$

Then $\Theta_{\omega, \nu}$ is the character of a unitary representation of $G$. Let $\Omega(P, \nu)$ denote the class of this representation. Then

1) $\Omega(P, \nu)$ is irreducible if $\varpi_P(\nu) \neq 0$,

2) $\Omega(P_1, \nu) = \Omega(P_2, \nu)$ for $P_1, P_2 \in \mathcal{P}(A)$.

The group $\mathcal{W} = \mathcal{W}(A)$ operates on $\mathcal{E}_2(M)$ as follows. Fix $s \in \mathcal{W}$ and $\omega \in \mathcal{E}_2(M)$. Choose a representative $y$ in $G$ for $s$ and a representation $\beta$ of $M$ in the class $\omega$. Put

$$\beta^y(m) = \beta(y^{-1}my) \qquad (m \in M) .$$

Since $y$ normalizes $M$, $\beta^y$ is a representation of $M$. We define $\omega^s$ to be the class of $\beta^y$. It is easy to see that $\omega^s \in \mathcal{E}_2(M)$ and it is independent of the choice of $y$ and $\beta$.

Lemma 12. Fix $\omega_1, \omega_2 \in \mathcal{E}_2(M)$ and $\nu_1, \nu_2 \in \mathfrak{a}^*$. Then

$$\Theta_{\omega_1, \nu_1} = \Theta_{\omega_2, \nu_2}$$

if and only if there exists an element $s \in \mathcal{W}$ such that

$$\omega_2 = \omega_1^s, \ \nu_2 = s\nu_1 .$$

The distributions $\theta_\omega$ and $\Theta_{\omega, \nu}$ are actually functions. We write[3)]

$$(\Theta_{\omega, \nu}, f) = \int_G \operatorname{conj} \Theta_{\omega, \nu} \cdot f dx \qquad (f \in \mathcal{C}(G)) ,$$

the integral being convergent. A similar notation is used also for $f \in \mathcal{C}(G, \tau)$.

Lemma 13. Let $F$ denote the projection in $V$ given by

$$Fv = \int_K \tau(k) v \tau(k^{-1}) dk \qquad (v \in V) .$$

Then

$$(\Theta_{\omega, \nu}, f) = F(\theta_\omega, f_\nu^{(P)}) \qquad (f \in \mathcal{C}(G, \tau))$$

for $\omega \in \mathcal{E}_2(M)$ and $\nu \in \mathfrak{a}^*$.

Here

$$(\theta_\omega, g) = \int_M \operatorname{conj} \theta_\omega . g dm$$

for $g$ in $\mathcal{C}(M)$ or $\mathcal{C}(M, \tau_M)$.

Fix $\nu_o \in \mathfrak{F}'$. Then $s\nu_o \neq \nu_o$ for $s \neq 1$ in $\mathfrak{W}$. Hence we can choose an open neighborhood $U$ of $\nu_o$ in $\mathfrak{F}'$ such that $U \cap sU = \emptyset$ for $s \neq 1$ in $\mathfrak{W}$.

Lemma 14. Put $L_\omega = \mathcal{C}_\omega(M, \tau_M)$ and suppose $a \in C_c^\infty(U)$ and $\psi \in L_\omega$. Fix $\omega' \in \mathcal{E}_2(M)$ and $\nu \in \mathfrak{F}$. Then

$$(\Theta_{\omega', \nu}, \phi_a) = 0$$

unless $\nu \in \bigcup_{s \in \mathfrak{W}} sU$. Now suppose $\nu \in U$. Then

$$(\Theta_{\omega', \nu}, \phi_a) = \begin{cases} \gamma(P) a(\nu) F(\theta_\omega, c_{P|\bar{P}}(1:\nu) c_{\bar{P}|P}(1:\nu)\psi) & \text{if } \omega' = \omega, \\ 0 & \text{otherwise.} \end{cases}$$

Here $\phi_a$ and $\gamma(P)$ have the same meaning as in Theorem 10.

## §12. The Plancherel measure $\mu_\omega$

Let $\mathcal{C}_A(G)$ denote the set of all $f \in \mathcal{C}(G)$ with the following property. If $P' = M'A'N'$ is any psgp of $G$, then $f^{P'} \sim 0$ (see [4(d), p. 538]) unless $P'$ is associated to $P$. For $f \in \mathcal{C}(G)$, put

$$\hat{f}(\omega : \nu) = (\Theta_{\omega, \nu}, f) \qquad (\omega \in \mathcal{E}_2(M), \nu \in \mathfrak{a}^*) .$$

We give $\mathcal{E}_2(M)$ the discrete topology.

Theorem 11.[4)] There exists a unique continuous function $\mu$ on $\mathcal{E}_2(M) \times \mathfrak{a}^*$ with the following properties:

1) $\mu(\omega^s : s\nu) = \mu(\omega : \nu)$ for $s \in \mathfrak{W}$.

2) For any $f \in \mathcal{C}(G)$, the series

$$\sum_{\omega \in \mathcal{E}_2(M)} d(\omega) \int_{\mathfrak{a}^*} |\hat{f}(\omega : \nu)\mu(\omega : \nu)| d\nu$$

is convergent.

3) $f(1) = \sum_{\omega \in \mathcal{E}_2(M)} d(\omega) \int_{\mathfrak{a}^*} \hat{f}(\omega : \nu)\mu(\omega : \nu) d\nu$

for all $f \in \mathcal{C}_A(G)$.

Fix $\omega \in \mathcal{E}_2(M)$ and put $\mu_\omega(\nu) = \mu(\omega : \nu)$.

Lemma 15. $\mu_\omega$ extends to a meromorphic function on $\mathfrak{a}_c^*$ which is holomorphic on $\mathfrak{a}^*$. Moreover $\mu_\omega(\nu) > 0$ on $\mathfrak{F}'$ and[5)]

$$[\mathfrak{W}]\mu_\omega(\nu).\, \gamma(P) c_{P|\bar{P}}(1 : \nu) c_{\bar{P}|P}(1 : \nu)\psi = \psi$$

for $\nu \in \mathfrak{F}'$ and $\psi \in L_\omega$.

## §13. Explicit determination of $\mu_\omega$

We now make $\mathfrak{W}$ operate on $L$ as follows. Fix $s \in \mathfrak{W}$ and $\psi \in L$ and let $y$ be a representative of $s$ in $K$. Then $s\psi$ is the function $m \longmapsto \tau(y)\psi(y^{-1}my)\tau(y^{-1})$ on $M$. We also put $P^s = P^y$.

Lemma 16. Let $P_1, P_2 \in \mathcal{P}(A)$ and $s, t \in \mathfrak{W}$. Then

$$s c_{P_2|P_1}(t : \nu) = c_{P_2^s|P_1}(st : \nu), \quad c_{P_2|P_1}(t : \nu)s^{-1} = c_{P_2|P_1^s}(ts^{-1} : s\nu)$$

and

$$s c^o_{P_2|P_1}(t : \nu) = c^o_{P_2^s|P_1}(st : \nu), \quad c^o_{P_2|P_1}(t : \nu)s^{-1} = c^o_{P_2|P_1^s}(ts^{-1} : s\nu)$$

for $\nu \in \mathfrak{a}_c^*$.

Fix $P = MAN$ in $\mathcal{P}(A)$ and for any $P_1 \in \mathcal{P}(A)$ $(P_1 = MAN_1)$ and $\nu \in \mathfrak{F}_c(P)$, define a linear transformation $J_{P_1|P}(\nu)$ on $L$ as follows.

$$(J_{P_1|P}(\nu)\psi)(m) = \int_{\bar{N}\cap N_1} \psi(\bar{n}m)e^{((-1)^{1/2}\nu-\rho)(H(\bar{n}))}d\bar{n} \qquad (\psi \in L,\ m \in M)$$

Here $d\bar{n}$ is the Haar measure on $\bar{N} \cap N_1$ which corresponds, under the exponential mapping, to the Euclidean measure on its Lie algebra. The function $\psi$ is extended on $G = KP$ as before by the rule

$$\psi(kman) = \tau(k)\psi(m) \qquad (k \in K,\ m \in M,\ a \in A,\ n \in N)\ .$$

The above integral is convergent when $\nu \in \mathfrak{F}_c(P)$. We note that from Lemma 9,

$$J_{P|P}(\nu) = 1\ , \quad J_{\bar{P}|P}(\nu) = \gamma(P)c_{\bar{P}|P}(1:\nu)\ ,$$

where $\gamma(P)$ has the same meaning as in Theorem 10.

Let $\Sigma$ be the set of all roots of $(P, A)$. A root $\alpha \in \Sigma$ is called reduced if $t\alpha \notin \Sigma$ for $0 < t < 1$ $(t \in \mathbb{R})$. Let $\Phi$ be the set of all reduced roots. For any $\alpha \in \Phi$, let $\Sigma(\alpha)$ denote the set of all roots in $\Sigma$ of the form $t\alpha$ $(t \geq 1)$. Put

$$\mathfrak{n}_\alpha = \sum_{\beta\in\Sigma(\alpha)} \mathfrak{n}(\beta)\ ,$$

where $\mathfrak{n}(\beta)$ is the set of all $X \in \mathfrak{n}$ such that $[H, X] = \beta(H)X$ for all $H \in \mathfrak{a}$. Put $N_\alpha = \exp \mathfrak{n}_\alpha$.

Let $\sigma_\alpha$ denote the hyperplane $\alpha = 0$ on $\mathfrak{a}$ and $Z_\alpha$ the centralizer of $\sigma_\alpha$ in $G$. Put $M_\alpha = {}^{0}(Z_\alpha)$, $A_\alpha = M_\alpha \cap A$ and $\bar{N}_\alpha = \theta(N_\alpha)$. Then

$${}^{*}P_\alpha = MA_\alpha N_\alpha\ , \quad {}^{*}\bar{P}_\alpha = MA_\alpha\bar{N}_\alpha$$

are maximal psgps of $M_\alpha$. Put $\rho_\alpha = \rho_{{}^*P_\alpha}$, $H_\alpha(y) = H_{{}^*P_\alpha}(y)$ $(y \in M_\alpha)$ and define

$$\gamma({}^{*}P_\alpha) = \int_{\bar{N}_\alpha} e^{-2\rho_\alpha(H_\alpha(\bar{n}))}d_\alpha\bar{n}\ ,$$

where $d_\alpha\bar{n}$ is the Haar measure on $\bar{N}_\alpha$ which corresponds to the Euclidean measure on its Lie algebra.

A point $H \in \mathfrak{a}$ is called regular if $\alpha(H) \neq 0$ for every $\alpha \in \Sigma$ and it is called semiregular if there is exactly one root $\alpha \in \Phi$ such that $\alpha(H) = 0$. Let $C$ be the set of all points $H \in \mathfrak{a}$ where $\alpha(H) > 0$ for all $\alpha \in \Sigma$. Then we can choose two points $H_o$, $H_1$ in $\mathfrak{a}$ such that the following conditions hold.

1) $H_o \in C$ and $-H_1 \in C$.

2) Put $H(t) = (1-t)H_o + tH_1$ $(0 \leq t \leq 1)$. Then $H(t)$ is either regular or semiregular.

Let $0 < t_1 < t_2 < \ldots < t_r < 1$ be all the values of $t$ such that $H(t)$ is semiregular. Let $\alpha_i$ be the root in $\Phi$ which vanishes at $H(t_i)$ $(1 \leq i \leq r)$. It is clear that $r = [\Phi]$. Put

$$c_i(\nu) = \gamma({}^*P_{\alpha_i})c_{{}^*\overline{P}_{\alpha_i}|{}^*P_{\alpha_i}}(1 : \nu_{\alpha_i}) \qquad (\nu \in \mathfrak{F}_c(P))$$

where $\nu_\alpha$ is the restriction of $\nu$ on $\mathfrak{a}_\alpha = \underset{\sim}{R}H_\alpha$ $(\alpha \in \Phi)$.

Lemma 17. $\gamma(P)c_{\overline{P}|P}(1 : \nu) = c_r(\nu)c_{r-1}(\nu) \ldots c_1(\nu) \quad (\nu \in \mathfrak{F}_c(P))$.

Since both sides are meromorphic on $\mathfrak{F}_c$, this relation must hold for all $\nu$. Lemma 17 may be regarded as a generalization of a result of Gindikin and Karpelevič [3].

Let $\mathcal{W}_\alpha$ denote the Weyl group of $A_\alpha$ in $M_\alpha$ and $\mu_{\omega,\alpha}$ the function on $\mathfrak{a}_\alpha^*$ which corresponds to $\mu_\omega$ when we replace $(G, P)$ by $(M_\alpha, {}^*P_\alpha)$. The following theorem is an immediate consequence of Lemmas 15 and 17.

Theorem 12. $\mu_\omega(\nu) = c \prod_{\alpha \in \Phi} \mu_{\omega,\alpha}(\nu_\alpha) \qquad (\omega \in \mathcal{E}_2(M),\ \nu \in \mathfrak{a}^*)$

<u>where</u>

$$c = \gamma(P)[\mathcal{W}]^{-1} \prod_{\alpha \in \Phi} [\mathcal{W}_\alpha]\gamma({}^*P_\alpha)^{-1} .$$

Observe that $\operatorname{prk} M_\alpha = 0$ and $\operatorname{prk} {}^*P_\alpha = 1$. Hence, in order to compute $\mu_\omega$, it is enough to consider the case when $\operatorname{prk} G = 0$ and $\operatorname{prk} P = 1$.

Then there are two possibilities.

1) $\mathcal{E}_2(G) = \emptyset$.

2) $\mathcal{E}_2(G) \neq \emptyset$.

The first case is easier and there one can show that $\mu_\omega$ is a polynomial function on $\mathfrak{a}^*$. On the other hand the second case can be dealt with by the method of [4(g), §24]. In this way one obtains an explicit formula for $\mu_\omega$.

## §14. Relation with the exceptional series

Now we assume that prk G = 0 and prk P = 1. Fix $\omega \in \mathcal{E}_2(M)$ and put $\mathfrak{W} = \mathfrak{W}(A)$.

Lemma 18.[6)] $\Theta_{\omega, o}$ is an irreducible character unless the following two conditions hold.

1) $\mu_\omega(0) > 0$, $[\mathfrak{W}] = 2$ and $\omega^s = \omega$ $(s \in \mathfrak{W})$.

2) $\mathcal{E}_2(G) \neq \emptyset$.

Actually 2) is a consequence of 1). Moreover it seems likely that $\Theta_{\omega, o}$ is the sum of two distinct irreducible characters when these conditions are fulfilled (cf. [4(b), Theorem 1] and [6(b)]).

Lemma 19. Fix $\omega \in \mathcal{E}_2(M)$ such that $\mu_\omega(0) = 0$. Then $[\mathfrak{W}] = 2$ and $\omega^s = \omega$ where $s$ is the element of $\mathfrak{W}$ other than 1. Put

$$C(\nu)\psi = -c^o_{P|P}(s : (-1)^{1/2}\nu)\psi \qquad (\nu \in \mathfrak{a}_c^*, \psi \in L_\omega) .$$

Then $C(\nu)$ is self-adjoint for $\nu \in \mathfrak{a}^*$ and

$$C(\nu)C(-\nu) = 1 .$$

Finally $C(0) = 1$.

Let $a$ denote the unique simple root of (P, A). We identify $\mathfrak{a}_c^*$ with $\underset{\sim}{C}$ by means of the mapping $\nu \mapsto \langle \nu, a \rangle$. Let $\delta_o > 0$ be the distance, from the origin, of the nearest pole of $C$ on the real axis. Then $C(\nu)$ is a positive-definite operator for $|\nu| < \delta_o$ $(\nu \in \mathbb{R})$. This enables us to prove the following theorem (cf. [6(a), Theorem 3.3]).

Theorem 13. Fix $\omega \in \mathcal{E}_2(M)$ such that $\mu_\omega(0) = 0$. Then we can choose $\delta > 0$ with the following property. Suppose $\nu \in (-1)^{1/2}\mathfrak{a}^*$, $|\langle \nu, a \rangle| < \delta$ and $\nu \neq 0$. Then $\Theta_{\omega,\nu}$ is the character of an irreducible unitary representation of $G$ belonging to the exceptional series.

It would be interesting to extend the above results to the case $\operatorname{prk} P > 1$.

## References

1. A. Borel and J. Tits, Groupes rèductifs, Inst. Hautes Études Sci. Publ. Math. No. 27 (1965), pp. 55-150.

2. I. M. Gelfand, Automorphic functions and the theory of representations, Proc. Internat. Congress Math. (1962), pp. 74-85.

3. S. G. Gindikin and F. I. Karpelevič, Plancherel measure of Riemannian symmetric spaces of nonpositive curvature, Sov. Math. vol. 3 (1962), pp. 962-965.

4. Harish-Chandra, (a) Automorphic forms on semisimple Lie groups, Lecture notes in Math., no. 62, Springer-Verlag (1968).
(b) Eisenstein Series over finite fields, Functional Analysis and Related Fields, pp. 76-88, Springer-Verlag (1970).
(c) Some applications of the Schwartz space of a semisimple Lie group, Lecture notes in Math., no. 140 (1970), pp. 1-7, Springer-Verlag.
(d) Harmonic Analysis on semisimple Lie groups, Bull. Amer. Math. Soc. vol. 78 (1970), pp. 529-551.
(e) Harmonic Analysis on reductive p-adic groups, Lecture notes in Math., no. 162 (1970), Springer-Verlag.
(f) Discrete series for semisimple Lie groups II, Acta Math. vol. 116 (1966), pp. 1-111.
(g) Two theorems on semisimple Lie groups, Ann. of Math. vol. 83 (1966), pp. 74-128.

5. S. Helgason, A duality theory for symmetric spaces with applications to group representations, Advances in Math. vol. 5 (1970), pp. 1-154.

6. A. W. Knapp and E. M. Stein, (a) Existence of complementary series, Problems in Analysis, pp. 249-259, Princeton University Press, 1970.
(b) Singular Integrals and the Principal Series II, Proc. Nat. Acad. Sci. U.S.A. vol. 66 (1970), pp. 13-17.

7. B. Kostant, On the existence and irreducibility of certain series of representations, Bull. Amer. Math. Soc. vol 75 (1969), pp. 627-642.

## Footnotes

1) For any finite-dimensional vector space $V$ over $\mathbb{R}$, we denote by $V^*$ its dual and by $V_c$ its complexification. Moreover $V_c^* = (V^*)_c$.

2) See Borel and Tits [1].

3) conj$c$ stands for the complex conjugate of $c \in \mathbb{C}$.

4) Cf. [4(d), §12].

5) $[F]$ denotes the number of elements in a finite set $F$.

6) The fact that there is a rather direct connection between questions of reducibility and the theory of the Eisenstein Integral was first pointed out to me by J. G. Arthur (cf. his thesis "Harmonic analysis of tempered distributions on semisimple Lie groups of real rank one," Yale, 1970). See also [5].

# HARMONIC ANALYSIS IN THE NON-EUCLIDEAN DISK

by

Sigurdur Helgason
Massachusetts Institute of Technology

Let $D$ denote the unit disk $x^2 + y^2 < 1$ with the Riemannian structure

$$ds^2 = \frac{dx^2+dy^2}{(1-x^2-y^2)^2},$$

which induces the customary non-Euclidean metric on $D$. In order to define a Fourier transform on $D$ let us observe a simple geometric characterization of the exponential functions on $\mathbb{R}^n$. Let us call a complex-valued $C^\infty$ function $f$ on $\mathbb{R}^n$ an <u>eigenwave</u> if

(i) $f$ is an eigenfunction of the Laplacian on $\mathbb{R}^n$,

(ii) $f$ is a plane wave, that is, there exists a unit vector $\omega \in \mathbb{R}^n$ such that $f$ is constant on each hyperplane perpendicular to $\omega$.

If $(,)$ denotes the inner product on $\mathbb{R}^n$, the eigenwaves are the linear combinations of $e^{\mu(x,\omega)}$ and $e^{-\mu(x,\omega)}$, where $\mu \in \mathbb{C}$ is arbitrary ($1$ and $(x,\omega)$ for $\mu = 0$), so roughly speaking Fourier analysis in $\mathbb{R}^n$ consists of a decomposition of "arbitrary" functions into eigenwaves.

This geometric viewpoint makes good sense for $D$ because the horocycles in $D$, being the orthogonal trajectories to a family of parallel geodesics, form the obvious analogs to hyperplanes in $\mathbb{R}^n$. The horocycles in $D$ are the circles tangential to the boundary $B$ of $D$; we refer to the point of contact as the <u>normal</u> to the horocycle. In analogy with (i) and (ii) we call a function $F$ on $D$ an eigenwave if it satisfies the conditions:

1) $F$ is an eigenfunction of the Laplace-Beltrami operator

$$\Delta = (1-(x^2+y^2))^2\left(\frac{\partial^2}{\partial x^2} + \frac{\partial^2}{\partial y^2}\right)$$

of $D$.

---

Supported in part by the National Science Foundation GP-22928.

2) There exists a point $b \in B$ such that $F$ is constant on each horocycle with normal $b$.

Denoting by $\langle z,b\rangle$ the signed distance from $0$ to the horocycle through $z \in D$ with normal $b \in B$, one finds that the eigenwaves are the linear combinations of $e^{(1+\mu)\langle z,b\rangle}$ and $e^{(1-\mu)\langle z,b\rangle}$, where $\mu \in \mathbb{C}$ is arbitrary ($e^{\langle z,b\rangle}$ and $\langle z,b\rangle e^{\langle z,b\rangle}$ for $\mu = 0$). By Fourier analysis on $D$ we therefore mean a decomposition of "arbitrary" functions on $D$ into functions of the form $z \to e^{(1+\mu)\langle z,b\rangle}$ $(\mu \in \mathbb{C}\ ,\ b \in B)$.

Theorem 1. *For* $f \in C_c^\infty(D)$ *let* $\tilde{f}$ *denote the "Fourier transform"*

$$\tilde{f}(\lambda,b) = \int_D f(z)e^{(-i\lambda+1)\langle z,b\rangle}dz \qquad (\lambda \in \mathbb{C},\ b \in B),$$

*where* $dz = (1-|z|^2)^{-2}dx\ dy$, *the non-Euclidean surface element on* $D$. *Then*

$$f(z) = \int_{\mathbb{R}}\int_B \tilde{f}(\lambda,b)e^{(i\lambda+1)\langle z,b\rangle}d\mu(\lambda,b), \tag{1}$$

*where*

$$d\mu = (2\pi)^{-2}\lambda\ \tanh(\tfrac{1}{2}\pi\lambda)d\lambda\ db,$$

$d\lambda$ *being the Euclidean measure on* $\mathbb{R}$, $db$ *the angular measure on* $B$. *Moreover if* $\mathbb{R}^+$ *denotes the set of positive reals the mapping* $f \to \tilde{f}$ *extends to an isometry of* $L^2(D,dz)$ *onto* $L^2(\mathbb{R}^+\times B, 2d\mu)$.

Formula (1), which is stated in [4] (for symmetric spaces), is proved in [5], Ch. I, by a simple reduction to the case of radial $f$, in which case it becomes the known inversion formula for the Fock-Mehler transform with the Legendre function as kernel (see [2]). The last part is proved in [6], p. 120.

It is a natural problem to try to find a Paley-Wiener type theorem for this Fourier transform, that is, to give an intrinsic characterization of the space $(C_c^\infty(D))^\sim$. In this context we consider the Radon transform

$$\hat{f}(\xi) = \int_{\xi} f(x)\,d\sigma(x),$$

where $\xi$ is any horocycle in $D$, $d\sigma$ the measure on $\xi$ induced by the Riemannian structure of $D$ and $f$ is any function on $D$ for which the integral exists. Writing $\hat{f}(t,b) = \hat{f}(\xi)$ if $\xi$ has normal $b$ and distance $t$ from the origin, it is easy to prove

$$\tilde{f}(\lambda,b) = \int_{-\infty}^{\infty} e^{(-i\lambda+1)t}\,\hat{f}(t,b)\,dt, \tag{2}$$

so one is led to the analogous problem of giving an intrinsic characterization of the space $(C_c^{\infty}(D))^{\wedge}$. For this we have the following partial result which can be proved by reducing the problem to the analogous one for the Radon transform on $\mathbb{R}^n$, settled in [3], Theorem 2.1.

<u>Theorem 2</u>. <u>Let</u> $K > 0$ <u>and assume</u> $f \in C^{\infty}(D)$ <u>satisfies the conditions</u>

(i) <u>For each integer</u> $m \geq 0$, <u>the function</u> $f(z)e^{md(o,z)}$ <u>is bounded</u>, $d$ <u>denoting distance</u>,

(ii) $\hat{f}(\xi) = 0$ <u>if</u> $d(o,\xi) > K$.

<u>Then</u>

$$f(x) = 0 \quad \underline{\text{if}} \quad d(o,x) > K.$$

This theorem can be used to prove that if $L \neq 0$ is any differential operator on $D$ which is invariant under all isometries, then

$$L\, C^{\infty}(D) = C^{\infty}(D); \tag{3}$$

that is, the differential equation $Lu = f$ has a solution $u$ for any $f \in C^{\infty}(D)$. However, since $L$ is necessarily a polynomial in the Laplace-Beltrami operator, (3) is contained in a more general theorem of Malgrange on elliptic operators ([9], p.341).

As mentioned earlier, if $\lambda \in \mathbf{C}$, $b \in B$, the function $z \to e^{\lambda\langle z,b\rangle}$ is an eigenfunction of $\Delta$. More precisely,

(4) $$\Delta_z(e^{\lambda\langle z,b\rangle}) = \lambda(\lambda-2)e^{\lambda\langle z,b\rangle},$$

so in particular, $e^{2\langle z,b\rangle}$ is harmonic; in fact it is identical to the Poisson kernel. Since the eigenvalue in (4) is independent of $b$, the function

(5) $$\int_B e^{\lambda\langle z,b\rangle}\, dm(b),$$

where $m$ is any measure in $B$, is still an eigenfunction of $\Delta$. According to Furstenberg [1], p.295, and Karpelevič [7], if $m$ is a positive measure and $\lambda \in \mathbb{R}$, the functions (5) constitute all the positive eigenfunctions of $\Delta$. Their methods are based on convexity theorems and seem unapplicable to the problem of finding all eigenfunctions. This turns out to involve analytic functionals on $B$, that is, the elements in the dual space of the topological vector space $\mathfrak{A}(B)$ of analytic functions on $B$.

Theorem 3. *The eigenfunctions of* $\Delta$ *are precisely the functions*

$$f(z) = \int_B e^{\lambda\langle z,b\rangle}\, dT(b)$$

*where* $\lambda \in \mathbf{C}$ *and* $T$ *is an analytic functional on* $B$.

This result was stated in [6], p.139, but without proof, so we shall indicate one here. We begin with a lemma.

Lemma. *Let* $n$ *be an integer. The eigenfunctions* $f$ *for* $\Delta$ *satisfying the homogeneity condition*

(6) $$f(e^{i\theta}z) \equiv e^{in\theta}f(z)$$

*are the constant multiples of the functions*

(7) $$f_n(z) = \frac{1}{2\pi} \int_B e^{\lambda\langle z,b\rangle} \alpha_n(b)\,db ,$$

where $\alpha_n(e^{i\phi}) = e^{in\phi}$.

If $r = d(o,z)$ then $z = (\tanh r)e^{i\theta}$, and $f(z) = e^{in\theta}\psi(r)$, where $\psi(r) = f(\tanh r)$. One finds easily that $\psi$ satisfies the ordinary differential equation

(8) $$\psi'' + 2\tanh^{-1}(2r)\psi' - 4n^2\sinh^{-2}(2r)\psi = c\psi,$$

where $c \in \mathbb{C}$. By analyticity, $\psi$ admits an expansion $\psi(r) = \sum_0^\infty a_k \sinh^k(2r)$. The differential equation (6) leads to the recursion formula

$$((k+2)^2 - n^2)a_{k+2} = (\frac{c}{4} - k(k+1))a_k.$$

If one considers separately the cases $n$ odd, $n$ even, this formula implies without difficulty that all solutions to (8) are proportional (although it is a second-order equation). This proves the lemma.

Now let $f$ be an arbitrary eigenfunction of $\Delta$. Select $\lambda \in \mathbb{C}$ such that $\Delta f = \lambda(\lambda-2)f$ and $\mathrm{Re}\,\lambda \geq 1$ (Re = real part). Using a Fourier series expansion on the function $\theta \to f(e^{i\theta}z)$ we obtain from the lemma an absolutely convergent expansion

(9) $$f(z) = \sum_n a_n f_n(z) \qquad a_n \in \mathbb{C} .$$

The function $f_n$ can be expressed in terms of a hypergeometric function; the crux of the proof is now that the absolute convergence of (9) implies

(10) $$\sum_n |a_n|\, r^{|n|} < \infty \qquad \text{for} \quad 0 \leq r < 1.$$

But it is essentially contained in [8], Satz 13, that (10) is the necessary and sufficient condition for the existence of an analytic functional $T$ with Fourier coefficients $(a_n)$:

$$T \sim \Sigma\, a_n e^{in\phi}.$$

Now (7) gives a convergent Fourier series

$$e^{\lambda\langle z,b\rangle} = \sum_n f_n(z)\alpha_{-n}(b).$$

This series actually converges in the topology of $\mathfrak{A}(B)$, so $T$ can be applied to it term-by-term, giving the representation of $f(z)$ stated in Theorem 3.

## References


[1] H. Furstenberg, Translation-invariant cones of functions on semisimple Lie groups, Bull. Amer. Math. Soc. 71 (1965), 271-326.

[2] R. Godement, Séminaire Bourbaki, 1957.

[3] S. Helgason, The Radon transform on Euclidean spaces, compact two-point homogeneous spaces and Grassmann manifolds, Acta Math. 113 (1965), 153-180.

[4] S. Helgason, Radon-Fourier transforms on symmetric spaces and related group representations, Bull. Amer. Math. Soc. 71 (1965), 757-763.

[5] S. Helgason, Lie groups and symmetric spaces, Battelle Rencontres 1967, 1-71, W. A. Benjamin, Inc. N.Y. 1968.

[6] S. Helgason, A duality for symmetric spaces with applications to group representations, Advances in Mathematics, 5 (1970), 1-154.

[7] F. I. Karpelevič, Non-negative eigenfunctions of the Beltrami-Laplace operator on symmetric spaces of non-positive curvature, Soviet Math. 4 (1963), 1180-1182.

[8] G. Köthe, Die Randverteilungen analytischer Funktionen, Math. Z. 57 (1952), 13-33.

[9] B. Malgrange, Existence et approximation des solutions des équations aux dérivées partielles et des équations de convolution, Ann. Inst. Fourier Grenoble, 6 (1955-56),271-355.

# PROBLEMS OF EXTRAPOLATION AND SPECTRAL SYNTHESIS ON GROUPS

by

C. Herz

Mc Gill University

Throughout this paper G denotes a locally compact group and H a closed subgroup. We write FS(G) for the representative ring of G; in general FS is a contravariant functor from locally compact groups to commutative Banach algebras with unity. We designate by A(G) the Fourier algebra of G, i.e. the space of representative functions of the (left or right makes no difference) regular representation of G; this is a closed ideal in FS(G). See Eymard's thesis [1] for the basic facts.

Although FS is a functor and hence the restriction map gives a Banach algebra morphism $\underline{Res}: FS(G) \to FS(H)$, an example given by Douady shows that this map need not be onto when G is non-commutative. In contrast the behavior of A, which is not a functor, is much better. We have [3]:

1. Theorem. Restriction of functions gives a quotient morphism of Banach algebras $\underline{Res}: A(G) \to A(H)$. In fact, for each $h \in A(H)$ there exists $g \in A(G)$ such that $\|g\|_A = \|h\|_A$ and $h = \underline{Res}\, g$.

Theorem 1 raises the question of whether A(H) is a Banach space coretract of A(G), i.e. does there exist a linear contraction $\Lambda: A(H) \to A(G)$ such that $\underline{Res} \circ \Lambda = \mathrm{Id}$ ? There are three trivial cases where the answer is yes.

Theorem. A(H) is a Banach space coretract of A(G) in the situations:

2. H is open in G; – define $\Lambda$ by $\Lambda h(x) = h(x)$ for $x \in H$ and $\Lambda h(x) = 0$ for $x \notin H$.

3. There is a semi-direct product decomposition $G = HK$ where K is a compact normal subgroup; – define $\Lambda$ by $\Lambda h(x) = h(y)$ where $x = yz$, $y \in H$ and $z \in K$.

4. H is compact and commutative

The proofs of Theorems 2 and 3 consist of straightforward verifications that the explicitly given maps $\Lambda$ are morphisms $A(H) \to A(G)$.

Proof of Theorem 4. Here we do not have an obvious choice for $\Lambda$, and I shall give a banal existential proof at this point. Since H is commutative we have that $A(H)$ is isomorphic to $L_1(\hat{H})$ by the Fourier transform. Since H is compact, each character $\chi \in \hat{H}$ is a member of $A(H)$; therefore, by Theorem 1, there exists $g_\chi \in A(G)$ with $\|g_\chi\|_A = \|\chi\|_A = 1$ such that $\underline{\text{Res}}\, g_\chi = \chi$. Having chosen the $g_\chi$, we define $\Lambda$ by $\Lambda h = \Sigma_{\chi \in \hat{G}} \hat{h}(\chi) g_\chi$, where $h = \Sigma\, \hat{h}(\chi)\chi$ is the Fourier series expansion of $h \in A(H)$; thus $\|h\|_A = \Sigma\, |\hat{h}(\chi)|$.

As we shall see later, the proof of Theorem 4 can be made more constructive, and we can drop the commutativity.

$4^*$. Theorem. If H is a compact subgroup of G then A(H) is a Banach space coretract of A(G).

It is tempting to try to generalize Theorem 4 by dropping the compactness assumption. If H is not compact, however, we have $\chi \notin A(H)$, and the Douady counterexample shows exactly that there need not exist $g_\chi \in FS(G)$ with $\underline{\text{Res}}\, g_\chi = \chi$ and $\|g_\chi\|_{FS} = 1$. Thus the proof of Theorem 4 cannot be generalized, and the best I know how to do is

5. Theorem. If H is a central subgroup of G then A(H) is a Banach space coretract of A(G), at least if H is $\sigma$-compact.

The dual Banach space to A(G) is denoted by PM(G), and the pairing is written $\int fT$ where $f \in A(G)$ and $T \in PM(G)$. The elements of PM(G) are called pseudomeasures, and every measure of bounded variation is a pseudomeasure with the norm in PM(G) dominated by the total variation. Let $CONV_2(G)$ designate the Banach space of bounded linear operators on $L_2(G)$ which commute with right-translations. The basic fact is that PM(G) is isometrically isomorphic to $CONV_2(G)$ by $T \mapsto \tilde{T}$ where $\tilde{T}u(x) = T*u(x) =_{df}$ $\int_G u(y^{-1}x)\, T(y)$. The expression on the right is defined by

$$\int_G (\tilde{T}u)(x)v(x)\,dx = \int f\,T, \text{ where}$$

$$f(y) = \int_G u(y^{-1}x)v(x)\,dx \in A(G), \quad \|f\|_A \le \|u\|_2\,\|v\|_2.$$

The dual version of Theorem 1 is the statement that if $S \in PM(H)$ then we get an operator $\tilde{S} \in CONV_2(G)$ by putting $\tilde{S}u(x) = \int_H u(\eta^{-1}x)\; T(\eta)$; this may be regarded as an isometric inclusion of PM(H) as a subspace of PM(G).

If $\Lambda: A(H) \to A(G)$ is a coretraction then the conjugate map $\Lambda': PM(G) \to PM(H)$ is a retraction. Thus the question we have been considering demands more than whether PM(H) is a retract of PM(G). This latter question is much easier.

6. Theorem. If H is commutative then PM(H) is a Banach space retract of PM(G).

Proof (by abstract nonsense). The Fourier transform gives an isomorphism of A(H) with $L_1(\hat{H})$. Thus, by duality, PM(H) is isomorphic to the dual of an $L_1$-space. Therefore PM(H) is an injective Banach space, i.e. a retract of every Banach space in which it is isometrically imbedded.

It is of considerable value to generalize the foregoing from $A(G) = A_2(G)$ to $A_p(G)$, $1 < p < \infty$. Let $L_p(G)$ be the Lebesgue space corresponding to the left-invariant Haar measure on G. Form the Banach space tensor product $L_p(G) \otimes L_{p'}(G)$; an element $t$ in this space can be represented in the form $t = \Sigma\, u_n \otimes v_n$ with $\|t\| \le \Sigma\, \|u_n\|_p\, \|v_n\|_{p'} < \|t\| + \epsilon$. A convenient notation is $t(x, y) = \Sigma\, u_n(x)\, v_n(y)$. The element $t$ can be viewed as the kernel of a trace-class operator on $L_p(G)$; the trace is $Tr(t) = \int_G t(x, x)dx$. More generally, given $\sigma \in G$ we put $(Pt)(\sigma) = \int_G t(\sigma^{-1}x, x)dx$. Then $Pt$ is a continuous function on G vanishing at $\infty$. The space of such functions $f = Pt$ with $\|f\| = \inf\{\|t\|: Pt = f\}$ constitutes the Banach space $A_p(G)$. If we write $PM_p(G)$ for the dual of $A_p(G)$

and $CONV_p(G)$ for the space of operators on $L_p(G)$ then $PM_p(G)$ may be identified with the smallest ultraweakly closed subspace of $CONV_p(G)$ containing the left translations. When $p = 2$ we have the identification of $PM_2(G)$ with $CONV_2(G)$ as given above. For $p \neq 2$ this identification persists, at least if G is an amenable group.

Theorems 1-5 above remain valid for $A_p$ except for the second sentence of Theorem 1; – all we can prove in complete generality is that given $h \in A_p(H)$ and $\epsilon > 0$ there exists $g \in A_p(G)$ with $\|g\| < \|h\| + \epsilon$ such that $h = \underline{\text{Res}}\, g$. The dual version of Theorem 1, namely the explicit isometric inclusion of $PM_p(H)$ in $PM_p(G)$, is a vast generalization of de Leeuw's theorem [ 7 ; Thm 4.5] on multipliers on quotient groups. From the point of view of convolution operators it is worthwhile to construct a variety of bounded linear transformations $\Lambda: A_p(H) \to A_p(G)$ without insisting on the condition $\underline{\text{Res}} \circ \Lambda = \text{Id}$. When G is separable metric the technique used to construct induced representations yields such maps $\Lambda$. When H is a central subgroup we do have $\underline{\text{Res}} \circ \Lambda = \text{Id}$, and this gives an alternate proof of Theorem 5. The procedure is, however, of general interest, and it yields the following result involving the Kunze-Stein phenomenon [6], [4].

7. Theorem. *Suppose* $G = KH$ *where* K *is a compact subgroup,* H *is an amenable subgroup, and* $K \cap H = \{1\}$. *Put* $\delta(x) = \Delta_H(y)\Delta_G^{-1}(y)$ *where* $x = \xi y$, $\xi \in K$, $y \in H$ *and* $\Delta_G$, $\Delta_H$ *are the modular functions of* G, H. *In order that a positive, bi-K-invariant, measurable function* **k** *on* G *be the kernel of a bounded convolution operator on* $L_p(G)$ *it is necessary and sufficient that*

$$\int_G \int_K k(x)\, \delta^{1/p}(x^{-1}\xi)\, d\xi\, dx < \infty.$$

The point is that when G is a non-compact, connected, semi-simple Lie group with finite center and K is a maximal compact subgroup, one can get information about the functions $\psi_p$, where $\psi_p(x) = \int_K \delta^{1/p}(x^{-1}\xi)\, d\xi$. It is worth noting that $\int_G k\psi_p\, dx$ is the exact operator norm.

The construction of a coretract $\Lambda: A(H) \to A(G)$ of the form given in the proof of Theorem 5 was the essential idea of my proof of spectral synthesis for the Cantor set [2]. The connection with spectral synthesis is not accidental. Consider the condition

(C). There exists a constant $c$ such that given $h \in A_p(H)$ of compact support there exists $g \in A_p(G)$ having compact support such that $\|g\| \leq c\|h\|$ and $h = \underline{\text{Res}}\, g$.

Remark. If condition (C) holds for some $c$ it holds for all $c > 1$.

Without the words "compact support" the statement would be a consequence of Theorem 1. On the other hand we have [5]

8. Theorem. Condition (C) holds iff H is a set of spectral synthesis for $A_p(G)$.

Local spectral synthesis always holds; condition (C) allows us to pass to global synthesis. That is, if $S \in PM_p(G)$ and $\underline{\text{supp}}\, S \subset H$ then we can conclude that S belongs to $PM_p(H)$, identified as a subset of $PM_p(G)$ by Theorem 1, when S has compact support; the restriction on $\underline{\text{supp}}\, S$ can be removed when (C) holds. The constructions used to prove Theorems 2, 3, $4^*$, and 5 all show that condition (C) holds under the hypotheses of those theorems. More generally condition (C) holds whenever H is amenable. For the case $G = SL_2(\mathbb{R})$ and $H = SL_2(\mathbb{Z})$, a discrete subgroup, I do not know how to prove spectral synthesis, and the existence of coretracts $\Lambda: A(H) \to A(G)$ in this case is in doubt.

Henceforth we shall assume that G is separable, metric, an assumption which may be weakened as indicated for the assertion of Theorems $4^*$ and 5. This hypothesis ensures the existence of a Borel map $\theta : G \to H$ such that $\theta(yx) = y\theta(x)$ for all $x \in G$, $y \in H$. We define $\vartheta : G \times G \to H$ by $\vartheta(\sigma, x) = \theta(x)[\theta(\sigma^{-1}x)]^{-1}$; observe that $\vartheta(\sigma, x) \equiv \sigma$ if $\sigma \in H$. Fix a function $s \in L_p(G) \otimes L_{p'}(G)$ with $\text{Tr}(s) = 1$; e.g. take $s(x_1, x_2) = u(x_1)v(x_2)$ where $u \in L_p(G)$, $v \in L_{p'}(G)$ and $\int u(x)v(x)\,dx = 1$. It can be verified, see [5], that given $t \in L_p(H) \otimes L_{p'}(H)$, the function g defined on G by

(a) $$g(\sigma) = \int_G \int_H s[y^{-1}\vartheta(\sigma, x)\sigma^{-1}x,\ y^{-1}x]\, t\,[\vartheta^{-1}(\sigma, x)y, y]\, dy\, dx$$

belongs to $A_p(G)$ and has norm $\|g\| \le \|s\|\|t\|$. Thus we have defined a bounded linear transformation

$$L_s : L_p(H) \otimes L_{p'}(H) \to A_p(G)$$

with $\|L_s\| \le \|s\|$. Moreover, if $P_H : L_p(H) \otimes L_{p'}(H) \to A_p(H)$ is the canonical morphism, i.e. $P_H t = h$ where $h(\tau) = \int_H t(\tau^{-1}y, y)dy$, then we have

$$\underline{\mathrm{Res}} \circ L_s = P_H .$$

Therefore $g = L_s t$ is an extrapolation of $h = P_H t$; in general, however, g is not determined by h since we could have $P_H t = 0$ and g non-trivial.

Proof of Theorem 4*. Suppose that H is compact. We can then choose s so that $s(y_1x_1, y_2x_2) = s(x_1, x_2)$ for all $x_1, x_2 \in G$ and $y_1, y_2 \in H$. In this case (a) becomes

($a_1$) $$g(\sigma) = \int_G s(\sigma^{-1}x, x)\, h[\vartheta(\sigma, x)]dx ,$$

where $h = P_H t$. Thus we have a factorization $L_s = \Lambda_s \circ P_H$ where $\Lambda_s : A(H) \to A(G)$ is a desired coretraction as long as $\|s\| = 1$. The map $\Lambda_s$ can be exhibited in a more appealing form. For each compact $E \subset H$ define

$$\mu(\sigma;E) = \int_G s(\sigma^{-1}x, x)\ e(\sigma, x)dx ,$$

where $e(\sigma, x) = 1$ if $\vartheta(\sigma, x) \in E$ and $e(\sigma, x) = 0$, otherwise. Then we have

(b) $g = \Lambda_s h$ is given by $g(\sigma) = \int_H h(y)\, \mu(\sigma;dy)$.

Observe that if $\tau \in H$ then $\vartheta(\sigma\tau, x) = \vartheta(\sigma, x)\tau$ which shows that $\mu(\sigma\tau;E\tau) = \mu(\sigma;E)$.

Proof of Theorem 5. Suppose now that H is a central subgroup. The change of variables $x \mapsto yx$ in (a) gives

($a_2$) $$g(\sigma) = \int_G s[\vartheta(\sigma, x)\sigma^{-1}x, x]\ h[\vartheta(\sigma, x)]dx.$$

This formula furnishes $\Lambda_s : A_p(H) \to A_p(G)$.

The most interesting instance of $(a_2)$ is the case where H is discrete and G/H is compact. Then there is a Borel fundamental domain B with compact closure; G has a unique product decomposition $G = HB$. Moreover B has Haar measure $b > 0$. Let s have the form $s(\sigma, x) = u(\sigma)$ for $x \in B$, $s(\sigma, x) = 0$ for $x \notin B$, where u is a positive function of integral 1 vanishing outside B. In this situation we get an extrapolation $\Lambda_u : A_p(H) \to A_p(G)$ given by

(c) $\quad g = \Lambda_u h$ where $g(\sigma) = \Sigma_{y \in H} h(y) \int_B u(y\sigma^{-1}x)dx.$

For fixed $\sigma$, only $y \in BB^{-1}\sigma$ comes into the picture. The bound on $\Lambda_u$ is estimated by $\|\Lambda_u\| \le \|u\|_p \, b^{1/p'}$.

If in formula (c) we take $u_0(x) = b^{-1}$ for $x \in B$, $u_0(x) = 0$ for $x \notin B$, we have exactly what I used to prove spectral synthesis for the Cantor set. To be very concrete, take $G = \mathbb{R}$, $H = b\mathbb{Z}$, and $B = (-b/2, b/2]$. Then $g = \Lambda_u h$ has the property that $g(\sigma)$ depends only on the values of h at lattice points whose distance from $\sigma$ is less than B. The function $u_0$ has the virtue that $\|\Lambda_{u_0}\| = 1$, but L. Carleson pointed out to me that there is a certain advantage in considering a more general choice of u. For example, given $\delta$ with $0 \le \delta < 1$ define

$$u_\delta(x) = b^{-1}(1-\delta)^{-1} \text{ for } -b(1-\delta) < 2x \le b(1-\delta), \; u_\delta = 0 \text{ elsewhere.}$$

Then $\|\Lambda_{u_\delta}\| = (1-\delta)^{-1/p'}$ which is $>1$ if $\delta > 0$, but, if there exists a lattice point $y_\sigma$ such that $|\sigma - y_\sigma| \le \frac{1}{2}b\delta$ then $g(\sigma)$ depends only on $h(y_\sigma)$.

Whenever we have an extrapolation $\Lambda : A_p(H) \to A_p(G)$ we get a projection $R : A_p(G) \to A_p(G)$ on putting $R = \Lambda \circ \underline{\mathrm{Res}}$. In the circumstances described above with $G = \mathbb{R}$, $H = b\mathbb{Z}$, and $\Lambda = \Lambda_{u_\delta}$ let $R_\delta^{(b)} : A_p(\mathbb{R}) \to A_p(\mathbb{R})$ be this projection. It is easy to see that

$$\lim_{b \to 0} \|f - R_\delta^{(b)} f\| = 0 \text{ for all } f \in A_p(\mathbb{R}).$$

Suppose $E$ is a closed subset of $\mathbb{R}$ with the property: there exists a sequence $\{b_n\}$ tending to $0$ such that each point of $b_n\mathbb{Z}$ is either in $E$ or at distance $\geq b_n(1-\delta/2)$ from $E$. Then $E$ is a set of spectral synthesis for $A_p(\mathbb{R})$. The proof consists in observing that if $f \in A_p(\mathbb{R})$ and $f = 0$ on $E$ then also $R_\delta^{(b_n)}f = 0$ on $E$, but $R_\delta^{(b_n)}f$ is easily seen to belong to the smallest closed ideal in $A_p(G)$ with zero-set $E$.

Let us change our point of view and consider the space of <u>left</u> cosets $\Gamma = G/H$. (In all that has gone before the space of right cosets was involved, albeit in a disguised form.) Let $\pi : G \to \Gamma$ be the projection map, and define the action of $G$ on $\Gamma$ by $\xi^\sigma = \pi(\sigma^{-1}x)$ where $\xi = \pi(x)$ and $\sigma \in G$. There exists a quasi-invariant measure on $\Gamma$ with multiplier $m$ such that if $\omega \in L_1(\Gamma)$ then

$$\int_\Gamma \omega(\xi^\sigma)\, m(\sigma, \xi)\, d\xi = \int_\Gamma \omega(\xi)\, d\xi .$$

There is also a Borel cross-section $\gamma : \Gamma \to G$ such that $\pi \circ \gamma = \mathrm{Id}$. Using $\gamma$ we define $\vartheta : G \times \Gamma \to H$ by

$$\sigma^{-1}\gamma(\xi) = \gamma(\xi^\sigma)[\vartheta(\sigma, \xi)]^{-1} \quad \text{for} \quad \sigma \in G,\ \xi \in \Gamma .$$

Now fix $s \in L_p(\Gamma) \otimes L_{p'}(\Gamma)$. It is not hard to prove that

(d) $\qquad g = \Lambda_s h$ where $g(\sigma) = \int_\Gamma s(\xi^\sigma, \xi)\, m^{1/p}(\sigma, \xi)\, h[\vartheta(\sigma, \xi)]\, d\xi$

defines a bounded linear transformation $\Lambda_s : A_p(H) \to A_p(G)$ with $\|\Lambda_s\| \leq \|s\|$.

When $H$ is a central subgroup formula (d) becomes a variant of $(a_2)$. Otherwise there is no reason to expect $\Lambda_s$ to furnish an extrapolation.

Suppose $H$ is an amenable subgroup, then we can choose functions $h \in A_p(H)$ with $\|h\| \leq 1$ which tend to 1 uniformly on compact sets. If $T \in PM_p(G)$ has compact support then we can pass to the limit and define

(e) $\qquad \langle T, s\rangle = \lim_{h \to 1} \int_G (\Lambda_s h)(\sigma)\, T(\sigma)$.

Put $UPM_p(G)$ for the norm closure of the elements of compact support in $PM_p(G)$.

Then we have

9. Theorem. If H is an amenable subgroup of the separable, metrizable group G then (e) defines a linear contraction

$$UPM_p(G) \to END(L_p(G/H)).$$

(Here END denotes the Banach space of endomorphisms.)

Theorem 9 may be viewed as a generalization of de Leeuw's results [7, §3] on restrictions of multipliers to subgroups. Note. If G is commutative then $UPM_2(G)$ is isomorphic to the space of bounded uniformly continuous functions on $\hat{G}$.

A very special corollary of Theorem 9 is this. Suppose that G/H has finite volume V. Put $\psi_p(\sigma) = V^{-1} \int_\Gamma m^{1/p}(\sigma, \xi)\, d\xi$. Then for $T \in UPM_p(G)$, $\int_G \psi_p(\sigma) T(\sigma)$ is defined and has modulus dominated by the norm of T in $PM_p$. In particular, for a positive measurable function k to give a bounded convolution kernel on $L_p(G)$ it is necessary that $\int_G k(x)\, \psi_p(x)\, dx < \infty$.

The necessary condition in Theorem 7 comes about by taking the Iwasawa decomposition $G = KAN$, putting $H = AN$, and identifying K with $\Gamma$. The sufficiency part of Theorem 7 results from a modification of the technique of [4]; the idea is that given $f \in A_p(G)$ and $\epsilon > 0$ there exists $s \in L_p(\Gamma) \otimes L_{p'}(\Gamma)$ with $\|s\| < \|f\| + \epsilon$ such that $|f| \le |\Lambda_s 1|$ where $\Lambda_s$ is described by (d).

## Bibliography


1. P. Eymard, L'algèbre de Fourier d'un groupe localement compact, Bull. Soc. Math. France 92 (1964), 181-236.

2. C. Herz, Spectral synthesis for the Cantor set, Proc. Nat. Acad. Sci. U.S.A. 42 (1956), 996-999.

3. C. Herz, Le rapport entre l'algèbre $A_p$ d'un groupe et d'un sous-groupe, C.R. Acad. Sci. Paris 271 (1970), 244-246.

4. C. Herz, Sur le phénomène de Kunze-Stein, C.R. Acad. Sci. Paris 271 (1970), 491-493.

5. C.Herz, Harmonic synthesis for subgroups, to appear.

6. R.A.Kunze and E.M.Stein, Uniformly bounded representations and harmonic analysis of the 2x2 real unimodular group, Amer.J.Math. 82 (1960), 1-62.

7. K.de Leeuw, On $L_p$ multipliers, Annals of Math. 81 (1965), 364-379.

# STRUCTURE OF INDUCED REPRESENTATIONS AND CHARACTERS OF IRREDUCIBLE REPRESENTATIONS OF COMPLEX SEMISIMPLE LIE GROUPS

by

Takeshi Hirai
Kyoto University

The present note is a sketch of results and proofs. Let $G$ be a connected complex semisimple Lie group, $H$ its Cartan subgroup and $B$ its Borel subgroup containing $H$. A character $\chi$ of $H$ is considered canonically as one of $B$. Inducing $\chi$ from $B$ to $G$, we obtain a representation $T^{\chi}$ of $G$ on a Hilbert space $\mathcal{H}(\chi)$ consisting of a certain functions on a maximal compact subgroup $K$ [3, I]. We consider also the representation $e(\chi)$ induced from $T^{\chi}$ on the subspace $\mathcal{D}(\chi)$ consisting of infinitely differentiable functions in $\mathcal{H}(\chi)$.

In this note, we investigate in detail the intertwining operators between $e(\chi)$'s, and clarify their invariant subspaces. Using its result, we obtain all irreducible constituents of $e(\chi)$ and of $T^{\chi}$, and obtain explicit formula of their characters under certain restrictive conditions that the character $\chi$ is of class A (for the definition, see § 8) and that, roughly speaking, the two different Zhelobenko's definitions of "the minimal representation" for $\chi$, the one in [6(a)] and the other in [6(c)], coincide. (Note that he denotes it by the same symbol $\mu(\chi)$ both in [6(a)] and [6(c)].) Moreover we can determine the constituents of the image and the kernel of certain intertwining operators. Any character $\chi$ is always of class A for all classical groups and the exceptional group of type $F_4$. We utilize essentially the important works of D. P. Zhelobenko in [6(a), (b), (c)] and our main idea is to describe things as far as possible in the terminology of Weyl groups. Many things are stated by using Bruhat ordering on a double coset space of a certain Weyl group.

Although this note is only a sketch of the results, the author expresses hereby his thanks to Prof. N. Iwahori who gives him much valuable information on the theory of Coxeter groups and to Prof. M. Tsuchikawa for his very encouraging discussions on the subjects.

## §1. Induced representations

In this section, we introduce some notations and definitions.

1.1. Since there is no loss of generality, we assume in this note that $G$ is simply connected. Let $\mathfrak{g}$ be the Lie algebra of $G$, $\mathfrak{h}$ a Cartan subalgebra of $H$, $\Delta$ the set of all non-zero roots of $(\mathfrak{g}, \mathfrak{h})$, $\Delta^+$ that of all positive roots with respect to a lexicographic order and $\Sigma$ that of all simple roots. Take a Weyl base $H_\alpha$ $(\alpha \in \Sigma)$, $E_\beta$, $E_{-\beta}$ $(\beta \in \Delta^+)$; then by definition,

$$(H, H_\alpha) = \alpha(H) \qquad (H \in \mathfrak{h}),$$

$$[E_\alpha, E_{-\alpha}] = H_\alpha, \qquad [E_\beta, E_{\beta'}] = N_{\beta,\beta'} \; E_{\beta+\beta'} .$$

Here $N_{\beta,\beta'} = N_{-\beta,-\beta'}$, and $(\cdot\,,\cdot)$ denotes the Killing form of $\mathfrak{g}$. Let $\mathfrak{g}_0$ be the real form of $\mathfrak{g}$ generated over $\mathbb{R}$ by $H_\alpha$, $E_\beta$, $E_{-\beta}$ $(\alpha \in \Sigma, \beta \in \Delta^+)$ and $\mathfrak{g}_u$ the compact form generated by $\sqrt{-1}\,H_\alpha$, $\sqrt{-1}\,(E_\beta + E_{-\beta})$, $E_\beta - E_{-\beta}$. The conjugation of $\mathfrak{g}$ with respect to $\mathfrak{g}_0$ is denoted by $X \to \bar{X}$ $(X \in \mathfrak{g})$. Let $\mathfrak{n}$ and $\mathfrak{n}_-$ be the nilpotent subalgebra generated over $\mathbb{C}$ by $E_\beta$'s and $E_{-\beta}$'s $(\beta > 0)$ respectively. Let $H$, $K$, $N$, $N_-$, $H_+$ and $H_-$ be the analytic subgroups of $G$ corresponding to the subalgebras $\mathfrak{h}$, $\mathfrak{g}_u$, $\mathfrak{n}$, $\mathfrak{n}_-$, $\mathfrak{h}_0 = \mathfrak{h} \cap \mathfrak{g}_0$, and $\sqrt{-1}\,\mathfrak{h}_0$ respectively. Put $B = HN_-$.

Any character $\chi$ of the Borel subgroup $B$ is trivial on $N_-$ and is expressed on $H$ uniquely as

$$\chi(\exp H) = \exp\,(p(H) + q(\bar{H})) \qquad (H \in \mathfrak{h}), \tag{1.1}$$

where $p, q \in \mathfrak{h}^*$. Let us denote this character by $\chi = (p, q)$. Introduce in $\mathfrak{h}^*$ the Killing form, and put for any $\beta \in \Delta$,

$$\tilde{\beta} = \frac{2\beta}{(\beta, \beta)} .$$

We say $p \in \mathfrak{h}^*$ is integral if $p_\beta = (p, \tilde{\beta})$ is integer for any $\beta \in \Delta$. A pair $(p, q)$ defines a character of $H$ by (1.1) iff $\nu = p - q$ is integral.

Let $d$ be the half sum of all $\beta \in \Delta^+$ and $\kappa$ be the character $(d, d)$ of

H. Let $D(\chi)$ be the space of all $f \in C^{\infty}(K)$ satisfying

$$f(h_-k) = \chi\kappa^{-1}(h_-)f(k) \qquad (h_- \in H_-,\ \ k \in K),$$

with the topology of uniform convergence of f and its derived functions. Complete $D(\chi)$ with respect to the norm

$$\|f\|^2 = \int_K |f(k)|^2 dk ,$$

where dk is a Haar measure on K. Then we obtain a Hilbert space $H(\chi)$. Define for $g \in G$ an operator $T_g^{\chi}$ as

$$T_g^{\chi} f(k) = \chi\kappa^{-1}(h_+)f(k_g) ,$$

where $kg = n_-h_+k_g$ is the decomposition of kg according to the Iwasawa decomposition $G = N_-H_+K$. Then $g \to T_g^{\chi}$ defines on $H(\chi)$ and on $D(\chi)$ representations $T^{\chi}$ and $e(\chi)$ of G.

Proposition 1.1. There exists canonical 1 - 1 correspondence between the closed invariant subspaces of $H(\chi)$ under $T^{\chi}$ and those of $D(\chi)$ under $e(\chi)$.

Let $H_1$ and $H_2$ be two different closed invariant subspaces of $H(\chi)$ such that $H_1 \subset H_2$ . If the representation on $H_2/H_1$ induced from $T^{\chi}$ is irreducible, it is called an irreducible constituent of $T^{\chi}$. We know that it is quasi-simple and therefore has character [3, III]. Let V be a topologically irreducible representation of G on a Hilbert space $H$. Let $\omega$ be an irreducible representation of K and $H_{\omega}$ the subspace consisting of all vectors in $H$ transformed according to $\omega$ under $V_k$ $(k \in K)$. If for some $\omega = \omega_0$, $0 < \dim H_{\omega_0} < \infty$ , then for any $\omega$, $\dim H_{\omega} \leq (\dim \omega)^2$ [8] and V is quasi-simple [4(a)]. Taking into account the result in [3, II], we obtain the following

Proposition 1.2. Any topologically irreducible representation on a Hilbert space $H$ such that $0 < \dim H_{\omega_0} < \infty$ for some $\omega_0$, is quasi-simple and infinitesimally equivalent to an irreducible constituent of $T^{\chi}$ on $H(\chi)$. And their characters are identical.

Hereafter we treat only $e(\chi)$ and not $T^{\chi}$.

## § 2. Intertwining operators

Let us consider intertwining operators between the representations $e(\chi)$.

2.1. The Weyl group $W$ of $G$ is generated by the set $S$ of all simple reflexions. A reflexion with respect to a simple root is called simple. The pair $(W, S)$ is a Coxeter group and we shall use the general properties of Coxeter groups without detailed reference (see [1]). For instance, for $\sigma \in W$, an expression

$$\sigma = s_{i_1} s_{i_2} \ldots s_{i_m} \qquad (s_i \in S)$$

is called reduced if the number $m$ is minimum in all expressions of $\sigma$ and then $m$ is called the length of $\sigma$ and denoted by $\ell(\sigma)$. The Weyl group $W$ operates on $\mathfrak{h}^*$ and we define for $\sigma \in W$ and $\chi = (p, q)$,

$$\sigma\chi = (\sigma p, \sigma q), \quad \bar{\sigma}\chi = (p, \sigma q), \quad \underline{\sigma}\chi = (\sigma p, q) .$$

2.2. For any simple reflexion $s_\alpha \in S$ $(\alpha \in \Sigma)$ and $\chi = (p, q)$, let us define an intertwining operator on $D(\chi)$ to $D(s_\alpha\chi)$. Put $\mathfrak{g}(\alpha) = \mathbb{C}H_\alpha + \mathbb{C}E_\alpha + \mathbb{C}E_{-\alpha}$ and let $K_\alpha$ be the analytic subgroup of $K$ corresponding to $\mathfrak{g}(\alpha) \cap \mathfrak{g}_u$. Consider a distribution $W_{\alpha,\chi}$ on $K_\alpha$ :

$$W_{\alpha,\chi}(k) = \frac{(\chi\kappa)^{-1}(ks_\alpha)}{\Gamma\left(\frac{|p_\alpha - q_\alpha| - p_\alpha - q_\alpha}{2}\right)} \qquad (k \in K_\alpha),$$

where $p_\alpha = (p, \tilde{\alpha})$, and for $g = n_-hn$ $(n_- \in N_-,\ h \in H,\ n \in N)$, we put $(\chi\kappa)^{-1}(g) = (\chi\kappa)^{-1}(h)$. This is a locally summable function if $\mathrm{Re}(p_\alpha)$ and $\mathrm{Re}(q_\alpha)$ are sufficiently small and, fixing $\nu_\alpha = p_\alpha - q_\alpha$, can be continued analytically in $c_\alpha = p_\alpha + q_\alpha \in \mathbb{C}$ to the whole complex plane. Define the intertwining operator

$A(s_\alpha, \chi)$ as the convolution operator with the distribution $W_{\alpha,\chi}$ :

$$A(s_\alpha, \chi)f = W_{\alpha,\chi} * f \qquad (f \in \mathcal{D}(\chi)) .$$

We know that

$$A(s_\alpha, \chi)\, e(\chi) = e(s_\alpha \chi)\, A(s_\alpha, \chi) .$$

2.3. For $\alpha \in \Sigma$, $\chi = (p, q)$, if $p_\alpha$ is an integer, so is also $q_\alpha$. In that case, if $p_\alpha \geq 0$ or $q_\alpha = 0$, we define an intertwining operator $B(s_\alpha, \chi)$ from $\mathcal{D}(\chi)$ onto $\mathcal{D}(\underline{s_\alpha}\chi)$ where $\underline{s_\alpha}\chi = (s_\alpha p, q)$. Considering $\mathfrak{g}$ as a complexification of $\mathfrak{g}_u$, every element of the enveloping algebra of $\mathfrak{g}$ is identified with a right invariant differential operator on K. When $p_\alpha \geq 0$, we put

$$B(s_\alpha, \chi) = E_{-\alpha}^{p_\alpha} .$$

The operator $B(s_\alpha, \chi)$ is always surjective and, when $p_\alpha \geq 0$ and $q_\alpha = 0$, it is bijective. Therefore when $p_\alpha < 0$ and $q_\alpha = 0$, we can define $B(s_\alpha, \chi)$ as

$$B(s_\alpha, \chi) = B(s_\alpha, \underline{s_\alpha}\chi)^{-1} .$$

When $q_\alpha = 0$, $\underline{s_\alpha}\chi = s_\alpha\chi$ and $B(s_\alpha, \chi)$ is a non-zero constant multiple of $A(s_\alpha, \chi)$.

Analogously if $q_\alpha$ is an integer and $q_\alpha \geq 0$ or $p_\alpha = 0$, we define an intertwining operator $C(s_\alpha, \chi)$ from $\mathcal{D}(\chi)$ onto $\mathcal{D}(\overline{s_\alpha}\chi)$, where $\overline{s_\alpha}\chi = (p, s_\alpha q)$. When $q_\alpha \geq 0$, put

$$C(s_\alpha, \chi) = E_\alpha^{q_\alpha} .$$

If $q_\alpha \geq 0$ and $p_\alpha = 0$, $C(s_\alpha, \chi)$ is bijective. Therefore when $q_\alpha < 0$ and $p_\alpha = 0$, we can put

$$C(s_\alpha, \chi) = C(s_\alpha, \overline{s_\alpha}\chi)^{-1} .$$

If $p_\alpha = 0$, $\overline{s_\alpha}\chi = s_\alpha\chi$ and $C(s_\alpha, \chi)$ is a non-zero constant multiple of $A(s_\alpha, \chi)$.

2.4. Let us give certain relations of these intertwining operators.

In the sequel, we express by $T \cong T'$ that two linear operators $T$ and $T'$ are constant multiples of one another. Let us say the signature $\chi = (p, q)$ of the representation $e(\chi)$ has positive (or negative) degeneracy along with $\beta \in \Delta$ if both $p_\beta$ and $q_\beta$ are positive (or negative) integers.

Proposition 2.1. For $\alpha \in \Sigma$ and $\chi = (p, q)$, assume that both $p_\alpha$ and $q_\alpha$ are positive integers or $p_\alpha q_\alpha = 0$. Then

(1) $A(s_\alpha, \chi) \cong B(s_\alpha, s_\alpha\chi)\ C(s_\alpha, \chi) \cong C(s_\alpha, s_\alpha\chi)\ B(s_\alpha, \chi)$ .

(2) $$A(s_\alpha, \underline{s_\alpha}\chi)\ B(s_\alpha, \chi) \cong C(s_\alpha, \chi) ,$$

$$A(s_\alpha, \overline{s_\alpha}\chi)\ C(s_\alpha, \chi) \cong B(s_\alpha, \chi) .$$

(3) The kernels of $A(s_\alpha, \chi)$, $B(s_\alpha, \chi)$ and $C(s_\alpha, \chi)$ are the same.

Proposition 2.2.

(1) If $\chi$ has degeneracy along with $\alpha \in \Sigma$ ,

$$A(s_\alpha, s_\alpha\chi)\ A(s_\alpha, \chi) = 0 ,$$

and otherwise,

$$A(s_\alpha, s_\alpha\chi)\ A(s_\alpha, \chi) \cong 1.$$

(2) For two different simple reflexions $s, s' \in S$, let $k$ be the order of $ss'$ . Then

$$\begin{aligned} &A(s, s\chi)\ A(s', s's\chi)\ A(s, ss's\chi)\ \ldots\ldots \\ &= A(s', s'\chi)\ A(s, ss'\chi)\ A(s', s'ss\chi)\ \ldots\ldots , \end{aligned} \tag{2.1}$$

where the both sides are products of k-terms.

The above equality was obtained essentially by R. A. Kunze and E. M. Stein. On the other hand, using the result of D. -N. Verma in [12(a)], we can prove the following

Proposition 2.3. *The following equalities mean that if one side is defined, then so is the other side and they are equal.*

(1) *For two different* $s, s' \in S$, *let* $k$ *be the order of* $ss'$. *Then*

$$B(s, \underline{s\chi})\ B(s', \underline{s's\chi})\ B(s, \underline{ss's\chi}) \cdots\cdots$$

$$= B(s', \underline{s'\chi})\ B(s, \underline{ss'\chi})\ B(s', \underline{s'ss'\chi}) \cdots\cdots ,$$

*where the both sides are products of* $k$*-terms.*

(2) *Analogously,*

$$C(s, \overline{s\chi})\ C(s', \overline{s's\chi})\ C(s, \overline{ss's\chi}) \cdots\cdots$$

$$= C(s', \overline{s'\chi})\ C(s, \overline{ss'\chi})\ C(s', \overline{s'ss'\chi}) \cdots\cdots .$$

(3) *For any* $s, s' \in S$,

$$C(s, \underline{s'\chi})\ B(s', \chi) = B(s', \overline{s}\chi)\ C(s, \chi) .$$

From the theory of Coxeter groups (e, g., see [5]), we obtain the following results. For any $\sigma \in W$, $\sigma \neq e$ (identity element), take a reduced expression $\sigma = s_{i_1} s_{i_2} \cdots\cdots s_{i_m}$ and put

$$A(\sigma, \chi) = A(s_{i_1}, s_{i_2} \cdots s_{i_m}\chi) \cdots\cdots A(s_{i_m}, \chi) ,$$

$$B(\sigma, \chi) = B(s_{i_1}, \underline{s}_{i_2} \cdots \underline{s}_{i_m}\chi) \cdots\cdots B(s_{i_m}, \chi) ,$$

$$C(\sigma, \chi) = C(s_{i_1}, \overline{s}_{i_2} \cdots \overline{s}_{i_m}\chi) \cdots\cdots C(s_{i_m}, \chi) .$$

Here $B(\sigma, \chi)$ and $C(\sigma, \chi)$ do not always exist.

Proposition 2.4. *The operator* $A(\sigma, \chi)$ *is uniquely determined not depending on the reduced expression of* $\sigma$. *If the product in the right hand side of* $B(\sigma, \chi)$ *is defined for some reduced expression, so is it for any other one and they determine* $B(\sigma, \chi)$ *uniquely. The operator* $C(\sigma, \chi)$ *is analogous.*

For $e \in W$, put $A(e, \chi) = B(e, \chi) = C(e, \chi) = 1$. Then,

Proposition 2.5.

(1) For any $\sigma, \tau \in W$, if the product $A(\sigma, \tau\chi)\, A(\tau, \chi)$ is not zero,

$$A(\sigma, \tau\chi)\ A(\tau, \chi) = A(\sigma\tau, \chi).$$

(2) If $B(\sigma, \tau\chi)$ and $B(\tau, \chi)$ are defined,

$$B(\sigma\ \tau\chi)\ B(\tau, \chi) = B(\sigma\tau, \chi).$$

Analogous fact holds also for $C(\sigma, \chi)$'s.

## § 3. The irreducible representation $\mu(\chi)$

Here let us explain a fundamental result of D. P. Zhelobenko in [6(b), (c)]. First note that for any $\sigma \in W$, $e(\chi)$ and $e(\sigma\chi)$ have the same set of irreducible constituents up to equivalence. We say after him that a signature $\chi = (p, q)$ is discretely positive if it has no negative degeneracy along with any $\beta \in \Delta^+$. Denote by $X_+$ the set of all discretely positive signatures. For any $\chi$, there exists $\sigma \in W$ such that $\sigma\chi \in X_+$. For the representation $e(\chi)$ on $D(\chi)$, consider the intersection $N(\chi)$ of all non-trivial kernels of $A(\sigma, \chi)$ $(\sigma \in W)$ on $D(\chi)$. Let us call $A(\sigma, \chi)$ monomial operator if it is expressed as $A(\sigma, \chi) = A(s, \tau\chi)\, A(\tau, \chi)$, where $s \in S$ and $A(\tau, \chi)$ is bijective but not $A(s, \tau\chi)$. Then $N(\chi)$ is the intersection of kernels of all monomial operators on $D(\chi)$. If there exists no monomial operator on $D(\chi)$, we put $N(\chi) = D(\chi)$. Moreover let us consider one more invariant subspace $M(\chi)$ of $D(\chi)$ for any $\chi \in X_+$. Let $w_0$ be the unique element of $W$ with the maximal length (see § 4), then $w_0^2 = 1$. Put $M(\chi) = A(w_0, w_0\chi)\, D(w_0\chi)$. Then $M(\chi)$ is closed and $M(\chi) \subset N(\chi)$. For $\chi \in X_+$, denote by $\mu'(\chi)$ and $\mu(\chi)$ the representations of $G$ induced from $e(\chi)$ on $N(\chi)$ and on $M(\chi)$ respectively. For $\chi = (p, q)$, put $\nu = p - q$ and let $|\nu|$ be a dominant element of the form $\sigma\nu$ $(\sigma \in W)$. The restriction of $e(\chi)$ on $K$ contains exactly once an irreducible representation $\omega(|\nu|)$ of $K$ with highest weight $|\nu|$.

---

Note that Mr. Zhelobenko uses the name "the minimal representation of $D(\chi)$" for $\mu'(\chi)$ on $N(\chi)$ in [6(a)] and for $\mu(\chi)$ on $M(\chi)$ in [6(c)] and denote them by the same symbol $\mu(\chi)$.

Proposition 3.1. (Zhelobenko [6(b), (c)]). *For any* $\chi \in X_+$, *let* $\mu(\chi)$ *be the representation of* $G$ *on* $M(\chi)$ *induced from* $e(\chi)$. *Then,* $\mu(\chi)$ *is completely irreducible and its restriction on* $K$ *contains* $\omega(|\nu|)$ *exactly once.*

We define $\mu(\chi)$ for $\chi \notin X_+$ as the irreducible constituent of $e(\chi)$ containing $\omega(|\nu|)$. Then $\mu(\chi)$ and $\mu(\sigma\chi)$ are equivalent for any $\sigma \in W$.

## § 4. Some properties of Weyl groups

4.1. Let $W_0$ be the Weyl group of a connected complex semisimple Lie group, $S_0$ the set of all simple reflexions and $R_0$ the set of all conjugates of elements of $S_0$. For $\sigma, \sigma' \in W_0$, we say $\sigma \to \sigma'$ iff there exist a reduced expression of $\sigma : \sigma = s_{i_1} s_{i_2} \cdots s_{i_m}$ $(s_{i_k} \in S_0)$, and $s \in S_0$ such that

$$\sigma' = s_{i_1} \cdots s_{i_k}\, s\, s_{i_{k+1}} \cdots s_{i_m} .$$

It is known that $\sigma \to \sigma'$ iff there exists $r \in R_0$ such that $\sigma' = r\sigma$ and $\ell(r\sigma) = \ell(\sigma) + 1$. Introduce an order in $W_0$ in such a way that $\sigma < \tau$ iff there exists a chain $\sigma \to \sigma_1 \to \sigma_2 \to \cdots \to \sigma_n \to \tau$, or equivalently, a reduced expression of $\sigma$ is a subexpression of some one of $\tau$ (see [2] and [11]).

Let X, Y be subsets of $S_0$ and $W_X = \langle X \rangle$, $W_Y = \langle Y \rangle$ be the subgroups of $W_0$ generated by X, Y respectively. Then the above order induces canonically an order in the double coset space $W_X \backslash W_0 / W_Y$. Denote $W_X \sigma W_Y$ by $[\sigma]$ and $[\sigma] \to [\tau]$ means that $[\sigma] < [\tau]$ and there is no element between $[\sigma]$ and $[\tau]$. We call the shortest element in a double coset (X, Y)-reduced as in [1]. When $X = \phi$, $\sigma W_Y \to \tau W_Y$ iff $(\phi, Y)$-reduced elements $\sigma_0$, $\tau_0$ of these cosets satisfy $\sigma_0 \to \tau_0$ [2]. In § 5, we use the following lemma and proposition.

Lemma 4.1. *For* $r \in R_0$, $\tau \in W_0$, *assume that* $\tau \to r\tau$, $r \notin W_X$, $\tau W_Y \neq r\tau W_Y$; *then* $[\tau] \neq [r\tau]$ *and hence* $[\tau] < [r\tau]$. (*But not necessarily* $[\tau] \to [r\tau]$.)

In the sequel, let us denote by $[\rho] \overset{\to}{\neq} [\rho']$ when $[\rho] \neq [\rho']$ and there exist $\tau \in [\rho]$, $\tau' \in [\rho']$ such that $\tau \to \tau'$.

<u>Proposition 4.2.</u> <u>For</u> $\sigma, \tau \in W_0$, $s \in S_0$, <u>assume that</u>

$$\ell(\sigma\tau) = \ell(\sigma) + \ell(\tau), \quad \ell(s\tau) = \ell(\tau) + 1 ,$$

$$\ell(\sigma s) = \ell(\sigma) + 1 ,$$

<u>then</u> $$\ell(\sigma s \tau) = \ell(\sigma) + 1 + \ell(\tau) .$$

This is obtained from the following

<u>Lemma 4.3.</u> <u>For</u> $\sigma_1, \sigma_2 \in W_0$, $s, s_1, s_2 \in S_0$, <u>assume that putting</u> $\rho = \sigma_1 s \sigma_2$, $\ell(\rho) = \ell(\rho_1) + 1 + \ell(\sigma_2)$, $\ell(s_1\rho) = \ell(\rho s_2) = \ell(\rho) + 1$ <u>and</u> $s_1\rho = \rho s_2$;

<u>then</u>

$$\ell(s_1\sigma_1\sigma_2 s_2) \leq \ell(\sigma_1) + \ell(\sigma_2) .$$

4.3. Now let $W$ and $S$ as before. Put $\Delta_p = \{\beta \in \Delta \; ; \; p_\beta$ is integer$\}$ and let $W_p$ be the subgroup of $W$ generated by $\{s_\beta \; ; \; \beta \in \Delta_p\}$. It is known that $\Delta_p$ is a root system and call $p$ (and $\chi = (p, q)$) <u>normal</u> if there exists $\sigma \in W$ such that $\sigma\Delta_p$ is generated by $\sigma\Delta_p \cap \Sigma$. Note that any character $\chi$ is normal for all classical groups of type A.

4.4. Assume that $\chi$ is normal; since $\sigma\Delta_p = \Delta_{p'}$ with $p' = \sigma^{-1}p$, and $e(\chi)$ and $e(\sigma^{-1}\chi)$ have the same irreducible constituents, it is sufficient to consider the case when $\Delta_p$ itself is generated by $\Delta_p \cap \Sigma = \Sigma_p$. Put $Z = \{s_\alpha \; ; \; \alpha \in \Sigma_p\}$, then $Z \subset S$ and $W_p = W_Z = \langle Z \rangle$. We denote $\Delta_p$ and $\Sigma_p$ also by $\Delta_Z$ and $\Sigma_Z$. Call $p$ Z-dominant if $p_\alpha \geq 0$ for any $\alpha \in \Sigma_Z$. We can assume also that $p$ is Z-dominant. There exist $\rho \in W_Z$ and Z-dominant $q_0$ such that $q = \rho q_0$. As is known [1], there exist $X, Y \subset Z$ such that

$$W_X = \{\sigma \in W_Z \; ; \; \sigma p = p\} , \quad W_Y = \{\sigma \in W_Z \; ; \; \sigma q_0 = q_0\} .$$

In the succeeding sections (§§ 5-7), we treat normal signature $\chi = (p, q)$. Then the order in $W_X \backslash W_Z / W_Y$ plays an essential role (in this case, $W_0 = W_Z$, $S_0 = Z$) .

4.5. Suppose $\chi$ is not normal. Introduce in $\Delta_p$ an order with respect to which $\Delta_p \cap \Delta^+ = \Delta_p^+$ is the set of positive roots. Let $\Sigma_p$ be the set of all simple roots of $\Delta_p$ with respect to this order and Z the set of reflexions $s_\gamma$ $(\gamma \in \Sigma_p)$. Although $Z \subset S$, we denote $W_p$, $\Delta_p$ and $\Sigma_p$ also by $W_Z$, $\Delta_Z$ and $\Sigma_Z$. Consider the pair $(W_Z, Z)$ as the one $(W_0, S_0)$ in § 4.1. Define Z-dominant analogously. Then we can assume p is Z-dominant and there exist $\rho \in W_Z$ and Z-dominant $q_0$ such that $q = \rho q_0$. Moreover there exist $X, Y \subset Z$ such that

$$W_X = \{\sigma \in W_Z \ ; \ \sigma p = p\} \ , \quad W_Y = \{\sigma \in W_Z \ ; \ \sigma q_0 = q_0\} \ .$$

In § 8, we treat abnormal $\chi$ and explain that Th's 1,2, and 3 in normal case hold also in abnormal case and that Th's 4, 5 and 6 hold also under a certain restrictive condition that $\chi$ is of class A (see § 8).

## § 5. Irreducible constituents of $e(\chi)$ (normal case)

Let the notations be as in § 4.4. Assume that $\chi = (p, q)$ is normal and p is Z-dominant. Then $\chi \in X_+$. Put $\chi_0 = (p, q_0)$ and $q = \rho q_0$, where $\rho \in W_Z$ and $q_0$ is Z-dominant. Let $\Omega(\chi_0)$ be the set of different $\overline{\tau}\chi_0 = (p, \tau q_0)$ with $\tau \in W_Z$. Then there exists 1-1 correspondence from the coset space $W_Z/W_Y$ onto the set $\Omega(\chi_0)$ as $\tau W_Y \to \overline{\tau}\chi_0$ .

5.1. Let us consider a monomial operator M on $D(\chi)$ :

$$M = A(s, \sigma\chi) \ \ A(\sigma, \chi) \ ,$$

where $s \in S$, $\sigma \in W_Z$, and $A(\sigma, \chi)$ is bijective but not $A(s, \sigma\chi)$. Using the properties of $A(\sigma', \chi')$ stated in § 2 and some properties of a root system and a Weyl group, we can prove that there exists such a standard monomial operator M', that it has the properties (i) and (ii) in the following proposition, and that $M \cong A(\sigma', \chi')M'$ . Therefore $\mathrm{Ker}(M) \supset \mathrm{Ker}(M')$ and we obtain

Propositon 5.1. To determine the invariant space $N(\chi)$ of $D(\chi)$ with

$\chi = \overline{\rho} x_0 = (p, \rho q_0)$, *it is sufficient to take the intersection of the kernels of monomial operators* $M$ *with the following properties* :

(5.1) $$M = A(s, \sigma_1 \sigma_2 \chi\,)\ \ A(\sigma_1, \sigma_2 \chi)\ \ A(\sigma_2, \chi)\ ,$$

*where* $A(\sigma_1, \sigma_2\chi)\ A(\sigma_2, \chi)$ *is bijective but not* $A(s, \sigma_1\sigma_2\chi)$.

(i) *Here* $s \in Z$, $\sigma_1 \in W_Z$, $\sigma_2 \in W_X$ ; *and* $\sigma_1$ *and* $s\sigma_1$ *is* $(\phi, X)$-*reduced.*

(ii) *Let* $\tau$ *be the longest element in the coset* $\sigma_2 \rho W_Y$; *then* $\sigma_2 \chi = (p, \sigma_2 q) = (p, \tau q_0) = \overline{\tau}\chi_0$ *and*

$$\ell(\sigma_1\tau) = \ell(\tau) - \ell(\sigma_1)\ ,$$

(5.2) $$\ell(s\sigma_1\tau) = \ell(\sigma_1\tau) + 1\ ,$$

$$\ell(s\sigma_1) = \ell(\sigma_1) + 1\ .$$

The proof is a little long and we omit it.

Using Prop. 4.2, it follows from (5.2) that $\tau \to r\tau$, where $r = \sigma_1^{-1} s \sigma_1 \in R_Z = \{s_\theta\ ;\ \theta \in \Delta_Z\}$ . From $s\sigma_1 p \neq \sigma_1 p$, $s\sigma_1\tau q_0 \neq \sigma_1 \tau q_0$, we see that $r \notin W_X$, $r\tau W_Y \neq \tau W_Y$. Hence by Lemma 4.1, $[\tau] \neq [r\tau]$ and $[\tau] \overset{\sim}{\to} [r\tau]$, where as before $[\tau] = W_X \tau W_Y$ .

5.2. Now we want to replace the (not always surjective) monomial operator $M$ by a surjective intertwining operator $D$ such that $\mathrm{Ker}(M) = \mathrm{Ker}(D)$. For the above $M$ in (5.1), put $\chi' = \sigma_2\chi = \overline{\tau}\chi_0$ . It follows from (5.2) that $A(\sigma_1, \overline{s\sigma_1}\chi')$ is bijective. Put

$$D(s, \sigma_1, \chi') = A(\sigma_1^{-1}, \overline{s\sigma_,}\chi')\ \ C(s, \sigma_1\chi')\ \ A(\sigma_1, \chi')\ ,$$

and

$$D = D(s, \sigma_1, \chi')\ \ A(\sigma_2, \chi)\ .$$

Then $\mathrm{Ker}(M) = \mathrm{Ker}(D)$ and $D$ is surjective from $\mathcal{D}(\chi)$ to $\mathcal{D}(\overline{r\sigma_2}\chi) = \mathcal{D}(\overline{r}\chi')$. Note that $\chi = \overline{\rho}\chi_0$, $\chi' = \overline{\tau}\chi_0$, $\overline{r}\chi' = \overline{r\tau}\chi_0$ , and $[\rho] = [\tau] \overset{\sim}{\to} [r\tau]$ .

5.3. Conversely if $[\rho] \tilde{\to} [\rho']$, we can find $\tau$ and $r \in R_Z$ such that $[\rho] = [\tau]$, $[\rho'] = [r\tau]$ and $\tau \to r\tau$. Hence $r \notin W_X$, $r\tau W_Y \neq \tau W_Y$. Moreover there exists $s \in Z$, $\sigma_1 \in W_Z$ which satisfy the relation (5.2) and $r = \sigma_1^{-1} s \sigma_1$. Put $\chi = \bar{\rho}\chi_0$ and $\chi_1 = \bar{\rho}'\chi_0$. Let $\tau W_Y = \sigma_2 \rho W_Y$, where $\sigma_2 \in W_X$. Then we can prove from the above facts that

$$M = A(s, \sigma_1\sigma_2\chi)\ \ A(\sigma_1, \sigma_2\chi)\ \ A(\sigma_2, \chi)$$

is a monomial operator on $D(\chi)$ satisfying (i) and (ii) in Prop. 5.1. Moreover if $\rho' W_Y = \sigma_3 r\tau W_Y$ with $\sigma_3 \in W_X$, the operator

$$D(\chi_1, \chi) = A(\sigma_3, \overline{r\tau}\chi_0)\ \ D(s, \sigma_1, \bar{\tau}\chi_0)\ \ A(\sigma_2, \bar{\rho}\chi_0)$$

maps $D(\chi) = D(\bar{\rho}\chi_0)$ onto $D(\chi_1) = D(\bar{\rho}'\chi_0)$.

Theorem 1.

(1) Take any $[\rho], [\rho'] \in W_X \backslash W_Z / W_Y$ such that $[\rho] \tilde{\to} [\rho']$. For any $\rho_1 W_Y \subset [\rho]$, $\rho_2 W_Y \subset [\rho']$, there exists at least one surjective but not injective intertwining operator $D(\overline{\rho_1}\chi_0, \overline{\rho_2}\chi_0)$ from $D(\overline{\rho_1}\chi_0)$ to $D(\overline{\rho_2}\chi_0)$.

(2) Put $\chi = \bar{\rho}\chi_0 = (p, \rho q_0)$. The invariant subspace $N(\chi)$ of $D(\chi)$ on which the representation $\mu'(\chi)$ is realized, is the intersection of the kernels of all $D(\bar{\rho}'\chi_0, \chi)$ such that $[\rho] \tilde{\to} [\rho']$.

We obtain from Th.1 the following

Theorem 2. Any irreducible constituent of $e(\bar{\rho}\chi_0)$ is equivalent to one of $\mu'(\bar{\rho}'\chi_0)$ on $N(\bar{\rho}'\chi_0)$ such that $[\rho] \leq [\rho']$. Conversely for any $[\rho']$ such that $[\rho] \leq [\rho']$, $\mu(\bar{\rho}'\chi_0)$ on $M(\bar{\rho}'\chi_0)$ is equivalent to some irreducible constituent of $e(\bar{\rho}\chi_0)$.

Let us denote by $\bar{e}(\chi)$ and $\bar{\mu}(\chi)$ the characters of $e(\chi)$ and $\mu(\chi)$ respectively. Then $\bar{e}(\bar{\rho}\chi_0)$ and $\bar{\mu}(\bar{\rho}\chi_0)$ are determined by the double coset $[\rho]$. Take a system $P$ of representatives of $W_X \backslash W_Z / W_Y$. We know that the characters $\bar{e}(\bar{\rho}\chi_0)$ $(\rho \in P)$ are linearly independent. On the other hand, Th.9 in [6(c)] says that for any two signatures $\chi$ and $\chi'$, $\mu(\chi)$ and $\mu(\chi')$ are equivalent iff there exists some $\sigma \in W$ such that $\chi' = \sigma\chi$. Therefore the characters $\bar{\mu}(\bar{\rho}\chi_0)$ $(\rho \in P)$ are also linearly independent. Moreover comparing the infinitesimal characters of the representations and utilizing the recent work of Naimark and Zhelobenko, we see that any irreducible constituent of $e(\bar{\rho}\chi_0)$ is equivalent to some $\mu(\bar{\rho}'\chi_0)$. Hence the character $\bar{\mu}(\bar{\rho}\chi_0)$ can be expressed as a linear combination of $\bar{e}(\bar{\rho}'\chi_0)$. Denote by $m([\rho], [\rho'])$ the multiplicity with which $\mu(\bar{\rho}'\chi_0)$ is contained in $e(\bar{\rho}\chi_0)$ as its irreducible constituent. It is determined by the double cosets $[\rho']$ and $[\rho]$. To obtain the explicit formula of $\bar{\mu}(\bar{\rho}\chi_0)$, we wish to determine $m([\rho], [\rho'])$. This is done under some condition in the following section. In general, rewriting a part of Th.2, we obtain the following theorem which of course must be sharpened. Let us remark that Prop.6.1. and Th.4 gives us important information on $m([\rho], [\rho'])$.

<u>Theorem 3.</u> $m([\rho], [\rho']) \geq 1$ if $[\rho] \leq [\rho']$.

## § 6. Multiplicity of irreducible constituents (normal case)

In this section we shall prove that under some condition on $\rho$,

$$m([\rho], [\rho']) = 1 \text{ if } [\rho] \leq [\rho'], \text{ and } = 0 \text{ otherwise.}$$

This follows from the following interesting properties of the intertwining operators $D(\bar{\rho}'\chi_0, \bar{\rho}\chi_0)$.

<u>Proposition 6.1.</u>

(1) <u>The intertwining operator</u> $D(\bar{\rho}'\chi_0, \bar{\rho}\chi_0)$ <u>for</u> $[\rho] \tilde{\rightarrow} [\rho']$ <u>is determined by</u> $\bar{\rho}\chi_0$ <u>and</u> $\bar{\rho}'\chi_0$ <u>to a non-zero constant factor, that is, does not depend on</u> $\tau$, $s$, $\sigma_1$, $\sigma_2$ <u>and</u> $\sigma_3$ <u>used to define</u> $D(\bar{\rho}'\chi_0, \bar{\rho}\chi_0)$ <u>in § 5.3</u>.

(2) <u>For any</u> $[\rho] < [\rho']$, <u>take a chain such that</u>

$$[\rho] \tilde{\rightarrow} [\rho_1] \tilde{\rightarrow} [\rho_2] \tilde{\rightarrow} \cdots\cdots \tilde{\rightarrow} [\rho_m] \tilde{\rightarrow} [\rho'] \qquad \underline{\text{and put}}$$

$$D(\bar{\rho}'\chi_0, \bar{\rho}\chi_0) = D(\bar{\rho}'\chi_0, \bar{\rho}_m\chi_0)\; D(\bar{\rho}_m\chi_0, \bar{\rho}_{m-1}\chi_0)\cdots\cdots$$
$$\cdots D(\bar{\rho}_2\chi_0, \bar{\rho}_1\chi_0)\; D(\bar{\rho}_1\chi_0, \bar{\rho}\chi_0).$$

Then $D(\bar{\rho}\chi_0, \bar{\rho}'\chi_0)$ is determined up to a non-zero constant factor, irrespective of a chain used to define it.

Now let us explain how (1) and (2) are proved. Here we omit the detailed calculation. For any $\tau \in W_Z$, the operator $C(\tau, \chi_0)$ is defined. Assume that $[\rho] \overset{\to}{\to} [\rho']$. Applying Prop's 2.1(2), 2.3(2), (3) and 2.4, we obtain

$$D(\overline{\rho}'\chi_0, \overline{\rho}\chi_0)\ \ C(\rho, \chi_0) \cong C(\rho', \chi_0).$$

It follows from this that there exists an intertwining operator T from $\mathcal{D}(\overline{\rho}\chi_0)$ onto $\mathcal{D}(\overline{\rho}'\chi_0)$ such that

$$T\ \ C(\rho, \chi_0) = C(\rho', \chi_0)\ .$$

Denote T by $[C(\rho', \chi_0) : C(\rho, \chi_0)]$. Then the above relation is written as

$$D(\overline{\rho}'\chi_0, \overline{\rho}\chi_0) \cong [C(\rho', \chi_0)\ :\ C(\rho, \chi_0)]\ . \tag{6.1}$$

This proves (1).

Moreover for any $[\rho] < [\rho']$, it follows from (6.1) that

$$D(\overline{\rho}'\chi_0, \overline{\rho}\chi_0) \cong [C(\rho', \chi_0)\ :\ C(\rho, \chi_0)]$$

This proves (2). Thus we obtain the following two theorems. Recall that we assumed $\chi_0 = (p, q_0)$ is normal.

Theorem 4. For any two elements $\rho, \rho'$ such that $[\rho] \leq [\rho']$, the operator

$$D(\overline{\rho}'\chi_0, \overline{\rho}\chi_0) \cong [C(\rho', \chi_0)\ :\ C(\rho, \chi_0)]$$

maps $\mathcal{D}(\overline{\rho}\chi_0)$ onto $\mathcal{D}(\overline{\rho}'\chi_0)$ and interwines $e(\overline{\rho}\chi_0)$ with $e(\overline{\rho}'\chi_0)$. It is not bijective iff $[\rho] < [\rho']$. The invariant subspace $\mathcal{N}(\overline{\rho}\chi_0)$ of $\mathcal{D}(\overline{\rho}\chi_0)$ on which $\mu'(\overline{\rho}\chi_0)$ is realized, is the intersection of all kernels of $D(\overline{\rho}'\chi_0, \overline{\rho}\chi_0)$ such that $[\rho] \to [\rho']$. (It is not necessary to take $[\rho] \overset{\to}{\to} [\rho']$ as in Th. 1.)

Theorem 5. The multiplicity $m([\rho], [\rho'])$ with which $\mu(\overline{\rho}'\chi_0)$ is contained in $e(\overline{\rho}\chi_0)$ is given as follows: if $\rho$ has the property that $\mu'(\overline{\rho}'\chi_0) = \mu(\overline{\rho}'\chi_0)$ or equivalently $N(\bar{\rho}'\chi_0) = M(\bar{\rho}'\chi_0)$ for any $\rho' \in W_Z$ such that $[\rho] \leq [\rho']$; then

$$m([\rho], [\rho']) = 1 \text{ if } [\rho] \leq [\rho']; \text{ and } = 0 \text{ otherwise.}$$

Let us make some remarks. Clearly Th's 4 and 5 are improvements of Th's 1 and 2. But as is explained is § 8, the latter hold also for abnormal signatures $\chi$ in general, whereas the former can be proved now for abnormal signatures of class A. Note that all signatures are of class A for any classical group and exceptional group of type $F_4$.

## § 7. Characters of irreducible representations

Here we give the explicit formula of the character of the irreducible representation $\mu(\chi)$ with $\chi = \bar{\rho}\chi_0$ under the same condition on $\rho$ as in Th. 5.

7.1. Put $\chi = \bar{\rho}\chi_0 = (p, \rho q_0)$ with $\rho \in W_Z$. Denote by $\bar{e}(\chi)$ and $\bar{\mu}(\chi)$ the characters of $e(\chi)$ and $\mu(\chi)$ respectively. Then it follows from Th.5 that

$$\bar{e}(\bar{\rho}\chi_0) = \sum_{[\rho] \leq [\rho']} \bar{\mu}(\bar{\rho}'x_0), \tag{7.1}$$

where $\rho'$ runs over a system of representatives of double cosets such that $[\rho] \leq [\rho']$.

7.2. Now let $\Omega$ be a partially ordered set (finite or infinite) such that every interval $[\rho, \sigma] = \{\tau ; \rho \leq \tau \leq \sigma\}$ $(\rho, \sigma \in \Omega, \rho \leq \sigma)$ contains only a finite number of elements. Let A be a matrix whose elements $a_{\rho,\sigma}$ with suffixes $(\rho, \sigma) \in \Omega \times \Omega$ are given as

$$a_{\rho,\sigma} = \begin{cases} 1 & \text{if } \rho \leq \sigma, \\ 0 & \text{otherwise.} \end{cases}$$

Then the inverse matrix B of A (AB = BA = 1) is given in [7] as follows. For any $\rho, \sigma \in \Omega$, let $a_{\rho,\sigma}(p)$ be the number of different chains of length p connecting $\rho$ and $\sigma$ in such a way that

$$\rho = \rho_1 < \rho_2 < \ldots < \rho_{p-1} < \rho_p = \sigma. \quad \text{Put}$$

$$(7.2) \qquad b(\rho, \sigma) = -\sum_{p \geq 1} (-1)^p \, a_{\rho,\sigma}(p).$$

Then $B = ( b(\rho, \sigma) )$.

Note that if $\rho \not\leq \sigma$, $b(\rho, \sigma)$ is always equal to zero, and that for $\rho = \sigma$, $b(\rho, \sigma) = 1$ ; for $\rho < \sigma$, $b(\rho, \sigma) = -\sum_{p \geq 2} (-1)^p \, a_{\rho,\sigma}(p)$. The element $b(\rho, \sigma)$ is completely determined by the type of order of the interval $[\rho,\sigma]$.

7.3. Now let us take $\Omega = W_X \setminus W_Z / W_Y$; then $A$ is the matrix whose element with suffix $([\rho''],[\rho'])$ is exactly the multiplicity $m([\rho''],[\rho'])$ if $[\rho''],[\rho'] \geq [\rho]$. Denote the elements of $B = A^{-1}$ by $b([\rho''],[\rho'])$ as above; then $b([\rho''],[\rho']) = 0$ if $[\rho''] \not\leq [\rho']$. Thus we obtain from (7.1) the following

<u>Theorem 6.</u> <u>The character $\overline{\mu}(\overline{\rho}\chi_0)$ of the irreducible representation $\mu(\overline{\rho}\chi_0)$ is given as</u>

$$(7.3) \qquad \overline{\mu}(\overline{\rho}\chi_0) = \sum_{[\rho] \leq [\rho']} b([\rho], [\rho']) \, \overline{e}(\overline{\rho}'\chi_0),$$

<u>where</u> $\rho'$ <u>runs over a system of representatives of double cosets such that</u> $[\rho] \leq [\rho']$, <u>under the assumption that</u> $\mu'(\overline{\rho}'\chi_0) = \mu(\rho'\chi_0)$ <u>for any</u> $[\rho'] \geq [\rho]$.

Note that we assume here that the signature $\chi$ is normal but as is explained in § 8 the above theorem holds for $\chi = \overline{\rho}\chi_0$ of class A. The character $\overline{e}(\chi)$ of the representation $e(\chi)$ is well-known.

7.4. Recently D. -N. Verma proved that in the case when $X = Y = \phi$ ,

$$(7.4) \qquad b(\rho, \sigma) = \operatorname{sign}(\rho^{-1}\sigma) \qquad \text{for } \rho \leq \sigma .$$

When $p$ is integral, i.e., $p_\beta$ is integer for any $\beta \in \Delta$, $Z = S$ and $W_Z = W$. Moreover if $p$ and $q_0$ are regular, $X = Y = \phi$. In this case the irreducible representation $\mu(\chi_0)$ is of finite dimension and taking into account (7.4), the character formula (7.3) will reproduce the Weyl's formula.

## § 8. Generalization of the results

Let us consider the case when $\chi = (p, q)$ is not normal. Let the notations be as in § 4.5. Assume that $p$ and $q_0$ are Z-dominant and put $\chi_0 = (p, q_0)$. Then $q = \rho q_0$ with $\rho \in W_Z$ and $\chi = \bar{\rho}\chi_0$. Put

$$\Omega(\chi_0) = \{\bar{\sigma}\chi_0 \;;\; \sigma \in W_Z\}, \qquad \Omega'(\chi_0) = \{\bar{\sigma}\underline{\tau}\chi_0 \;;\; \sigma,\ \tau \in W_Z\} .$$

Let the set $\Sigma$ of all simple roots of $\Delta$ be $\{\alpha_1, \alpha_2, \ldots, \alpha_n\}$ and denote by $s_i$ the simple reflexion $s_{\alpha_i}$.

We need the following

*Lemma 8.1. A root $\gamma \in \Delta_Z^+$ belongs to $\Sigma_Z$ iff there exist $s_{i_1}, s_{i_2}, \ldots, s_{i_m} \in S$ and $\alpha \in \Sigma$ such that*

(1) $\gamma = s_{i_1} s_{i_2} \cdots s_{i_m} \alpha$ *and* $\sigma = s_{i_1} s_{i_2} \cdots s_{i_m}$ *is reduced,*

(2) $\theta_k = s_{i_1} s_{i_2} \cdots s_{i_{k-1}} \alpha_{i_k} \notin \Delta_Z$ for $1 \leq k \leq m$.

We omit the proof. Using this lemma we obtain

*Lemma 8.2. For any $\gamma \in \Sigma_Z$, take $\sigma$ and $\alpha$ satisfying the above conditions (1) and (2). Put $s = s_\alpha \in S$. Then for any $\chi' \in \Omega'(\chi_0)$, the intertwining operators*

$$A(\sigma^{-1}, \chi'),\quad A(\sigma, s\alpha^{-1}\chi'),\quad A(\sigma, \bar{s}\sigma^{-1}\chi'),\quad A(\sigma, \underline{s}\sigma^{-1}\chi')$$

*are all bijective.*

Let us fix once for all for every $\gamma \in \Sigma_Z$, one pair $(\sigma, \alpha)$ as above. Since $\gamma = \sigma\alpha$, we see $s_\gamma = \sigma s \sigma^{-1}$. Define for $\chi' \in \Omega'(\chi_0)$ and a simple reflexion $s_\gamma \in Z$ with respect to the pair $(W_Z, Z)$ the following operators :

$$A'(s_\gamma, \chi') = A(\sigma, s\sigma^{-1}\chi') \;\; A(s, \sigma^{-1}\chi') \;\; A(\sigma^{-1}, \chi')$$

$$B'(s_\gamma, \chi') = A(\sigma, \underline{s}\sigma^{-1}\chi') \;\; B(s, \sigma^{-1}\chi') \;\; A(\sigma^{-1}, \chi')$$

$$C'(s_\gamma, \chi') = A(\sigma, \overline{s}\sigma^{-1}\chi') \;\; C(s, \sigma^{-1}\chi') \;\; A(\sigma^{-1}, \chi') .$$

Here, of course, $B'(s_\gamma, \chi')$ is defined iff so is $B(s, \sigma^{-1}\chi')$. We know from § 2 that $A'(s_\gamma, \chi') \cong A(s_\gamma, \chi')$ .

In case $\chi$ is not normal, we wish to use these operators instead of $A(s', \chi')$, $B(s', \chi')$, $C(s', \chi')$ $(s' \in Z)$ in the normal case. To guarantee the success of this idea, the analogies of Prop's 2.1, 2.2 and 2.3 must be established for newly defined $A'(s_\gamma, \chi')$, $B'(s_\gamma, \chi')$ and $C'(s_\gamma, \chi')$.

First let us state those necessary to prove Th's 1.2 and 3. We omit the word " the analogy of (Prop. 2.1 etc) " for simplicity.

<u>Proposition 8.3.</u>

(1) <u>Prop. 2.1 holds also for</u> $A'(s_\gamma, \chi')$, $B'(s_\gamma, \chi')$ <u>and</u> $C'(s_\gamma, \chi')$.

(2) <u>Prop. 2.2(1) holds also for</u> $A'(s_\gamma, \chi')$ $(\gamma \in \Sigma_Z)$.

(3) <u>Prop. 2.2(2) holds under the following restrictive condition</u> : <u>for two different</u> $s, s' \in Z$, <u>at least one side of the equality (2.1) is bijective or both sides are not zero</u>. <u>(Moreover the equality sign</u> " = " <u>must be replaced by</u> " $\cong$ ".)

Carefully checking the discussions to obtain Th's 1, 2 and 3, we see that they also go well for abnormal $\chi$, only using Prop. 8.3. Thus it is proved that Th's 1, 2 and 3 hold in general even when $\chi$ is not normal.

To repeat the discussions sketched in § § 6-7 to obtain Th's 4, 5 and 6, we need the following properties of $B'(s_\gamma, \chi')$ and $C'(s_\gamma, \chi')$.

Proposition 8.4. For $s = s_\gamma$, $s' = s_{\gamma'} \in Z$, Prop. 2.3 holds for newly defined $B'(\cdot, \cdot)$ and $C'(\cdot, \cdot)$ in one of the following two cases :

(i) When we can take $\sigma = s_{i_1} s_{i_2} \ldots s_{i_m}$ (reduced), $\alpha, \alpha' \in \Sigma$ such that simultaneously $\gamma = \sigma\alpha$, $\gamma' = \sigma\alpha'$ and that $\sigma$ has the property (2) of Lemma 8.1.

(ii) When simple reflexions $s$ and $s'$ commute and moreover the intertwining operator in one (or equivalently both) side of the equality in question is from $e(\chi')$ on $\mathcal{D}(\chi')$ with discretely positive $\chi'$ (i.e., $\chi' \in X_+$) to another.

(Also " = " must be replaced by " $\cong$ ".)

To prove the above proposition in the case (ii), we need the following

Lemma 8.5. For any $\chi' \in X_+$, a continuous intertwining operator from $e(\chi')$ into itself is always a constant multiple of the identity operator.

To obtain this lemma, we use essentially Lemma 11.1 in [6(c)].

We omit detailed discussions and now give a definition to state the last conclusion.

Definition : Any normal signature $\chi = (p, q)$ is called of class A. An abnormal $\chi$ is called of class A if for any not commuting $s_\gamma$ and $s_{\gamma'}$ in Z, we can find $\sigma \in W$, $\alpha, \alpha' \in \Sigma$ which satisfy the condition (i) in Prop. 8.4.

Conclusion : Th's 1, 2 and 3 hold in general. Th's 4, 5 and 6 hold for signatures of class A.

## § 9. Kernels and images of intertwining operators

Assume that $\chi_0 = (p, q_0)$ is of class A and that $p$, $q_0$ are Z-dominant. Put $\chi = (\sigma p, \tau q_0)$ with $\sigma, \tau \in W_Z$. The set of its irreducible constituents is equal to that of $e(p, \sigma^{-1}\tau q_0)$. Hence if we assume that $\mu'(\bar{\rho}\chi_0) = \mu(\bar{\rho}\chi_0)$ for any $\rho$ such that $[\rho] \geq [\sigma^{-1}\tau]$, then it is the set of

(9.1) $\mu(\bar{\eta}\chi_0)$ such that $[\eta] \geq [\sigma^{-1}\tau]$ ,

where $\eta$ runs over a system of representatives of double cosets of $W_X \backslash W_Z / W_Y$.

Now for any $\kappa \in W = W_S$, consider $A(\kappa, \chi)$ on $e(\chi)$ with $\chi = (\sigma p, \tau q_0)$. Take a reduced expression of $\kappa : \kappa = s_{j_1} s_{j_2} \ldots s_{j_m}$, where $s_j = s_{\alpha_j} \in S$ $(\alpha_j \in \Sigma)$. Put for $1 \leq k \leq m$,

$$\theta_k = s_{j_m} \ldots s_{j_{k+1}} \alpha_{j_k} .$$

Then $\theta_k \in \Delta^+$. Let $\Theta^+$ (or $\Theta^-$) be the set of $\theta_k$ such that $(\sigma p, \tilde{\theta}_k)$ and $(\tau q_0, \tilde{\theta}_k)$ are both positive (or negative resp.) integers. We can prove the following theorem under the assumption stated just above.

Theorem 7. The set of irreducible constituents of the image of $A(\kappa, \chi)$ of $e(\chi)$ with $\chi = (\sigma p, \tau q_0)$ is $\mu(\bar{\eta}\chi_0)$ such that

$$[\eta] \geq [\sigma^{-1}\tau],$$

$$[\eta] \geq [\sigma^{-1} s_{\theta_k} \tau] \quad \text{for} \quad \theta_k \in \Theta^+ ,$$

$$[\eta] \not\geq [\sigma^{-1} s_{\theta_k} \tau] \quad \text{for} \quad \theta_k \in \Theta^- .$$

That of the kernel of $A(\kappa, \chi)$ is the complement of the above set in the set (9.1).

Analogous results can be obtained for any intertwining operators expressed as product of $A(\cdot , \cdot)$, $B(\cdot , \cdot)$ and $C(\cdot , \cdot)$.

## References


[1] N. Bourbaki, Groupes et algèbres de Lie, Chap. 4, 5 et 6, Hermann, Paris, 1968.

[2] C. Chevalley, Sur les décompositions cellulaires des espaces G/B, preprint.

[3] Harish-Chandra, Representations of semisimple Lie groups, I, II and III, Trans. Amer. Math. Soc., 74(1953), 185-243, 76(1954), 26-65 and 234-253.

[4] T. Hirai,
(a) The characters of some induced representations of semisimple Lie groups, J. Math. Kyoto Univ., 8(1968), 313-363.

(b) Invariant eigendistributions on real simple Lie groups, I, Japan. J. Math., 40(1970). 1-68.

[5] N. Iwahori, On the structure of a Hecke ring of a Chevalley group over a finite field, J. Fac. Sci. Univ. Tokyo, 10(1963-'64), 215-236.

[6] D. P. Zhelobenko,
(a) Symmetry in the class of elementary representations of a semisimple complex Lie group (in Russian), Functional Analysis and its Applications, 1(1967), 15-38.
(b) Analysis on the irreducibility in the class of elementary representations of a semisimple complex Lie group, Izv. AN SSSR, Ser. Math., 32(1968), 108-133.
(c) Operational calculus on a semisimple complex Lie group, ibid., 33(1969), 931-973.

[7] G. -C. Rota, On the foundations of combinatorial theory I, Theory of Möbius functions, Zeits. für Wahrsheinlichkeitstheorie, 2(1964), 340-368.

[8] H. Shin'ya, Spherical functions on locally compact groups, to appear.

[9] G. Shiffmann, Intégrales d'entrelacement, C. R. Acad. Sci. Paris, Ser. A, 226(1968), 47-49.

[10] A. M. Knapp and E. M. Stein, Intertwining operators for semisimple groups, preprint.

[11] R. Sternberg, Lectures on Chevalley groups, Lecture note of Yale Univ., 1967.

[12] D. -N. Verma,
(a) Structure of certain induced representations of complex semisimple Lie algebras, Bull. Amer. Math. Soc., 74(1968), 160-166.
(b) Möbius inversion for the Bruhat ordering on a Weyl group, preprint.

[13] D. P. Zhelobenko, M.A. Naimark, Description of completely irreducible representations of a semisimple complex Lie group (in Russian), Izv. AN SSSR, Ser. Math., 34(1970), 57-82.

[14] D. P. Zhelobenko, Classification of extremely irreducible and normally irreducible representations of a semisimple complex Lie group, ibid, 35(1971), 573-599.

# HAUSDORFF DIMENSION, LACUNARY SERIES, AND SOME PROBLEMS ON EXCEPTIONAL SETS

by

Robert Kaufman

University of Illinois at Urbana-Champaign and University of Washington

0. Let $E$ be a compact linear set and $\Lambda = (\lambda_k)_1^\infty$ an increasing sequence of positive real numbers. The following relations between $E$ and $\Lambda$ are in sharp contradiction:

(A) There is a subsequence $\Lambda = (\lambda_k^1)$ so that $\lim \cos 2\pi \lambda_k^1 x = 1$ uniformly in $E$.

(B) There are numbers $k_0$ and $\delta > 0$ so that

$$\sup_E \left| \sum_{k \geq k_0} a_k e^{2\pi i \lambda_k x} \right| \geq \delta \sum_{k \geq k_0} |a_k|$$

for every trigonometric polynomial with exponents from $(\lambda_k)_{k_0}^\infty$.

Theorem 1. Suppose that $E$ has Hausdorff dimension $> \eta$, and suppose that $\Lambda$ has Hadamard gaps: $\lambda_{k+1} > q\lambda_k$, for a certain $q > 1$. Then the sequence $t \cdot \Lambda = (t\lambda_k)_1^\infty$ fulfills condition (B) relative to $E$, for all $t > 0$ except a set of dimension $\leq 1 - \eta$.

Theorem 2. To each $\eta$ in $(0,1)$ there is a sequence $\Lambda$, a set $E$ of dimension $\eta$ and a set $F$ of dimension $1-\eta$ so that

$$\lim \cos 2\pi t\lambda_k x = 1 \quad \text{uniformly for } t \text{ in } F, \ x \text{ in } E.$$

The basic work on this subject is due to Helson and Kahane [2], who introduced property (B) under a different name, and used classical techniques of Riesz and Banach, presented by Zygmund [6, V]. Connections with probability theory and singular measures were obtained in [3,4] and also by Takahashi [5].

1. In the proof of Theorem 1 we use the abbreviations

$$e(u) \equiv e^{2\pi i u} \quad \text{and} \quad \hat{\mu}(u) \equiv \int e(-ux)\mu(dx) \quad \text{for measures } \mu.$$

We now review the method of Helson and Kahane, introducing a minor refinement that is the key to our argument. By $\Lambda^*$ we denote a certain set of linear forms in the elements of $\Lambda$, whose exact nature depends on $q$.

Suppose that $\mu$ is a probability in $E$ and that for some $\alpha > 0$ and $u_0 > 0$,

$$|\hat{\mu}(tu)| \leq |u|^{-\alpha} \quad \text{if} \quad u \in \Lambda^*, \quad |u| > u_0.$$

Then (B) holds for $t \cdot \Lambda$ and $E$.

To enumerate the linear forms we choose an integer $p$ so that $q^p > 2$ and $q^p > r(q-1)^{-1}$. Let $\xi$ stand for $-1,0,1$. We then have linear forms

$$\text{(i)} \qquad \lambda_{pk+j} + \sum_{m=1}^{k-1} \xi_m \lambda_{pm+j} \qquad (k = 1,2,3,\ldots,j=0,\ldots,p-1).$$

Writing $L$ for the form (i) we find

$$|\lambda_{pk+j} - L| \leq 2q^{-p}\lambda_{pk+j} < \tfrac{1}{2}(q-1)\lambda_{pk+j}.$$

Thus in particular $L \geq C\lambda_{pk+j}$ while the <u>number</u> of forms (i) involving $\lambda_r$ as leading term does not exceed $3^{rp^{-1}}$.

Next there are forms

$$\text{(ii)} \qquad \lambda_{pk+j} + \sum_{m=1}^{k-1} \xi_m \lambda_{pm+j} + w\lambda_r, \qquad w = \pm 1.$$

When $r = pk+j$ and $w = 1$, then this form is a variant of the form (i), subject to the same analysis. When $r = -1$, a form of type (i) or its negative is obtained. When $r > pk+j$, then the form obtained exceeds in modulus $C\lambda_r$ by the previous calculation, and the number of forms having a certain $\lambda_r$ as leading coefficient does not exceed $2p3^{r/p}$. When $r < pk+j$ the form exceeds $C\lambda_{pk+j}$ and the number corresponding to a given $s = pk+j$ is at most $2s3^{sp^{-1}}$.

2. As the dimension of $E$ exceeds $\eta$ there is a measure $\mu$ in $E$ satisfying a Lipschitz condition to exponent $\eta$: $\mu(a,a+h) \leq Ch^{\eta}$ for all intervals of length $h$. It is well known that then

$$\int_0^T |\hat{\mu}(u)|^2 du \leq CT^{1-\eta} \quad \text{for} \quad T \geq 1.$$

Let us fix a small $\alpha > 0$, and study the set

$$\{0 \leq u \leq T, \quad |\hat{\mu}(u)|^2 > T^{-\alpha}\}.$$

Because $\mu$ has compact support $E$, $|\hat{\mu}(u)|^2$ has a derivative bounded in modulus by some $C_1$. Now $[0,T]$ can be covered by adjacent intervals of length $(2C_1)^{-1}T^{-\alpha}$. Whenever $|\hat{\mu}(u)|^2 > T^{-\alpha}$ for some $u$ in $I$, then $|\hat{\mu}(u)|^2 > C_2T^{-\alpha}$ throughout $I$, so that the exceptional numbers $u$ are contained in $CT^{1-\eta+\alpha}$ intervals of length $< 1$. This yields immediately a basic lemma:

The set defined by the inequalities

$$0 \leq t \leq 1, \quad |\hat{\mu}(tT)|^2 > T^{-\alpha},$$

is contained in $CT^{1-\eta+\alpha}$ intervals of length $T^{-1}$.

3. Let $D > 1-\eta$ and let $\alpha > 0$ and $p$ be chosen successively so that

$$1 - \eta + \alpha + \log 3/p \log q < D.$$

Let $L_r$ represent any form in $\Lambda^*$, with leading term $\lambda_r$. Then

$|L_r| > C\lambda_r$, and the number of forms $L_r$ does not exceed $Cr3^{rp^{-1}}$ while $\lambda_r > Cq^r$. Each set

$$0 \leq t \leq 1, \quad |\hat{\mu}(tL_r)| > |L_r|^{-\alpha},$$

is contained in $C\lambda_r^{1-\eta+\alpha}$ intervals of length $\lambda_r^{-1}$. The natural estimate for D-dimensional measure, summed over all forms $L_r$, has magnitude

$$r\lambda_r^{1-\eta+\alpha}\cdot\lambda_r^{-D}\cdot\lambda_r^{s}, \qquad s = \log 3/p \log q.$$

As $1-\eta+\alpha-D+s < 0$, the sum on all $r \geq 1$ is convergent; it is plain that outside a set of D-dimensional measure zero, the requirement in §1 is fulfilled. Of course the same argument holds for every t-interval and the proof of Theorem 1 is complete.

4. In the proof of Theorem 2 we denote by $\|x\|$ the distance between a real number $x$ and the nearest integer. To each $b > 0$ we can construct $E$ so that $\dim E = (b+1)^{-1}$, while there is a sequence $Q = (q_k)_1^\infty$ of positive integers tending to infinity so that

$$\|qx\| \leq q^{-b} \qquad (q \in Q, \quad x \in E).$$

For example, let $M = (m_j)_1^\infty$ be an increasing sequence of positive integers such that $\limsup m_j/j \leq (b+1)^{-1}$, while $m_{j+1} \geq 1+(b+1)m_j$ infinitely often. The first condition implies that the set $E$ of all sums $\sum \xi_j 2^{-m_j}$ has dimension at least $(b+1)^{-1}$. Suppose now that $q = 2^{m_j}$, where $m_j$ is specified in the second requirement. Then

$$\|2^{m_j} \sum \xi_k 2^{-m_k}\| \leq \sum_{j+1}^{\infty} 2^{m_j - m_k}$$

$$\leq 2\cdot 2^{m_j} 2^{-m_{j+1}} \leq 2^{-bm_j} = q^{-b}.$$

We suppose that $Q$ has geometric growth and take for $\Lambda$ a certain subsequence of the sequence $\lambda_k = k^{-1}\lambda_k^{b+1}$. To define $F$ we consider the sets $S_k$ defined by the inequalities

$$|t| \leq 2, \qquad \|q_k^{-1}\lambda_k t\| \leq k^{-1}q_k^{-1}.$$

To estimate $\|t\lambda_k x\|$, for $x$ in $E$ and $t$ in $S_k$, we note that $x = q_k^{-1}n + yq_k^{-b-1}$, with $n$ integral and $|y| \leq 1$. Then

$$\|t\lambda_k x\| \leq |n|\cdot\|\lambda_k q_k^{-1}x\| + 2\lambda_k q_k^{-b-1} \leq Ck^{-1} + 2k^{-1}.$$

Thus the proof of Theorem 2 can be completed by proving that there is an infinite set $S$ such that $\bigcap_{k\in S} S_k$ has dimension $1-\eta$.

Now $S_k$ contains all numbers $mq_k\lambda_k^{-1} + y(k\lambda_k)^{-1}$ wherein $|y| \leq 1$, $m$ is an integer and $|m| \leq q_k^{-1}\lambda_k$. Suppose $D < 1-\eta$, and let us observe that the natural estimate for D-dimensional measure, applied to $S_k$, exceeds

$$q_k^{-1}\lambda_k\cdot(k\lambda_k)^{-D} = k^{-D}q_k^{-1}\lambda_k^{1-D}.$$

This tends to infinity because $(1-D)(b+1) = (1-D)\eta^{-1} > 1$. Now by the method of Eggleston [1] we see that $\bigcap_{k\in S} S_k$ has dimension at least $1-\eta$ for an infinite sequence $S$.

Theorem 2 follows.

5. The proof of Theorem 1 suggests a problem on distribution *modulo* 1; that can be treated by a cognate method.

*Lemma*. *Let* $\phi$ *be a trigonometric polynomial and*

$$\hat{\phi}(0) = 0, \qquad |\hat{\phi}(n)| \leq |n|^{-1}, \quad n \neq 0.$$

*Then*

$$\int_0^1 \left|\int \phi(tTx)\mu(dx)\right|^2 dt \leq CT^{-\eta}$$

*where* $C$ *depends only on* $\mu$.

The integral is in fact

$$\sum\sum \hat{\phi}(n)\hat{\phi}(m)\iiint e(ntTx - mtTy)\mu(dx)\mu(dy)dt ,$$

and this is smaller in modulus than

$$2\sum\sum |n|^{-1}|m|^{-1}\iint \inf(1, |nTx - mTy|^{-1})\mu(dx)\mu(dy).$$

To estimate the inner integral we suppose that $|n| \geq |m|$ and observe that the set $(|nTx - mTy| < h)$ has product measure $< C|n|^{-\eta}T^{-\eta}h^{\eta}$. Hence over the set $(2^r \leq |nTx - mTy| < 2^{r+1})$ the integral $<$ $C|n|^{-\eta}T^{-\eta}2^{r(\eta-1)}$. We sum this for integers $r \geq 0$, obtaining $C|n|^{-\eta}T^{-\eta}$, while the integral over $(|nTx - mTy| \leq 1)$ is $C|n|^{-\eta}T^{-\eta}$ also. Thus the integral is at most

$$CT^{-\eta}\sum\sum |n|^{-1}|m|^{-1}(|n|+|m|)^{-\eta} = CT^{-\eta}.$$

Now suppose $0 < a < b < 1$ and define $\phi(x) = 1 - b + a$ if $a < x \leq b$ modulo 1, $\phi(x) = a - b$ otherwise. Then $\hat{\phi}(0) = 0$ and $|\hat{\phi}(n)| \leq |n|^{-1}$, and the lemma shows that the function of $t$ defined by

$$\mu\{x : a < tTx \leq b \text{ modulo } 1\} + a - b$$

has $L^2$-norm $CT^{-\frac{1}{2}\eta}$ on $[0 \leq t \leq 1]$. This gives a bound for the Lebesgue measure of certain sets; fractional length can be handled by the following method. Suppose that $r > T^{-1}$ and for a given $t$

(i) $$\mu\{x : a < tTx \leq b \text{ modulo } 1\} > b - a + r.$$

Then, if $|t-s| < \frac{1}{4}T^{-1}r$,

$$\mu\{x : a - \tfrac{1}{4}r < sTx \leq b + \tfrac{1}{4}r\} > b - a + r,$$

and the latter inequality defines an s-set of measure $CT^{-\eta}(\frac{1}{2}r)^{-2} =$ $= Cr^{-2}T^{-\eta}$. Hence the set defined in (i) is contained in $Cr^{-1}T^{1-\eta}$

intervals of length $T^{-1}r$.

## References

[1] H. G. Eggleston, Sets of fractional dimensions in number theory, Proc. London Math. Soc. 54(2) (1951), 42-93.

[2] H. Helson and J. P. Kahane, A fourier method in Diophantine problems, J. D'Analyse Math. 15 (1965), 245-262.

[3] R. Kaufman, A random method for lacunary series, J. D'Analyse Math. 22 (1969), 171-175.

[4] R. Kaufman, Lacunary series and probability, Pacific J. Math. 36 (1971), 195-200.

[5] S. Takahashi, Tohôku Math. J. 22 (1970), 502-510.

[6] A. Zygmund, Trigonometric Series, Cambridge 1959 and 1968.

## Supplementary References

[1] A. S. Besicovitch, Sets of fractional dimensions (IV): On rational approximation to real numbers, J. London Math. Soc. 9 1934, 126-131.

[2] P. Erdos and S. J. Taylor, On the set of points of convergence of a lacunary trigonometric series... , Proc. London Math. Soc. 7 (3) (1957), 598-615.

[3] P. Erdos and S. J. Taylor, The Hausdorff measure of the intersection of positive Lebesgue measure, Mathematika 10 1963, 1-9.

[4] R. Kaufman, Probability, Hausdorff dimension, and fractional distribution, Mathematika 17 1970, 57-62.

[5] R. Kaufman, A remark on Sidon sets; Hausdorff measure and Sidon sets. (To appear.)

[6] V. Jarník, Über einen Satz von A. Khintchine II, Acta Arithmetica 2 (1937), 1-22.

# IRREDUCIBILITY THEOREMS FOR THE PRINCIPAL SERIES

by

A.W. Knapp* and E.M. Stein+

Cornell University and Princeton University

## 1. Introduction

In earlier work [6] we have developed a class of intertwining integrals for semisimple Lie groups. These operators exhibit various members of the principal series of representations as unitarily equivalent in a way that mirrors the action of the Weyl group. Where members of the Weyl group act with fixed points, the operators give self-equivalences of representations of the principal series and thereby provide information about reducibility. One of the main results of the present announcement is that for (at least) some of these groups, the operators actually give complete information about reducibility of principal series representations.

To be more specific, let $G$ be a connected semisimple Lie group of matrices and Let $MAN$ be a minimal parabolic subgroup. Here $M$ is compact, $A$ is a vector group, and $N$ is nilpotent. (For details of the notation, see §6 of [6].) The principal series consists of those representations $U(\sigma,\lambda)$ of $G$ obtained by inducing from $MAN$ the finite-dimensional representation $man \to \lambda(a)\sigma(m)$, where $\sigma$ is an irreducible unitary representation of $M$ and $\lambda$ is a unitary character of $A$.

Let $W = M'/M$ be the Weyl group relative to $A$. The members $w$ of $M'$ act on representations of $M$ and characters of $A$ by $w\sigma(m) = \sigma(w^{-1}mw)$ and $w\lambda(m) = \lambda(w^{-1}mw)$. A central result of [6] is that, corresponding to each triple $(w,\sigma,\lambda)$, there is a unitary

*Supported partly by a research associateship at Princeton University and partly by NSF Grant GP 28251 at Cornell University.
+National Science Foundation fellow.

operator $\mathcal{A}(w,\sigma,\lambda)$ with the property that

$$U(w\sigma, w\lambda)\,\mathcal{A}(w,\sigma,\lambda) = \mathcal{A}(w,\sigma,\lambda)U(\sigma,\lambda). \tag{1.1}$$

The dependence of these operators on $\lambda$ is holomorphic (in a neighborhood of $\lambda$ unitary), and they satisfy a cocycle relation

$$\mathcal{A}(w_1w_2,\sigma,\lambda) = \mathcal{A}(w_1, w_2\sigma, w_2\lambda)\,\mathcal{A}(w_2,\sigma,\lambda). \tag{1.2}$$

Fix $(w,\sigma,\lambda)$ and suppose that $w\sigma$ is equivalent with $\sigma$ and that $w\lambda = \lambda$. Then it is possible to extend $\sigma$ to a representation of the subgroup of $M'$ generated by $M$ and $w$. (The enlarged $\sigma$ operates on the same vector space as before.) With $\sigma(w)$ defined in this way, (1.1) yields

$$U(\sigma,\lambda)[\sigma(w)\,\mathcal{A}(w,\sigma,\lambda)] = [\sigma(w)\,\mathcal{A}(w,\sigma,\lambda)]U(\sigma,\lambda). \tag{1.3}$$

If $\sigma(w)\,\mathcal{A}(w,\sigma,\lambda)$ is not scalar, then (1.3) exhibits $U(\sigma,\lambda)$ as reducible. With $\sigma$ and $\lambda$ fixed, we shall call the set of all such operators $\sigma(w)\,\mathcal{A}(w,\sigma,\lambda)$ the <u>set of intertwining operators for $(\sigma,\lambda)$</u>. This paper deals with the following two problems:

(1) Normally many of the operators $\sigma(w)\,\mathcal{A}(w,\sigma,\lambda)$ coincide. Give an explicit description of the distinct operators in the set.

(2) Decide whether the linear span of the set of intertwining operators for $(\sigma,\lambda)$ is the entire set of bounded operators $L$ such that $U(\sigma,\lambda)L = LU(\sigma,\lambda)$.

For $G$ of real-rank one, problem 1 is solved by Theorem 5 of [6], and the question raised in problem 2 is answered affirmatively by the proof of Proposition 20. For $G$ of higher real-rank, the two problems ostensibly are independent. However, progress by our methods on the second problem for a given $G$ has occurred only after the first problem was solved for $G$. Our main conjectures are as follows:

<u>Conjecture 1</u>. <u>Let</u> $W_{\sigma,\lambda}$ <u>be the subgroup of elements</u> $w$ <u>of</u> $W$

*such that* $w\sigma$ *is equivalent with* $\sigma$ *and* $w\lambda = \lambda$. *There exist subgroups* $W'$ *and* $R$ *of* $W_{\sigma,\lambda}$ *such that* $W'$ *is abstractly isomorphic to a Weyl group,* $R$ *is a direct sum of copies of* $\mathbb{Z}_2$, $W'$ *is the subgroup on which* $\sigma(w)\mathcal{A}(w,\sigma,\lambda)$ *is scalar,* $W_{\sigma,\lambda}$ *is the semidirect product* $W_{\sigma,\lambda} = W'R$ *with* $W'$ *normal, and the set of all operators* $\sigma(r)\mathcal{A}(r,\sigma,\lambda)$ *for* $r$ *in* $R$ *is linearly independent.*

*Conjecture 2.* *In every case the intertwining operators for* $(\sigma,\lambda)$ *do span the space of bounded operators* $L$ *such that* $U(\sigma,\lambda)L = LU(\sigma,\lambda)$.

The first conjecture if true is a sufficiently precise answer to problem (1) provided the subgroups $W'$ and $R$ are defined explicitly enough.

As evidence for these conjectures we have the following new results:

(1a) a proof of Conjecture 1, together with an explicit description of $W'$ and $R$, for the case that $G$ is split over $\mathbb{R}$ and $\sigma$ and $\lambda$ are arbitrary. See §2.

(1b) a proof of part of Conjecture 1 for general $G$. In this case the subgroups $W'$ and $R$ are not defined explicitly enough to provide a useful solution to problem (1). See the end of §2.

(2) a proof of Conjecture 2 when $G$ has real-rank 2 and when $G = SL(n,\mathbb{R})$. The method in these cases applies to other groups as well; no group is known for which it fails. However, we as yet do not have an argument that works simultaneously for all $G$. See §3 and §4.

We should mention that the solution (1b) shows that the basic intertwining operators $\sigma(r)\mathcal{A}(r,\sigma,\lambda)$ are those whose normalizing factors (see §18 of [6]) are regular at $\lambda$. If Conjecture 2 is true, it would appear that the order of $R$ (and hence the decision between reducibility and irreducibility) could be expressed in terms of the Plancherel measure and similar quantities (cf. Theorem 5 of [6]).

Other authors have worked on the problem of deciding irreducibility of principal series. In addition to [6], one should consult Gelfand-Graev [4], Bruhat [3], Kostant [7], Helgason [5], Zelobenko [12], and Wallach [11].

## 2. Operators when $G$ is split over $\mathbb{R}$

In this section we assume that the group $G$, satisfying the conditions of §1, has the further property of being split over $\mathbb{R}$. Then the Lie algebra $\mathfrak{A}$ of $A$ is a Cartan subalgebra, $\mathfrak{m}$ is $0$, and $M$ is a finite abelian group. To simplify the exposition, we shall assume that $G$ is simple, so that $\mathfrak{G}$ is completely determined by one of the standard Dynkin diagrams.

From work of Satake [8, p. 93], for example, $M$ is completely understood. For each root $\alpha$ let $H_\alpha$ be the member of $\mathfrak{A}$ corresponding to $\alpha$ and let $H'_\alpha = 2\langle\alpha,\alpha\rangle^{-1}H_\alpha$. Set $\gamma_\alpha = \exp \pi i H'_\alpha$. The element $\gamma_\alpha$ is in $M$ and has order 1 or 2. Let $\epsilon_1,\ldots,\epsilon_n$ be the simple roots. Then the elements $\gamma_{\epsilon_i}$, $1 \leq i \leq n$, generate $M$.

Let $\sigma$ be an irreducible unitary representation of $M$. Then $\sigma$ is one-dimensional and $\sigma(\gamma_\alpha) = \pm 1$ for each root $\alpha$. In view of the remarks above, $\sigma$ is completely determined by specifying which simple roots $\epsilon_i$ satisfy $\sigma(\gamma_{\epsilon_i}) = -1$.

As in §1, the group $M'$ acts on the representations $\sigma$ and characters $\lambda$. Since $\sigma$ is one-dimensional, equivalence becomes identity, and the action of $M'$ reduces to an action of $W$. Then our concern is with the subgroup $W_{\sigma,\lambda}$ of elements of $W$ that leave $\sigma$ and $\lambda$ fixed. Again since $\sigma$ is one-dimensional, we can disregard $\sigma(w)$ in (1.3). By (1.3) the operators $\mathcal{A}(w,\sigma,\lambda)$ for $w$ in $W_{\sigma,\lambda}$ commute with the principal series representation $U(\sigma,\lambda,x)$. Normally many of these operators coincide. Our problem in this section is to give an explicit description of the distinct operators in the set. The case that $\sigma$ is trivial is of no interest for the problem since $\mathcal{A}(w,1,\lambda)$ is easily seen to be scalar if $w$ is in $W_{1,\lambda}$.

For simplicity we shall assume until after Theorem 1 that the character $\lambda$ of $A$ is trivial. Let $W_\sigma = W_{\sigma,1}$.

We say that the representation $\sigma$ of $M$ is fundamental if there is exactly one simple root $\epsilon_k$ such that $\sigma(\gamma_{\epsilon_k}) = -1$. In this case we write $\sigma = \sigma_k$. The condition that $\sigma$ be fundamental for $G$ is a mod 2 analog of the condition for $G^{\mathbb{C}}$ that an integral form be dominant.

Proposition 2.1. Each nontrivial representation $\sigma$ of $M$ is equivalent under $W$ with a fundamental representation.

That is, there is a $p$ in $W$ and there is a $k$ such that $p\sigma = \sigma_k$. Now it is shown in [6] that $\mathcal{A}(p,\sigma,1)$ is a unitary equivalence of $U(\sigma,1)$ and $U(\sigma_k,1)$, and it is easy to see from Theorem 7 of [6] that $\mathcal{A}(p,\sigma,1)$ conjugates the intertwining operators $\{\mathcal{A}(w,\sigma,1) \mid w \in W_\sigma\}$ into

$$\{\mathcal{A}(w',\sigma_k,1) \mid w' \in W_{\sigma_k} = p\, W_\sigma\, p^{-1}\}.$$

Thus if we characterize the intertwining operators for $\sigma_k$, we have characterized them for $\sigma$. So for the rest of the section we assume $\sigma$ is fundamental. Say $\sigma = \sigma_k$.

Let $\Delta$ and $\Pi$ be, respectively, the roots and simple roots for $G$. Consider the following conditions on a positive non-simple root $\alpha$:

(i) $\sigma(\gamma_\alpha) = 1$.

(ii) $\sigma(\gamma_\beta) = -1$ for every $\beta > 0$ different from $\alpha$ such that $p_\alpha\beta < 0$.

(ii') $\langle \alpha, \epsilon_i \rangle \leq 0$ for $i \neq k$.

Here (ii) implies (ii'). [In fact, if $\langle\alpha,\epsilon_i\rangle > 0$, then $p_\alpha \epsilon_i < 0$ and (ii) gives $\sigma(\gamma_{\epsilon_i}) = -1$. But $\sigma = \sigma_k$ and $i \neq k$ imply $\sigma(\gamma_{\epsilon_i}) = 1$.]

Lemma 2.2. *There is at most one positive non-simple root* $\alpha$ *satisfying both (i) and (ii).*

We shall define a new root system $\Delta'$ in terms of $\alpha$. If there is no $\alpha$ in Lemma 1, let $\Pi'$ consist of the $\epsilon_i$ for $i \neq k$. If $\alpha$ does exist, let $\Pi'$ consist of $\alpha$ and the $\epsilon_i$ for $i \neq k$. Let $\Delta'$ be the subset of $\Delta$ generated by $\Pi'$ and the Weyl group reflections corresponding to members of $\Pi'$. By (ii') $\Delta'$ is a root system in which $\Pi'$ can be taken as the set of simple roots. The Dynkin diagram of $\Pi'$ we shall call the *$\alpha$-diagram* of G and $\sigma_k$. Ordinarily the $\alpha$-diagram is not connected.

Computation of $\alpha$ in examples is simplified by the following lemma.

Lemma 2.3. *The least positive* $\alpha$ *in* $\Delta$ *satisfying (i) and (ii') satisfies (ii).*

Let $W(\Delta')$ be the Weyl group for the root system $\Delta'$. One can show that $W(\Delta') \subseteq W_\sigma$. Let $R_\sigma$ be the subgroup of members $w$ of $W_\sigma$ such that $w(\Pi') \subseteq \Pi'$. Each element of $R_\sigma$ defines an automorphism of the $\alpha$-diagram of G and $\sigma$. If $\alpha$ exists, distinct members of $R_\sigma$ lead to distinct automorphisms. In this case, in particular, if the $\alpha$-diagram admits no nontrivial automorphism, then $R_\sigma = \{1\}$.

Theorem 1. *$W_\sigma$ is the semidirect product $W_\sigma = W(\Delta')R_\sigma$ with $W(\Delta')$ normal. $W(\Delta')$ is the subgroup of $W_\sigma$ on which $\mathcal{A}(w,\sigma,1)$ is scalar. Consequently if $w = w_1 r$ is the decomposition of a member of $W_\sigma$ according to the semidirect product, then $\mathcal{A}(w,\sigma,1) = c\,\mathcal{A}(r,\sigma,1)$ for a scalar $c$ of modulus one. Moreover, the set of all operators $\mathcal{A}(r,\sigma,1)$ for $r$ in $R_\sigma$ is linearly independent.*

With a case-by-case argument, one can check that $|R_\sigma| = 1, 2$, or 4. The case $|R_\sigma| = 4$ occurs only for $G$ of type $D_n$ with $n$ even, and when $|R_\sigma| = 4$, $R_\sigma$ is $\mathbb{Z}_2 \oplus \mathbb{Z}_2$. If $G$ is of type $A_n$, $|R_\sigma| = 1$ or 2; this case is discussed in §4. For $G$ of type $B_n$ or $C_n$, $|R_\sigma|$ can be 1 or 2. But in $F_4$ and $G_2$, $|R_\sigma| = 1$ for all $\sigma$. In $E_7$ there is a fundamental $\sigma$ for which $|R_\sigma| = 2$.

For an explicit example, take $G$ of type $C_n$. In standard notation the simple roots are

$$\epsilon_1 = e_1 - e_2,\ \epsilon_2 = e_2 - e_3,\ \ldots,\ \epsilon_{n-1} = e_{n-1} - e_n,\ \epsilon_n = 2e_n.$$

Choose $\sigma = \sigma_n$. Then $\alpha = e_{n-1} + e_n$, and

$$\Pi' = \{\epsilon_1, \epsilon_2, \ldots, \epsilon_{n-1}, \alpha\}.$$

The $\alpha$-diagram of $\sigma$ is of type $D_n$, and $R_\sigma = \{1, p_n\}$.

We pass to the case of general $\sigma$ and $\lambda$. If $(\sigma, \lambda)$ is given, we first apply to $(\sigma, \lambda)$ a member of $W$ that makes $\lambda$ dominant. After this change it is quite easy to check that $W_{1,\lambda}$ is the Weyl group generated by the simple reflections that fix $\lambda$. These simple reflections correspond to a Dynkin diagram and fall into components corresponding to the components of the Dynkin diagram. We then operate with a second member of $W$, this one in $W_{1,\lambda}$, to make $\sigma$ fundamental on each component. Let $\Delta'$ be the roots corresponding to the union of the $\alpha$-diagrams for the components, and let $\Pi'$ be the corresponding simple system. Let

$$R_{\sigma,\lambda} = \{w \in W_{\sigma,\lambda} \mid w(\Pi') \subseteq \Pi'\}.$$

<u>Theorem 2</u>. $W_{\sigma,\lambda}$ <u>is the semidirect product</u> $W_{\sigma,\lambda} = W(\Delta')R_{\sigma,\lambda}$ <u>with</u> $W(\Delta')$ <u>normal.</u> $W(\Delta')$ <u>is the subgroup of</u> $W_{\sigma,\lambda}$ <u>on which</u> $\mathcal{A}(w,\sigma,\lambda)$ <u>is scalar. Consequently if</u> $w = w_1 r$ <u>is the decomposition of a member of</u> $W_\sigma$ <u>according to the semidirect product, then</u> $\mathcal{A}(w,\sigma,\lambda) = c\,\mathcal{A}(r,\sigma,1)$ <u>for a scalar</u> $c$ <u>of modulus one. Moreover,</u>

<u>the set of all operators</u> $\mathcal{A}(r,\sigma,\lambda)$ <u>for</u> $r$ <u>in</u> $R_{\sigma,\lambda}$ <u>is linearly independent.</u>

An example of this situation is in §4. In any event, one can use the form of $R_\sigma = R_{\sigma,1}$ to show that $R_{\sigma,\lambda}$ is a subgroup of a (perhaps large) direct sum of copies of $\mathbb{Z}_2$ and therefore itself is a direct sum of copies of $\mathbb{Z}_2$. This completely settles Conjecture 1 for $G$ split over $\mathbb{R}$.

For the case of a general $G$ not necessarily split over $\mathbb{R}$, one can prove an analog of Theorem 2 but without a satisfactory description of $\Delta'$ and $R_{\sigma,\lambda}$. To do so, let

$$\Delta' = \{\beta \mid p_\beta \in W_{\sigma,\lambda} \text{ and } \sigma(p_\beta)\mathcal{A}(p_\beta,\sigma,\lambda) \text{ is scalar}\},$$

where $p_\beta$ is the reflection relative to $\beta$, and let

$$R_{\sigma,\lambda} = \{p \in W_{\sigma,\lambda} \mid p\beta > 0 \text{ for every } \beta > 0 \text{ in } \Delta'\}.$$

It is easy to see that $\Delta'$ is a root system and thus has a Weyl group $W(\Delta')$. Once again we have the semidirect product decomposition $W_{\sigma,\lambda} = W(\Delta')R_{\sigma,\lambda}$ with $W(\Delta')$ the normal subgroup corresponding to trivial operators and with the operators $\sigma(r)\mathcal{A}(r,\sigma,\lambda)$ for $r$ in $R_{\sigma,\lambda}$ independent. This result is considerably easier to prove than Theorems 1 and 2. However, it has the shortcoming that $\Delta'$ is defined in such a way that one can prove rather little about $R_{\sigma,\lambda}$. These facts make clearer the thrust of Theorems 1 and 2, namely the possibility of conjugating $\sigma$ and $\lambda$ suitably so that the simple roots of $\Delta'$ can be expressed readily in terms of $\Pi$ and special roots $\alpha$.

## 3. Completeness theorem for G of real-rank two

Return to the notation of §1. Fix an irreducible unitary representation $\sigma$ of $M$ and a unitary character $\lambda$ of $A$. If $w$ is a member of $M'$ for which $\sigma$ is equivalent with $w\sigma$, then the operator $\sigma(w)\mathcal{A}(w,\sigma,\lambda)$ is defined. Here $\sigma(w)$ is determined up to a scalar factor; once a choice is made for the scalar, the operator depends only on the coset of $w$ in $W = M'/M$. We make such a choice of scalars for each member of $W_{\sigma,\lambda}$ without imposing any consistency conditions on the different choices. Then we can speak unambiguously of the linear span of the operators $\sigma(w)\mathcal{A}(w,\sigma,\lambda)$ for $w$ in $W_{\sigma,\lambda}$.

Theorem 3. *Let G be of real-rank two. For any $(\sigma,\lambda)$, the linear span of the operators $\sigma(w)\mathcal{A}(w,\sigma,\lambda)$ for $w$ in $W_{\sigma,\lambda}$ is the set of all bounded linear operators $L$ such that $U(\sigma,\lambda)L = LU(\sigma,\lambda)$.*

Consequently $U(\sigma,\lambda)$ is irreducible if and only if all the operators $\sigma(w)\mathcal{A}(w,\sigma,\lambda)$ are scalar. The detailed proof of the theorem shows exactly which operators $\sigma(w)\mathcal{A}(w,\sigma,\lambda)$ are scalar.

The exposition will be simpler if we sketch the proof for a particular case and then describe the extent to which the general case differs from the special case. Temporarily take $G$ to be the real symplectic group $Sp(2,\mathbb{R})$. Let $\epsilon_1$ be the shorter simple root and $\epsilon_2$ be the longer one. In the notation of §2, let $\sigma = \sigma_2$ and $\lambda = 1$. Then $W_{\sigma,\lambda} = W_\sigma$ is all of $W$ and has order 8. So

$$W_\sigma = \{1,\ p_1,\ p_2,\ p_1p_2,\ p_2p_1,\ p_1p_2p_1,\ p_2p_1p_2,\ p_1p_2p_1p_2\}.$$

Recall the Bruhat decomposition of $G$: The MAN double cosets are in one-to-one correspondence with the elements of $W$, and $G = \bigcup MANwMAN$ with the union over a system of representatives of the cosets of $M'$ modulo $M$. Let $C(w)$ be the double coset corresponding to $w$.

Let $\sigma$ operate in the space $E^\sigma$, let $C^{\sigma,1}$ be the subspace of functions in $C^\infty(G,E^\sigma)$ that lie in the representation space for $U(\sigma,1)$, and let $\pi_{\sigma,1}$ be the standard mapping of $C^\infty_{com}(G,E^\sigma)$ onto $C^{\sigma,1}$, given by the case $\lambda = 1$ of

$$(3.1)\quad (\pi_{\sigma,\lambda}f)(x) = \int_{MAN} e^{-\rho H(\xi)}\lambda(\exp H(\xi))^{-1}\sigma(m(\xi))^{-1}f(\xi x)d_r\xi.$$

If $L$ is a bounded operator commuting with $U(\sigma,1)$, then the mapping $\delta_L$ given by

$$(3.2)\qquad f \to L(\pi_{\sigma,1}f)(1)$$

is an $End(E_\sigma)$-valued distribution on $G$. (See [1].)

Bruhat [3] examined $\delta_L$ and found that it satisfies a functional equation under translation on the right and left by MAN. For $w$ in $W$ he showed essentially that if

(i) $\delta_L$ vanishes on $C(w')$ for each $w' \neq w$ such that $\overline{C(w')} \supseteq C(w)$ and

(ii) $\delta_L$ does not vanish on $C(w)$,

then

(i') $w$ is in $W_{\sigma,1}$ (here this conclusion is empty)

(ii') the component of $\delta_L$ transverse to $C(w)$ vanishes

(iii') the restriction of $\delta_L$ to $C(w)$ is a multiple of a distribution $\delta_w$ that is independent of $L$.

The idea of the proof is to construct a "pseudo-operator" $T(w)$ for each $w$ in $W_\sigma$ such that the main part of the distribution corresponding to $T(w)$ is $\delta_w$. Subtracting from $L$ a suitable linear combination of the $T(w)$, we obtain an operator whose distribution vanishes on certain double cosets and satisfies a functional equation reflecting properties of both $L$ and the $T(w)$'s. This functional equation will have no solutions unless each $T(w)$ that was subtracted from L is already one of the operators $\sigma(w)\mathcal{A}(w,\sigma,\lambda)$. Consequently $L$ is in the span of the operators $\sigma(w)\mathcal{A}(w,\sigma,\lambda)$.

We shall describe parts of the proof in more detail, but first we must define the pseudo-operators $T(w)$. Let $w$ be in $W_\sigma$, and let $\Delta_w$ be the set of all positive roots $\alpha$ such that $w\alpha < 0$ and

$\sigma(\gamma_\alpha) = +1$. If we regard $\lambda$ as a variable on $A$ or its Lie algebra, we can speak of differentiation $D_\alpha$ of a function of $\lambda$ with respect to the vector $\alpha$. Define the <u>pseudo-operator</u> $T(w)$ by

$$T(w) = \Big(\prod_{\alpha\in\Delta_w} D_\alpha\Big)\, \mathcal{A}(w,\sigma,\lambda)\,\Big|_{\lambda=1}.$$

This operator is well-defined if we use the compact picture for the induced representation. It has three main properties:

(1) It maps $C^{\sigma,1}$ into itself.

(2) It satisfies an obvious functional equation. This equation comes by applying the differential operator $\Pi D_\alpha$ to both sides of (1.1), using the Liebritz rule for differentiating products and using a formula relating $U(\sigma,\lambda)$ to $U(\sigma,1)$ in the compact picture. The result is of the form

$$U(\sigma,1,x)T(w) = T(w)U(\sigma,1,x) + \text{remainder terms.} \tag{3.3}$$

(3) Its distribution, defined in analogy with (3.2), coincides with a nonzero multiple of $\delta_w$ on the union of all double cosets of dimension $\geq \dim C(w)$.

There is a helpful (though slightly inaccurate) notation for dealing with these operators. We write formally

$T(p_1) = T_1$ $\qquad\qquad$ $T(p_1p_2p_1) = T_1H_2T_1$

$T(p_2) = H_2$ $\qquad\qquad$ $T(p_2p_1p_2) = H_2T_1H_2$

$T(p_1p_2) = T_1H_2$ $\qquad\qquad$ $T(p_1p_2p_1p_2) = T_1H_2T_1H_2$.

$T(p_2p_1) = H_2T_1$

Each $T(w)$ is written as a product of $T$'s and $H$'s, with the subscripts matching those in a minimal decomposition of $w$ into the product of simple reflections. We use $T$ or $H$ in the kth factor according as the kth operator $\mathcal{A}$ in the expansion of $\mathcal{A}(w,\sigma,\lambda)$ by (1.2) is or is not, respectively, scalar for $\lambda = 1$. [The notation here is meant to suggest that $T(w)$ is formally the product of rank-one pseudo-operators, $T$ being a rank-one pseudo-identity and $H$ being a rank-one Hilbert transform. Actually the product of these operators is not exactly equal to $T(w)$, but it is equal in first approximation, in the sense that the main part of the distribution for the product operator is $\delta_w$.]

Form $\delta_L$ as in equations (3.2) and (3.1), and consider the open double coset $C(p_1p_2p_1p_2)$. In view of property (3) of $T(w)$, we can find a constant $c$ such that the distribution $\nu$ of $L - cT(p_1p_2p_1p_2)$ vanishes on $C(p_1p_2p_1p_2)$. Combining (3.3) and the commutativity of $L$ with $U(\sigma,\lambda)$, we see that $L - cT(p_1p_2p_1p_2)$ satisfies an analog of (3.3). It follows that $\nu$ satisfies a functional equation under right translation by $MAN$. Also property (1) of $T(w)$ implies that $\nu$ satisfies another functional equation under left translation by $MAN$.

It turns out that the main contribution to $\nu$ is on $C(p_2p_1p_2)$, and one can show from the functional equation that $\nu$ has no transverse derivatives to this double coset. It follows readily that the restriction of $\nu$ to $C(p_2p_1p_2)$ makes sense and is a function. Evaluating this function at a representative $w'$ of $p_2p_1p_2$ and using the functional equation for right translation by $a$ in $A$ and left translation by $wa^{-1}w^{-1}$, we are led to the equality of a bounded expression and $c\alpha(\log a)$ for a certain root $\alpha$. Consequently $c = 0$ and $\delta_L$ vanishes on $C(p_1p_2p_1p_2)$.

Next we attempt to show that $\delta_L$ vanishes on $C(p_1p_2p_1)$ and $C(p_2p_1p_2)$. (The transverse derivatives to these double cosets must vanish, according to [3].) We form

$$L - c_1T(p_1p_2p_1) - c_2T(p_2p_1p_2)$$

and argue similarly. The conclusion $c_1 = 0$ will come immediately from considering $C(p_1p_2)$, but the conclusion $c_2 = 0$ will come only later by considering $C(1)$ after $L$ has been handled on the double cosets that lie between $C(p_2p_1p_2)$ and $C(1)$.

The argument continues in this way, with $L$ adjusted on double cosets of lower and lower dimension. The details are cumbersome to list, but we can say the following. For each $w$ we push $L$ off $C(w)$ by using $T(w)$. The contradiction that eliminates $T(w)$ comes from the double coset corresponding to the formal expansion of $T(w)$ in $H$'s and $T$'s, but with one $T$ deleted. The root in the final equation that gives the contradiction is obtained as follows: If the factor $p_{i_k}$ is deleted from $p_{i_1} \cdots p_{i_n}$, then the root is $p_{i_n} \cdots p_{i_{k+1}} \varepsilon_{i_k}$. It is possible for a linear combination of as many as two roots to appear in the final equation, but these roots will be distinct and hence independent.

In the end, $L$ will be expressed as a linear combination of $T(p_2)$ and $T(1)$, that is, of $H_2$ and $I$. These are the operators $\mathcal{A}(w,\sigma,1)$ for $w = p_2$ and $w = 1$, with no differentiations, and the theorem is proved for this special $G$ and $\sigma$.

Now consider the case of general $G$ of real-rank two. In view of the results in [6] concerning the real-rank one case, we may assume that $G$ is simple. We are given $\sigma$ and $\lambda$, but the same kind of argument as after Proposition 2.1 shows that there is no loss of generality in taking $\lambda$ dominant. Observe that $W_{\sigma,\lambda} = W_{\sigma,1} \cap W_{1,\lambda}$. With $\lambda$ dominant, $W_{1,\lambda}$ is generated by the simple reflections that it contains. If there are no simple reflections in $W_{1,\lambda}$, then $W_{1,\lambda} = \{1\}$, and the the theorem follows from Bruhat's results [3]. If there is one simple reflection, the problem is substantially a rank-one problem and is handled by a simpler version of the argument to follow. If both simple reflections are in $W_{1,\lambda}$, then $\lambda = 1$; this is the only hard case.

Thus suppose $\lambda = 1$. Call a restricted root $\alpha > 0$ <u>primitive</u> if $\alpha/2$ is not a restricted root. Fix $w$ in $W_{\sigma,1}$ and consider the primitive restricted roots $\alpha > 0$ such that $w\alpha < 0$. Recall from §18 of [6] that the restriction $\sigma|_{M_\alpha}$ is equivalent with a multiple of a single irreducible representation $\sigma_\alpha$ of $M_\alpha$. We shall say that a primitive $\alpha$ with $w\alpha < 0$ is in $\Delta_w$ if the rank-one intertwining operator $\sigma_\alpha(p_\alpha)\mathcal{A}_\alpha(p_\alpha,\sigma_\alpha,1)$ is scalar. (A necessary and sufficient condition for this is given in Theorem 5 of [6].) Define the <u>pseudo-operator</u> $T(w)$ by

$$T(w) = \Big(\prod_{\alpha\in\Delta_w} D_\alpha\Big)\sigma(w)\mathcal{A}(w,\sigma,\lambda)\Big|_{\lambda=1},$$

with notation as in the case of $Sp(2,\mathbb{R})$. Just as in $Sp(2,\mathbb{R})$ it is convenient to have symbolic notation for $T(w)$. If $w$ decomposes minimally as $w = p_{i_1}\cdots p_{i_n}$, we write $T(w)$ formally as a product of T's and H's, using $T_{i_j}$ at the jth stage if the associated root $p_{i_n}\cdots p_{i_{j+1}}\epsilon_{i_j}$ is in $\Delta_w$ and using $H_{i_j}$ otherwise. (Again we are thinking of $T(w)$ as a product of pseudo-identities T and nontrivial rank-one operators H, and again this is only an approximation. In this general case, the symbol $H_i$ is standing for both an operator and its inverse, and each H in a string must be interpreted suitably.)

With this notation we describe $W_{\sigma,1}$ and the associated pseudo-operators. A case-by-case check using the results of [10] shows that we can conjugate $\sigma$ by a member of $M'$ in order to arrive at one of the following situations:

(1) $A_2$ as restricted root diagram. Here $|W| = 6$. $W_\sigma$ can be any of $\{1\}$, $\{1,p_1\}$, $\{1,p_2\}$, W. In all cases each $T(w)$ is formally a product of T's.

(2) $G_2$ as restricted root diagram. Here $|W| = 12$. $W_\sigma$ can be $\{1\}$, $\{1,p_1\}$, $\{1,p_2\}$, or W with each $T(w)$ a product of T's.

Alternatively $W_\sigma$ can be $\{1, p_1, p_2p_1p_2p_1p_2, p_1p_2p_1p_2p_1p_2\}$ with formally $T(p_1) = H_1$, $T(p_2p_1p_2p_1p_2) = H_2H_1T_2H_1H_2$, and the other $T(w)$ given as the product.

(3) $BC_2$ as restricted root diagram. Here $|W| = 8$. Let $\epsilon_1$ and $\epsilon_2$ be the simple restricted roots, with $\epsilon_1$ longer than $\epsilon_2$. Then $2\epsilon_2$ is a restricted root. There are two possibilities for $W_\sigma$.

(a) $W_\sigma = W$. Then $T(p_1) = T_1$, $T(p_2) = H_2$ or $T_2$, and the other $T(w)$'s are given as products.

(b) $W_\sigma = \{1, p_2, p_1p_2p_1, p_2p_1p_2p_1\}$. Then $T(p_2) = H_2$ or $T_2$ and, independently, $T(p_1p_2p_1) = H_1T_2H_1$ or $H_1H_2H_1$. The other $T(w)$ is given as the product.

(4) $B_2$ as restricted root diagram. Here $|W| = 8$. Let $\epsilon_1$ and $\epsilon_2$ be the simple restricted roots, with $\epsilon_1$ shorter then $\epsilon_2$. Then $W_\sigma$ and the pseudo-operators can be as in (3a) and (3b) above, or else $W_\sigma$ can be $\{1\}$, $\{1,p_1\}$, or $\{1,p_2\}$ with only T's occurring.

We need one more fact. This is a result due to Steinberg [9, p. 127] for Chevalley groups, and to Borel and Tits [2] in the general case. Namely let $w = p_{i_1} \cdots p_{i_n}$ be a minimal decomposition into simple reflections. Then the closure of $C(w)$ is the union of the double cosets $C(w')$ as $w'$ runs over all products (in order) of subsets of the $p_{i_j}$.

Putting this fact and the detailed description of the pseudo-operators together, one sees that a simple modification of the completeness argument given for $Sp(2,\mathbb{R})$ and the special $\sigma$ works for all $G$ of real-rank two and all $\sigma$.

## 4. Completeness theorem for $SL(n,\mathbb{R})$

For $SL(n,\mathbb{R})$ the completeness theorem is as follows.

Theorem 4. Let $G = SL(n,\mathbb{R})$. For any $(\sigma,\lambda)$ the linear span

<u>of the operators</u> $\mathcal{A}(w,\sigma,\lambda)$ <u>for</u> $w$ <u>in</u> $W_{\sigma,\lambda}$ <u>is the set of all bounded linear operators</u> $L$ <u>such that</u> $U(\sigma,\lambda)L = LU(\sigma,\lambda)$.

As we shall see, this linear span has dimension 1 or 2. Dimension 1 is necessary and sufficient for irreducibility. Partial results on irreducibility for $SL(n,\mathbb{R})$ were known already. Gelfand and Graev [4] settled $n$ odd, and Wallach [11] proved the irreducibility of $U(\sigma,\lambda)$ when all $\mathcal{A}(w,\sigma,1)$ for $w$ in $W_{\sigma,1}$ are scalar.

Before commenting on the proof, we introduce notation. In $SL(n,\mathbb{R})$, we shall take $M$ to be diagonal matrices with $\pm 1$ in the diagonal entries and $A$ to be diagonal matrices with positive diagonal entries. The Weyl group $W$ is the permutation group on $n$ letters, and it operates by permuting the diagonal entries.

Turning to the proof, we may without loss of generality deal with a convenient image of $(\sigma,\lambda)$ under the operation of $W$. First conjugate $(\sigma,\lambda)$ so that $\lambda$ is dominant. The effect of this is to decompose $\{1,\ldots,n\}$ into disjoint strings of consecutive integers in such a way that the members of $W_{1,\lambda}$ are exactly the permutations that leave each string stable. Next, $\sigma$ is given as a product of certain signs of diagonal entries, and we make a further conjugation leaving each string stable so that the signs that are used within each string occur consecutively at the beginning of the string.

With $(\sigma,\lambda)$ in this form, one can check that there are only two possibilities:

(1) $W_{\sigma,\lambda}$ is a direct product of smaller permutation groups, and each $T(w)$ for $w$ in $W_{\sigma,\lambda}$ is formally the product of $T$ operators only.

(2) $n$ is even, and exactly the first half of the entries in each $\lambda$-string obtained above is used in computing $\sigma$. In this case $W_{\sigma,\lambda}$ has a subgroup $W'$ of index 2 that is a direct product of

smaller permutation groups. Write $W_{\sigma,\lambda} = W' \cup w_0 W'$, where $w_0$ is the element of shortest length in the nontrivial coset. Then $T(w_0)$ is a product of $H$ operators only, and therefore $\mathcal{A}(w_0,\sigma,\lambda)$ is not scalar by property (iii) of $T(w_0)$. Say, $\mathcal{A}(w_0,\sigma,\lambda) = H_0$. In addition, each $T(w)$ for $w$ in $W'$ is a product of $T$'s, and each $T(w)$ for $w$ in $w_0 W'$ is the product of $H_0$ by a product of $T$'s.

The rest of the proof proceeds along the lines of §3. The only additional thing that is needed is an algebraic result to ensure that if Theorem 3 fails, then the functional equation satisfied by the difference of $L$ and its first approximation is actually contradictory. This result is given as Proposition 4.2.

Let $W$ be a Weyl group, and let $\ell(w)$ be the length of the element $w$ of $W$. If $p$ and $q$ are members of $W$, we shall say that $p$ is a parent of the child $q$ if $\ell(p) = \ell(q)+1$ and if $q$ can be obtained from some (or any, in view of Steinberg's result mentioned in §3) minimal decomposition of $p$ by striking out one of the simple-reflection factors. In this case it is a simple matter to see that $p = qw_\alpha$ for some root-reflection $w_\alpha$. We write $\alpha = \alpha_{p,q}$. For fixed $q$, let $P_q$ be the set of parents of $q$.

Lemma 4.1. *Let $W$ be any Weyl group, let $p$ and $q$ be members of $W$ with $\ell(p) = \ell(q)+1$, and suppose $p = qw_\alpha$ for some $\alpha$. Then $p$ is a parent of the child $q$.*

Proposition 4.2. *Let $W$ be the Weyl group of $SL(n,\mathbb{R})$, and fix a length $\ell$. Suppose that to each element $p$ in $W$ of length $\ell$ is associated a complex number $c_p$ in such a way that the set $\{c_p\}$ satisfies*

$$\sum_{p \in P_q} c_p \alpha_{p,q} = 0$$

*for all $q$ in $W$ of length $\ell-1$. Then all the $c_p$ are equal to 0.*

## References

[1] R. J. Blattner, On induced representations, Amer. J. Math. 83 (1961), 79-98.

[2] A. Borel and J. Tits, Compléments à l'article: Groupes reducitifs, preprint.

[3] F. Bruhat, Sur les représentations induites des groupes de Lie, Bull. Soc. Math. France 84 (1956), 97-205.

[4] I. M. Gelfand and M. I. Graev, Unitary representations of the real unimodular group (principal nondegenerate series), Translations Amer. Math. Soc. (2) 2 (1956), 147-205.

[5] S. Helgason, A duality for symmetric spaces with applications to group representations, Advances in Math. 5 (1970), 1-154.

[6] A. W. Knapp and E. M. Stein, Intertwining operators for semi-simple groups, Annals of Math. 93 (1971), 489-578.

[7] B. Kostant, On the existence and irreducibility of certain series of representations, Bull. Amer. Math. Soc. 75 (1969), 627-642.

[8] I. Satake, On representations and compactifications of symmetric Riemannian spaces, Annals of Math. 71 (1960), 77-110.

[9] R. Steinberg, Lectures on Chevalley groups, Yale University, 1967.

[10] J. Tits, Classification of algebraic semisimple groups, Proc. Symposia in Pure Math. IX, Algebraic Groups and Discontinuous Subgroups, Amer. Math. Soc., Providence, 1966, 33-62.

[11] N. Wallach, Cyclic vectors and irreducibility for principal series representations, Trans. Amer. Math. Soc. 158 (1971), 107-113.

[12] D. Zelobenko, The analysis of irreducibility in a class of elementary representations of complex semi-simple Lie groups, Isvestia 2 (1968), 108-133.

# CALCUL SYMBOLIQUE SUR L'ALGÈBRE DE WIENER CENTRALE DE  SU(3)

par

Marie-Paule et Paul Malliavin

University of Paris

Le calcul symbolique sur l'algèbre de Wiener et sur des algèbres dérivées a joué dans les quinze dernières années un rôle de premier plan dans le développement de l'analyse harmonique commutative [4,6]. On se propose dans ce travail de montrer que dans le cas semi-simple central les résultats sont profondément différents.  Un résultat de Noël Leblanc [5] avait déjà révélé ce fait pour la structure algébrique duale de celle utilisée ici.

## 1.  Notations et résultats

G désigne un groupe compact, $D_G$ les classes de représentations irréductibles de  G.  Si  $\lambda \in D_G$  on note $U^\lambda$  la représentation correspondante,  $d_\lambda$  la dimension de l'espace où la représentation  $U^\lambda$  est définie,  $\chi_\lambda$  le caractère de la représentation  $\lambda$.  Pour  $f \in L^1(G)$ on pose

$$c_\lambda(f) \ = \ \int U^\lambda_{g^{-1}} f(g) \ dg$$

$$A(G) \ = \ \{f; \ \textstyle\sum_\lambda d_\lambda |c_\lambda(f)|\}$$

où

$$|a| \ = \ \text{Trace } (a^* a)^{1/2}.$$

Alors  $A(G)$  est une algèbre pour la multiplication des fonctions (cf. [2]).  On note  $A_c(G)$  la sous-algèbre de  $A(G)$  constituée par les fonctions centrales.  Alors

$$A_c(G) \ = \ \{f = \textstyle\sum_\lambda a_\lambda \chi_\lambda \ \text{ où } \ \sum d_\lambda |a_\lambda| < \infty\}.$$

Si  F  est une fonction définie sur un intervalle  I  de  $\underline{R}$  à valeurs dans  $\underline{C}$, on dit que  F  opère sur  $A(G)$  si  $F \circ f \in A(G)$  pour toute  $f \in A(G)$  telle que  $f(G) \subset I$.  On a alors:

1.1.  **Théorème**. _Si_  G  _est un groupe de Lie et si_  F  _opère sur_  $A(G)$ _alors_  F  _est une fonction analytique au voisinage de_  I.

Nous démontrerons dans ce travail le résultat:

1.2.  **Théorème**.  _Prenons_  G = SU(3);  _si_  F  _est de classe_ $C^k$, _alors_ F _opère sur_  $A_c(G)$  _dès que_  $k \geq 6$.

Ce résultat reste vrai pour tout groupe de Lie semi-simple compact en prenant k assez grand. La démonstration dans le cas de SU(2) est triviale. Le cas de SU(3) est le cas le plus simple où l'on rencontre les difficultés qui apparaissent dans le cas général.

La démonstration du théorème 1.1. résulte du résultat suivant de Carl S. Herz [1]:

Si H est un sous-groupe de G alors

$$A(H) \simeq A(G)/J$$

(où J est l'idéal des fonctions nulles sur H).

Prenons alors pour H un tore à une dimension, le résultat 1.1. résulte alors du même résultat pour $\underline{\underline{T}}$ (Y. Katznelson [4]).

La démonstration du théorème 1.2. dépendra de deux parties. D'abord de techniques de développment Taylorien incomplet introduites dans le calcul symbolique par Noël Leblanc [5], ensuite d'estimations quantitatives des coefficients de Clebsch-Gordan de SU(3) qui seront calculés à partir des formules de Freudenthal et Kostant.

La structure multiplicative de $A_c(G)$ est déterminée par les entiers constituant les coefficients de Clebsch-Gordan $c_\mu(\lambda,\lambda')$ définis par l'identité

$$\chi_\lambda \chi_{\lambda'} = \sum_\mu c_\mu(\lambda,\lambda')\chi_\mu .$$

On note

$$\lambda \circ \lambda' = \{\mu \in D_G;\ c_\mu(\lambda,\lambda') \neq 0\} .$$

Si G est abélien alors, $\lambda \circ \lambda'$ ne compte qu'un seul point; à l'opposé la semi-simplicité de G se traduira par le fait que le cardinal de $\lambda \circ \lambda'$ sera grand, phénomène que l'on pourrait appeler la "diffusion" du produit; cette diffusion sera exploitée dans le paragraphe suivant en termes d'inégalités $L^2$.

Nous introduirons enfin sur $D_G$ une relation de préordre notée < relation qui sera définie dans la section suivante. Si $\lambda \in D_G$ on pose

$$B_\lambda = \{\mu \in D_G;\ \mu < \lambda\}.$$

## 2. Formules de Taylor incomplètes et calcul symbolique

Posons

$$p(\lambda_1,\lambda_2,\lambda_3) = \|\chi_{\lambda_1}\chi_{\lambda_2}\chi_{\lambda_3}\|_{L^2(G)} (d_{\lambda_1} d_{\lambda_2} d_{\lambda_3})^{-1}$$

$$q_s(\lambda_1,\lambda_2,\lambda_3) = \left(\sum_{\mu\in\lambda_1\circ\lambda_2\circ B^s_{\lambda_3}} d_\mu^2\right)^{1/2} ,$$

où on pose

$$B_{\lambda_3}^{s} = B_{\lambda_3} \circ B_{\lambda_3}^{s-1}.$$

2.1. Théorème. Supposons qu'il existe deux constantes $c_1$ et $c_2$ telles que quels que soient $s > 1$ et $\lambda_1, \lambda_2, \lambda_3 \in D_G$, $\lambda_1 > \lambda_2 > \lambda_3$, on ait

$$(2.1) \qquad p(\lambda_1,\lambda_2,\lambda_3)\, q_s(\lambda_1,\lambda_2,\lambda_3) < c_1 (1+s)^{c_2}.$$

Alors quelle que soit $f \in A_c(G)$, $f$ réelle, on a

$$(2.2) \qquad \|e^{if}\|_{A(G)} \leq c_1 (1 + \|f\|)^{3+c_2}.$$

Remarque: Si $\hat{F}$ est la transformée de Fourier de $F$ alors la formule d'inversion permet d'écrire

$$F \circ f = \int_{\mathbb{R}} e^{itf} \hat{F}(t)\, dt,$$

et la majoration (2.2) montre que si $F \in C^k$ avec $k > 6$ cette intégrale est convergente comme intégrale à valeur vectorielles dans $A(G)$.

Tout revient donc à démontrer le théorème 2.2.

2.2. Lemme. Soit $Q$ un polynôme d'une variable tel que

$$Q(0) = Q'(0) = Q''(0) = 0.$$

Soient $f_1, \cdots, f_n$, $n$ indéterminées. Posons

$$s_0 = 0$$

$$s_k = s_{k-1} + f_k \qquad 1 \leq k \leq n.$$

Alors on a l'identité

$$Q(s_n) = A_1 + A_2 + A_3 + A_4$$

où

$$A_1 = \sum_{q=1}^{n} [Q(s_q) - Q(s_{q-1}) - f_q Q'(s_{q-1}) - \frac{f_q^2}{2} Q''(s_{q-1})]$$

$$A_2 = \sum_{q=1}^{n} \frac{f_q^2}{2} \sum_{r=1}^{q-1} [Q''(s_r) - Q''(s_{r-1})]$$

$$A_3 = \sum_{q=1}^{n} f_q \sum_{r=1}^{q-1} [Q'(s_r) - Q'(s_{r-1}) - f_r Q''(s_{r-1})]$$

$$A_4 = \sum_{q=1}^{n} f_q \sum_{r=1}^{q-1} f_r \sum_{\ell=1}^{r-1} [Q''(s_\ell) - Q''(s_{\ell-1})].$$

Preuve: Vérification formelle.

2.3. *Lemme. Les hypothèses et notations étant celles de 2.2, supposons les* $f_k$ *réels. Posons*

$$M = \max|Q'''(\xi)|, \qquad -\delta < \xi < \delta$$

*où*

$$\delta = \sum_{k=1}^{n} |f_k| .$$

*Alors on a*

$$A_1 = \sum_{q=1}^{n} f_q^3 K_3(s_{q-1}, f_q)$$

$$A_2 = \sum_{q>r} f_q^2 f_r K_1(s_{r-1}, f_r)$$

$$A_3 = \sum_{q>r} f_q f_r^2 K_2(s_{r-1}, f_r)$$

$$A_4 = \sum_{q>r>\ell} f_q f_r f_\ell K_1(s_{\ell-1}, f_\ell)$$

*où* $K_j$ $(j = 1,2,3)$ *sont des polynômes de deux indéterminées de degré total inférieur au degré de* $Q$ *et où*

$$|K_j(s_{q-1}, f_q)| \leq M.$$

Preuve: On pose

$$K_3(u,v) = v^{-3}[Q(u+v) - Q(u) - vQ'(u) - \frac{v^2}{2} Q''(u)]$$

et des expressions analogues pour $K_1$ et $K_2$.

2.4. Démonstration du Théorème

Soit $f = A_c(G)$. Si $\|f\| \leq R$, posons

$$Q(f) = \sum_{2<p<R} \frac{(if)^p}{p!} .$$

Alors

$$\|e^{if}\| \leq \|Q(f)\| + 2 + \|f\| + \|\tfrac{f}{2}\|^2 .$$

Soit $\overline{\lambda}$ la représentation définie par

$$\chi_{\overline{\lambda}} = \overline{\chi_\lambda} .$$

Choisissons une suite $\lambda_1 < \lambda_2 < \cdots < \lambda_n < \cdots$.
On vérifie dans la section suivante que

$$\bar\lambda_1 < \bar\lambda_2 < \cdots < \bar\lambda_n \cdots$$

On a alors, si $f \in A_c(G)$, $f$ réelle,

$$f = \sum_q f_{\lambda_q}$$

où

$$f_{\lambda_q} = a_q(\chi_{\lambda_q} + \chi_{\bar\lambda_q}) + b_q i(\chi_{\lambda_q} - \chi_{\bar\lambda_q})$$

où les $a_q$ et $b_q$ sont réels et où

$$\|f\| = \sum 2d_{\lambda_q}(|a_q| + |b_q|) .$$

Utilisons le lemme 2.3. on obtient en remarquant que

$$|Q'''(\xi)| < 2 \quad \text{si} \quad -R < \xi < R$$

$$\|f_q^3 K_3\|_{L^2(G)} < 2 \|f_q^3\|_{L^2(G)} .$$

D'où

$$\|f_q^3 K_3\|_{L^2(G)} \leq 2[p(\lambda_q,\lambda_q,\lambda_q) + \cdots] \|f_q\|_A^3 ,$$

les pointillés dénotant cinq autres termes obtenus en remplaçant l'un des $\lambda_q$ par $\bar\lambda_q$.

De même

$$\|f_q^2 f_r K_1\|_{L^2(G)} \leq 2[p(\lambda_q,\lambda_q,\lambda_r) + \cdots] \|f_q\|_A^2 \|f_r\|_A$$

$$\|f_q f_r^2 K_2\|_{L^2(G)} \leq 2[p(\lambda_q,\lambda_r,\lambda_r) + \cdots] \|f_q\|_A \|f_r\|_A^2$$

$$\|f_q f_r f_\ell K_1\|_{L^2(G)} \leq 2[p(\lambda_q,\lambda_r,\lambda_\ell) + \cdots] \|f_q\|_A \|f_r\|_A \|f_\ell\|_A$$

Si on note par $\mathrm{Supp}(h)$ l'ensemble des $\lambda \in D_G$ tels que le coefficient $c_\lambda(h)$ de Fourier correspondant soit non nul, alors l'inégalité de Schwarz donne

$$\|h\|_{A(G)} = (\sum |c_\lambda(h)| d_\lambda) < (\sum_{\lambda \in \mathrm{Supp}(h)} d_\lambda^2)^{1/2} \|h\|_{L^2(G)}$$

Remarquons que

$$\mathrm{Supp}(f_q^3 K_3) \subset (\lambda_q \cup \bar\lambda_q)^3 \circ B^R_{\lambda_q}$$

$$\mathrm{Supp}(f_q^2 f_r K_2) \subset (\lambda_q \cup \bar\lambda_q)^2 \circ (\lambda_r \cup \bar\lambda_r) \circ B^R_{\lambda_r}$$

$$\mathrm{Supp}(f_q f_r f_\ell K_1) \subset (\lambda_q \cup \bar\lambda_q) \circ (\lambda_r \circ \bar\lambda_r) \circ (\lambda_\ell \circ \bar\lambda_\ell) \circ B^R_{\lambda_\ell}$$

On obtient utilisant (2.1) que

$$\|f_q^3 K_3\|_A \leq C_1(1+R)^{C_2} \|f_q\|_A^3$$

et les trois inégalités analogues d'où

$$\|Q(f)\|_A \leq C_1(1+R)^{C_2} (\textstyle\sum_q \|f_q\|)_A^3 = C_1(1+R)^{C_2} \|f\|^3$$

Ce qui établit (2.2).

## 3. Structure multiplicative de $A_c(G)$: Calcul des coefficients de Clebsch-Gordan

### 3.1. Notations et rappels

Le tore maximal de SU(3) étant de dimension 2, les caractères de SU(3) pourront s'écrire comme des polynômes trigonométriques usuels de deux variables.

Dans un plan euclidien on choisit deux vecteurs unitaires $\alpha_1$, $\alpha_2$, faisant un angle de $2\pi/3$ et on pose $\alpha_3 = \alpha_1 + \alpha_2$. Soit $S_1$, la symétrie définie par

$$S_1(\alpha_1) = -\alpha_1$$
$$S_1(\alpha_2) = +\alpha_3 .$$

$S_1$ est la symétrie par rapport à la bissectrice de $(\alpha_2, +\alpha_3)$.
De même soient $S_2$, $S_3$ les symétries vérifiant

$$S_2(\alpha_2) = -\alpha_2 \qquad S_3(\alpha_3) = -\alpha_3$$
$$S_2(\alpha_3) = +\alpha_1 \qquad S_3(\alpha_1) = -\alpha_2 .$$

On a

$$S_3 = S_1 S_2 S_1 = S_2 S_1 S_2 .$$

Notons par W le groupe engendré par $S_1$ $S_2$; W est isomorphe au groupe symétrique opérant sur 3 lettres. Si $\sigma \in W$ on note par $\varepsilon_\sigma$ le déterminant de l'isométrie $\sigma$ $(\varepsilon_\sigma = \pm 1)$.

On note $R$ le groupe additif engendré par $\alpha_1$ et $\alpha_2$,

$$R = Z\alpha_1 + Z\alpha_2 .$$

Comme $\alpha_3 = +(\alpha_1 + \alpha_2)$, on peut engendrer $R$ par $(\alpha_1, \alpha_3)$ par suite W opère sur $R$. On prend comme domaine fondamental $R_0$

$$R_0 = \{m \in R ; |\text{angle } (m, \alpha_3)| \leq \tfrac{\pi}{6}\} .$$

Alors l'orbite par W de tout point de $R$ rencontre $R_0$ en un point unique.

On note par $C$ le groupe compact dual de $R$; $C$ peut être identifié au tore maximal de SU(3). Par suite les caractères de SU(3) s'identifient à des polynômes trigonométriques sur $C$:

$$\chi_\lambda(\theta) = \sum_{r\in R} c_r \exp(ir.\theta),$$

où les $c_r$ sont des coefficients entiers. Posant

$$\chi_\lambda^\sigma(\theta) = \sum_{r\in R} c_r \exp(i\sigma(r)\theta),$$

alors on a $\chi_\lambda^\sigma = \chi_\lambda$ pour tout $\sigma \in W$.
Si $\lambda \in R_0$, posons

$$p_\lambda(\theta) = \sum_{\sigma\in W} \varepsilon_\sigma \exp(i\sigma(\lambda).\theta).$$

On a alors

_Théorème de Weyl._ $D_G$ _est en bijection avec_ $R_0$. _Si_ $\lambda \in R_0$, _le caractère correspondant est donné par_

$$\chi_\lambda(\theta) = \frac{p_{\lambda+\delta}(\theta)}{p_\delta(\theta)}$$

_où_ $\delta = \alpha_3$.

Un exemple trivial est SU(2). Le tore maximal est de dimension 1, les caractères sont ainsi des polynômes trigonométriques d'une variable. Le groupe de Weyl est réduit à 2 éléments; l'identité et la symétrie par rapport à l'origine. On a $\delta = 1/2$; la formule de Weyl donne alors

$$\chi_\lambda(\theta) = \frac{e^{i(\lambda+\frac{1}{2}\theta)} - e^{-i(\lambda+\frac{1}{2}\theta)}}{e^{\frac{i\theta}{2}} - e^{-\frac{i\theta}{2}}}$$

et on retrouve le noyau de Dirichlet

$$\chi_n(\theta) = \sum_{q=-n}^{n} e^{iq\theta}$$

3.2. Methode utilisée dans le cas de SU(3)

On déterminera les coefficients de Fourier de $\chi_\lambda$:

(3.1) $$\chi_\lambda(\theta) = \sum m_\lambda(\mu) e^{i\mu.\theta},$$

où $m_\lambda(\mu)$ est un entier positif, appelé la multiplicité du poids $\mu$ dans la représentation $\lambda$. Le calcul des $m_\lambda(\mu)$ conduit ainsi à une formule assez analogue à celle des noyau de Dirichlet.

3.3. Calcul des caractères de SU(3)

A tout $\lambda \in R_0$ on associe l'orbite de $\lambda$ sous l'action de W, on note par $\mathcal{H}_\lambda$ l'hexagone convex obtenu en prenant la borne convex de cette orbite. Cet hexagone a ses côtés parallèles à $\alpha_1, \alpha_2, \alpha_3$. On note $\partial\mathcal{H}_q$ les cotés de cet hexagone.

On note par $\lambda_1^0, \lambda_2^0$ les deux points de $R_0$ situés sur les deux droites $\arg -\lambda = \pi/6$ et $\arg \lambda = -\pi/6$ tels que

$$\lambda_1^0 + \lambda_2^0 = k\alpha_3 , \quad \text{où } k > 0 \text{ et tels qu'ils engendrent } R_0 .$$

Alors tout $\lambda \in R_0$ s'écrit

$$\lambda = m_1\lambda_1^0 + m_2\lambda_2^0 , \qquad m_1, m_2 \in \mathbb{N}.$$

On note $\overline{\lambda}$ le symétrique de $\lambda$ par rapport à $\alpha_3$.

$$\overline{\lambda} = m_2\lambda_1^0 + m_1\lambda_2^0$$

On pose enfin

$$\lambda_1 < \lambda_2$$

si on a l'inégalité suivante entre les produits scalaires euclidiens

$$(\lambda_1|\alpha_3) < (\lambda_2|\alpha_3).$$

Si $m_2 \leq m_1$ on note

$$\mu_q = \lambda - q\alpha_3 \qquad 0 \leq q \leq m_2 .$$

Alors $\mu_q \in R_0$.
On pose

$$K_\lambda = \mathcal{H}_{\mu_{m_2}} , \qquad \mathcal{H}_q^\lambda = \mathcal{H}_{\mu_q} .$$

On voit alors que $K_\lambda$ est un triangle équilatéral dont les cotés sont parallèles à $\alpha_1, \alpha_2, \alpha_3$. On a

3.3.1. <u>Théorème</u>. <u>Soit</u> $\lambda = m_1\lambda_1 + m_2\lambda_2$ <u>avec</u> $m_1 \geq m_2$. <u>Alors on a</u>

$$n_\lambda(\sigma) = m_2 + 1 \quad \underline{\text{si}} \quad \sigma \in K_\lambda$$

$$n_\lambda(\sigma) = q+1 \quad \underline{\text{si}} \quad \sigma \in \partial(\mathcal{H}_q^\lambda) \quad \underline{\text{avec}} \quad 0 \leq q \leq m_2$$

$$n_\lambda(\sigma) = 0 \quad \underline{\text{si}} \quad \sigma \notin \mathcal{H}_\lambda$$

3.3.2. <u>Corollaire</u>. $\overline{\chi_\lambda} = \chi_{\overline{\lambda}}$.

3.3.3. <u>Lemme</u>. <u>Sur chaque coté de l'hexagone</u> $\mathcal{H}_\lambda$ <u>la multiplicité vaut</u> 1.

Preuve: Prenons par exemple un poids situé sur le coté $\lambda$, $S_1\lambda$. Il s'écrit

$$\mu = \lambda - k\alpha_1, \quad 0 \leq k \leq m_1 .$$

On raisonne par récurrence sur $k$, sachant le lemme vrai pour $k = 0$. On applique la formule de récurrence de Freudenthal: qui s'écrit alors

$$[(\lambda+\delta,\lambda+\delta) - (\mu+\delta,\mu+\delta)]n_\mu = 2\sum_{h-1}^{\infty} n_{\mu+h\alpha_1}(\mu+h\alpha_1,\alpha_1),$$

où $\delta = \alpha_3 = \alpha_1 + \alpha_2$ est la demi-somme des racines positives. On a par l'hypothèse de récurrence:

$$n_{\mu+\alpha_1} = n_{\lambda-(k-1)\alpha_1} = 1$$

$$\cdots\cdots\cdots\cdots\cdots\cdots\cdots\cdots$$

$$n_{\mu+k\alpha_1} = n_\lambda = 1$$

d'où

$$\begin{aligned}[(\lambda+\delta,\lambda+\delta) - (\mu+\delta,\mu+\delta)]n_\mu &= 2\sum_{h=1}^{k} (\mu+h\alpha_1,\alpha_1)\\ &= 2\sum_{h=1}^{k} (\lambda+(h-k)\alpha_1,\alpha_1)\\ &= 2k(\lambda,\alpha_1)2(\alpha_1,\alpha_1)\frac{(k-1)k}{2}\end{aligned}$$

D'où

$$n_\mu[2k(\lambda+\delta,\alpha_1) - k^2(\alpha_1,\alpha_1)] = 2k(\lambda,\alpha_1) - k(k-1)(\alpha_1,\alpha_1)$$

$$n_\mu[2(\lambda,\alpha_1) + 2(\delta,\alpha_1) - k(\alpha_1,\alpha_1)] = 2(\lambda,\alpha_1) - (k-1)(\alpha_1,\alpha_1)$$

$$2\left[\frac{(\lambda,\alpha_1)}{(\alpha_1,\alpha_1)} + (2-k) + 2\frac{(\alpha_2,\alpha_1)}{(\alpha_1,\alpha_1)}\right]n_\mu = 2\frac{(\lambda,\alpha_1)}{(\alpha_1,\alpha_1)} - (k-1)$$

Comme

$$2\frac{(\alpha_2,\alpha_1)}{(\alpha_1,\alpha_1)} = -1$$

on obtient

$$2\left[\frac{(\lambda,\alpha_1)}{(\alpha_1,\alpha_1)} - (k-1)\right] n_\mu = 2\,\frac{(\lambda,\alpha_1)}{(\alpha_1,\alpha_1)} - (k-1)$$

qui n'est pas nul car

$$2\,\frac{(\lambda,\alpha_1)}{(\alpha_1,\alpha_1)}\,\alpha_1 = \lambda - S_1(\lambda) = m_1\alpha_1,$$

et on a supposé $k \leq m_1$.

3.3.4. <u>Lemme</u>. <u>Lorsque le poids</u> $\mu$ <u>parcourt la corde passant par</u> $\lambda$ <u>parallèlement à</u> $\alpha_3$, <u>jusqu'au poids</u> $\lambda'$ <u>la croissance de la multiplicité</u> $n_\mu$ <u>est linéaire et constante sur les parallèles aux côtés</u> $\lambda$, $S_1\lambda$ <u>et</u> $\lambda$, $S_2\lambda$ <u>et elle vaut</u> $k+1$.

Preuve: Il s'agit d'étudier

$$\mu = \lambda - k\alpha_3 \quad \text{pour} \quad 0 \leq k \leq m_2,$$

sachant que le lemme 3.2.4 est vrai à l'aide du lemme 3.3.3 pour $k = 0$. Nous poserons $n_k = n_\mu$ si $\mu = \lambda - k\alpha_3$. Les poids de la représentation qui sont de la forme

$$\mu' = \mu + h\alpha_1 \qquad \mu'' = \mu + h\alpha_2 \qquad h \geq 1$$

sont tels que $1 \leq h \leq k$. Mais on peut écrire:

$$\begin{aligned} \mu' &= \lambda - k\alpha_3 + h\alpha_1 \\ &= \lambda - (k-h)\alpha_3 - h\alpha_2 . \end{aligned}$$

Comme $h \geq 1$ on a $n_{\mu'} = n_{(k-h)}$ par l'hypothèse de récurrence et on a aussi

$$n_{\mu''} = n_{(k-h)} .$$

Dans ce cas la formule de Freudenthal s'écrit:

$$\begin{aligned} &[(\lambda+\alpha_3,\lambda+\alpha_3) - (\mu+\alpha_3,\mu+\alpha_3)]n_h = \\ &= 2\sum_{h-1}^{k} n_{k-h}(\mu+h\alpha_3,\alpha_3) + 2\sum_{h-1}^{k} n_{k-h}(\mu+h\alpha_1,\alpha_1) \\ &\qquad + 2\sum_{h-1}^{k} n_{k-h}(\mu+h\alpha_2,\alpha_2) \\ &= k[2(\lambda,\alpha_3) + (2-k)(\alpha_3,\alpha_3)]n_k . \end{aligned}$$

On vérifie facilement que

$$2\frac{(\lambda,\alpha_3)}{(\alpha_3,\alpha_3)} = m_1 + m_2$$

d'où

$$k[m_1+m_2+2-k]n_k = 2\sum_{h=1}^{k} n_{k-h}[m_1+m_2-2k+3h] .$$

Comme $k-h$ est $< k$ on applique l'hypothèse de récurrence et on a $n_{k-h} = k - h + 1$. D'où le résultat en appliquant $\sum_{h=1}^{k} h^2 = \frac{k(k+1)(2k+1)}{6}$ ,

$$n_k = k + 1.$$

Il reste à prouver que si

$$\mu = -\lambda - k\alpha_3 \quad \text{où} \quad 0 \leq k \leq m_2$$

et si $\mu' = \mu - h\alpha_1$ est un poids situé dans le trapèze $\lambda$, $\lambda'$, $S_1(\lambda)$, $S_1(\lambda')$, alors $n_\mu = n_{\mu'}$. Pour cela on appliquera la formule de Kostant

$$n_\mu = \sum_{S\in W} (\det S)P[S(\lambda+\delta) - (\mu+\delta)],$$

où $\delta = \alpha_1 + \alpha_2 = \alpha_3$ est la demi-somme des racines positives, et où $P$ est la fonction de partition de Kostant qui vérifie la formule de récurrence:

$$P(-\mu) = \sum_{\substack{S\in W\\ S\neq 1}} (\det S)P(\mu-[\delta-S\delta])$$

On obtient ainsi:

$$n_\mu = n_{\mu+\alpha_3-\alpha_2} + n_{\mu+\alpha_3-\alpha_1} - n_{\mu+\alpha_3+\alpha_1} - n_{\mu+\alpha_3+\alpha_2} + n_{\mu+2\alpha_3}$$

Prenons le cas $\mu' = \mu - h\alpha_1$, $1 \leq h \leq m_1-k$.

On a

$$S_1(\mu) = \lambda - m_1\alpha_1 - k\alpha_2$$
$$= \mu - (m_1-k)\alpha_1$$

et on a

$$n_\mu = n_{S_1(\mu)} = k + 1 .$$

Donc il suffit de regarder ce qui se passe pour

$$\mu' = \mu - h\alpha_1 \qquad 1 \leq h \leq m_1-k-1$$

D'abord pour $h = 1$. Appliquons la formule de Kostant. On a

$$n_\mu = n_{\mu+\alpha_3-\alpha_2} + n_{\mu+\alpha_3-\alpha_2} - n_{\mu+\alpha_3+\alpha_1} - n_{\mu+\alpha_3+\alpha_2} + n_{\mu+2\alpha_3}.$$

Alors

$$n_{\mu-\alpha_1} = n_\mu + n_{\mu+\alpha_3-2\alpha_1} - n_{\mu+\alpha_3} - n_{\mu+\alpha_3+\alpha_2-\alpha_1} + n_{\mu+2\alpha_3-\alpha_1}$$
$$= (k+1) + k - k - (k-1) + k - 1 = k + 1.$$

Supposons que

$$n_{\mu-(h-1)\alpha_1} = k + 1 \quad \text{et} \quad 2 \leq h \leq m_1-k-1.$$

Alors

$$n_{\mu-h\alpha_1} = k + 1 + n_{\mu-(h)\alpha_1+\alpha_2} - n_{\mu-(h-1)\alpha_1+\alpha_3}$$
$$-n_{\mu-h\alpha_1+\alpha_3+\alpha_2} + n_{\mu-h\alpha_1+\alpha_3+\alpha_2}$$
$$= k +1 + k - k - (k-1) + (k-1) = k + 1.$$

En particulier pour $\mu = \lambda - m_2\alpha_1$ on a $n_\mu = m_2 + 1$.

3.3.5. Lemme. A l'intérieur du triangle $K_\lambda$ de sommets $\lambda'$, $S_1(\lambda')$, $S_2S_1(\lambda')$ la multiplicité vaut $m_2+1$.

Preuve: On sait que en chaque point du côté $\lambda'$, $S_2S_1(\lambda')$ la multiplicité vaut $m_2+1$. Il suffit de prouver que cette multiplicité vaut $m_2+1$ en chaque point d'une parallèle à $\lambda'$, $S_1\lambda'$ passant par ce point. La démonstration est la même que celle du lemme 3.3.4.

3.3.6. Proposition. Soit $\lambda_1,\lambda_2 \in R_0$, $\lambda_2 < \lambda_1$. On a

$$\chi_{\lambda_1}\chi_{\lambda_2} = \textstyle\sum_{\mu\in R_0} c_\mu(\lambda_1,\lambda_2)\chi_\mu$$

Posons

$$\phi(\mu) = n_{\mu-\lambda_1}(\lambda_2)$$
$$\phi^*(\mu) = \phi(\mu) - \phi(\tilde{S}_1\mu) - \phi(\tilde{S}_2\mu),$$

où $\tilde{S}_1$ (resp. $\tilde{S}_2$) denote la symétrie d'axe parallèle à $S_1$ (resp. $S_2$) envoyant $-\alpha_3$ sur zéro.
Alors

$$\phi^*(\mu) = c_\mu .$$

3.3.7. Corollaire. Si $\mathcal{H}_{\lambda_2} + \lambda_1 \subset R_0$ alors $c_\mu = \phi(\mu)$.

Preuve: Écrivons la formule de Weyl.

$$\chi_{\lambda_1}\chi_{\lambda_2} = p_{\lambda_1+\delta}p_{\lambda_2+\delta}p_\delta^{-2}$$

$$\chi_\mu = p_{\mu+\delta}p_\delta^{-1}$$

d'où l'identité

$$p_{\lambda_1+\delta}\frac{p_{\lambda_2+\delta}}{p_\delta} = \sum_{\mu\in R_0} c_\mu p_{\mu+\delta}$$

d'où

$$p_{\lambda_1+\delta}\sum_{\mu\in R} n_{\tilde{\mu}}(\lambda_2)e^{i\tilde{\mu}.\theta} = \sum_{\mu\in R_0} e_\mu p_{\mu+\delta}.$$

Remarquons que chaque $p_{\mu+\delta}$ contient une seule exponentielle dans $R_0$ à savoir $\mu+\delta$. On obtient

$$\sum_{\tilde{\mu}} \det(S)n_{\tilde{\mu}}(\lambda_2) = c_\mu$$

où

$$\tilde{\mu} + S(\lambda_1+\delta) = \mu + \delta$$

d'où

$$c_\mu = \sum_{S\in W} \det(S)n_{\mu+\delta-S(\lambda_1+\delta)}(\lambda_2). \tag{3.2}$$

L'hypothèse $\lambda_2 < \lambda_1$, qui n'a pas été utilisée pour établir (3.2), implique que dans cette somme les termes non nuls ne peuvent que provenir de $S_0$ = identité, $S_1$ ou $S_2$.

Comme n est invariant par W,

$$n_{\mu+\delta-S(\lambda_1+\delta)} = n_{S(\mu)+S(\delta)-\lambda_1+\delta},$$

d'où le résultat.

3.3.8. <u>Exemple</u>. Soit $\lambda = m_1\lambda_1^0$; alors

$$c_\mu(\lambda,\lambda) \equiv 1 \quad \text{si} \quad \mu = 2\lambda - q\alpha_1, \quad 0 \le q \le 2m_1 = 0 \quad \text{sinon}.$$

De même

$$c_\mu(\lambda,\overline{\lambda}) = 1, \quad \mu = k\alpha_3, \quad 0 \le k \le 2m_1.$$

3.4. Démonstration des estimations $L^2$

3.4.1. <u>Lemme</u>. <u>Si</u> $d_\lambda$ <u>est la dimension de la représentation irréductible de poids dominant</u> $\lambda = m_1\lambda_1^0 \ell m_2\lambda_2^0$, <u>on a</u>:

$$d_\lambda = (m_1+1)(m_2+1)\frac{m_1+m_2+2}{2}$$

Preuve: On a $d_\lambda = \sum_\mu n_\mu(\lambda)$. On calcule somme à l'aide du Théorème 3.3.1.

Le polynôme $d_\lambda$ a son degré qui s'abaise d'une unité si $m_1$ ou $m_2 = 0$. Cette circonstance est défavorable du point de vue des estimations $L_2$ dont nous avons besoin. Mais le produit des Clebsch-Gordan de deux représentations de ce type pénètre profondément dans l'intérieur (cf. exemple 3.3.8.).

Si $\lambda_1 > \lambda_2 > \lambda_3$, alors le produit $\lambda_1 \circ \lambda_2$ pénètre assez dans $R_0$ pour que le produit avec $\lambda_3$ soit abtenu pour sa plus grande partie en untilisant le corollaire 3.3.7.

Ce fait combiné avec les estimations données en 3.3.6. et en 3.3.1. permet d'achever la démonstration du théorème 1.2.

## BIBLIOGRAPHIE


(1) Carl S. Herz. Article dans ce même volume.

(2) E. Hewitt et Kenneth A. Ross. Abstract Harmonic Analysis, Springer-Verlag. Band 152.

(3) N. Jacobson. Lie Algebras, Interscience, Tracts in Pure and Applied Mathematics, N° 10.

(4) Y. Katznelson. An Introduction to Harmonic Analysis, John Wiley, New York.

(5) Noël Leblanc. Calcul symbolique dans le centre d'un algèbre de groupe, Annales de I'Institut Fourier, Tome XIX, Fasc I, 1969, p. 109.

(6) W. Rudin. Fourier Analysis on Groups, Interscience, Tracts in Pure and Applied Mathematics, N° 12.

# HELSON SETS IN $T^n$

by

O. Carruth McGehee
Louisiana State University

It is still not known whether every Helson set obeys harmonic synthesis. This problem is the reason we have been interested in finding Helson sets in new sizes and shapes. The work reported here does not solve the problem, but the results are interesting in themselves. The main result is an explicit construction of a continuous curve in $T^n$, for $n \geq 3$, whose Helson constant is no greater than $\frac{n}{n-2}$. We begin with definitions and background.

Let $G$ be a locally compact abelian group. Let $A = A(G)$ denote the Fourier representation of the convolution algebra $L^1(\hat{G})$, where $\hat{G}$ is the dual group of $G$. When $E$ is a compact subset of $G$, $A(E)$ denotes the quotient algebra $A/I(E)$, where $I(E)$ is the closed ideal consisting of the functions in $A$ that vanish on $E$; it may be thought of as the algebra of restrictions to $E$ of functions in $A$. A set $E$ is a <u>Helson set</u> if every continuous function on $E$ may be extended to a function in $A$, that is, if $C(E) = A(E)$; or, equivalently, if the quantity

$$\alpha(E) = \sup \left\{ \frac{\|f\|_{A(E)}}{\|f\|_{C(E)}} : f \in A(E) , f \neq 0 \right\}$$

is finite. The inclusion map $A(E) \subseteq C(E)$ is always univalent and norm-decreasing; $1 \leq \alpha(E) \leq \infty$ for every $E$. The adjoint inclusion map takes $C(E)^*$ into $A(E)^*$. Now $C(E)^*$ is the space of all Borel measures with support in $E$; and $A(E)^*$ is the annihilator of $I(E)$ in $PM = A^*$. Therefore

$$\alpha(E) = \sup \left\{ \frac{\|\mu\|_M}{\|\mu\|_{PM}} : \mu \in M(E) , \mu \neq 0 \right\},$$

where $\|\mu\|_M$ denotes the total variation norm of $\mu$, and $\|\mu\|_{PM} = \sup \{|\hat{\mu}(\gamma)| : \gamma \in \hat{G}\}$. The <u>Sidon constant of</u> a set $E$ is given by

$$\alpha_S(E) = \sup\left\{\frac{\|\mu\|_M}{\|\mu\|_{PM}} : \mu \in M(E)\ , \mu \neq 0\ , \mu \text{ discrete}\right\};$$

and $E$ is a <u>Sidon set</u> if $\alpha_S(E) < \infty$. Evidently $1 \le \alpha_S(E) \le \alpha(E) \le \infty$ for all $E$; every Helson set is a Sidon set; and $\alpha_S(E) = \alpha(E)$ when $G$ is discrete.

There exist Sidon sets that are not Helson sets. The standard method of proving it is to construct a set $E$, say in $R^n$, which is independent (considering $R^n$ as a module over the rational numbers) but which supports a nonzero measure $\mu$ such that $\lim\sup_{\gamma\to\infty} \hat{\mu}(\gamma) = 0$. The first condition implies that $E$ is a Sidon set, with $\alpha_S(E) = 1$, by Kronecker's Theorem (see [15, Chapter 5] or [1, p. 53]). The second condition implies that $E$ is not a Helson set (see [3]; for Helson's original proof, see [5]; for Helson's technique generalized, see [15, 5.6.10]; for an interesting probabilistic proof, see [11, Chapter XI, Theorem 5]). The first construction of such a set was done by Rudin [14]. A different and more beautiful technique for doing the same thing was found recently by Körner [12], and Varopoulos [17] has obtained like results for more general groups. But we are more concerned here with recent constructions of Stegeman [16] and Varopoulos [18], of which the following is a sample.

<u>Theorem</u> (Stegeman [16, p. 99]). <u>Let</u> $0 < \beta \le 1$. <u>In the metric space of Lipschitz-</u> $\beta$ <u>functions</u> $f$ <u>on</u> $R^n$, <u>there is a set of the first Baire category, such that for all</u> $f$ <u>in its complement,</u> $f^{-1}(0)$ <u>is the union of at most</u> $[\frac{n}{\beta}]$ <u>independent subsets</u>.

It is not difficult to prove that a finite union of independent sets is a Sidon set; this is a special case of a theorem which, in its most refined form, is due to Rider [13]. We know from more recent work (see [7, Corollary to Theorem 3]) that Stegeman's Theorem implies, for example, that in fact $\alpha_S(f^{-1}(0)) \le 3^{3/2}$ for all $f$ in $\Lambda_1(R^2)$ outside a set of category one. On the other hand, the Sidon sets obtained by the Varopoulos-Stegeman procedures are not in general Helson sets (see [18, Section 3] and [6]).

Let $T$ denote the circle group, considered as the real numbers modulo $2\pi$.

We shall state and prove our results in $T^n$ instead of $R^n$ ; the choice is purely technical. We construct curves in $T^n$ that are Helson sets, making the Helson constant as small as we know how. The existence of Helson curves in $T^n$ has been proved by Kahane (see [9], or [10, Section VII.9]; compare [8]). Whereas he used Baire category arguments, we give an explicit construction; but the basic idea is the same.

Theorem I. For $n \geq 3$ , there is a continuous curve $E \subset T^n$ with $\alpha(E) \leq \frac{n}{n-2}$.

Proof. The curve $E$ will be parametrized by homeomorphisms $h_1,\cdots,h_n$ from $T$ onto $T$ . For each $m$ , $1 \leq m \leq n$ , $h_m$ will be the uniform limit of a sequence $\{h_{mk}\}_{k=0}^{\infty}$ , to be constructed inductively. Let $\{N_k\}_{k=1}^{\infty}$ be a sequence of positive integers such that $N_k$ is always an integral multiple of $N_{k-1}$ ; $N_k = 3^k$ will do for the present, although as we shall explain later, if we want the $h_m$'s to obey certain smoothness conditions, then it will be useful to require faster growth of the $N_k$'s .

Let $h_{m,o}(t) = t$ for $0 \leq t \leq 2\pi$ . Choose a finite set $F_{m,1} = \{x_j = x(m,1,j):j = 0,1,\cdots,N_1 - 1\}$ such that $x_o = 0$ , $\alpha(F_{m,1}) = 1$ , and $|x_j - h_{m,o}(\frac{2\pi j}{N_1})| < 10^{-N_1}$ for each $j$ . Then there is a finite set of integers $U_{m,1}$ such that for every function $f$ on $F_{m,1}$ with $|f| = 1$ , there is an integer $p \in U_{m,1}$ and a real number $\theta$ such that $|e^{i(px_j+\theta)} - f(x_j)| < 1/10$ for every $j$ . There is a number $d_{m,1} > 0$ such that $p \in U_{m,1}$ , $0 \leq j < N_1$ , $|x-x_j| < d_{m,1} \Rightarrow |e^{ipx} - e^{ipx_j}| < 1/10$ . The function $h_{m,1}$ will be strictly increasing and will map $[0,2\pi]$ onto itself. First we specify its value at each of $N_1 + 1$ points:

$$h_{m,1}(\frac{2\pi j}{N_1}) = x_j \quad \text{for } 0 \leq j < N_1 \; ; \quad h_{m,1}(2\pi) = 2\pi .$$

Then we require that $h_{m,1}$ increase linearly by the small amount $d_{m,1}/2$ on each of the $N_1 + 1$ intervals

$$I_j = [\ \frac{2\pi((j-1)n+m)}{nN_1}\ ,\ \frac{2\pi(jn+m-1)}{nN_1}\ ] \quad (0 \leq j \leq N_1) .$$

This defines $h_{m,1}$ on some points that lie outside $[0,2\pi]$, so we restrict it now to $\bigcup_{j=0}^{N_1}([0,2\pi] \cap I_j)$. Finally, we complete the definition by making $h_{m,1}$ linear on each of the $N_1$ intervals that make up the complement of $\bigcup_{j=0}^{N_1} I_j$ in $[0,2\pi]$. Thus most of the increase takes place on these $N_1$ intervals, each of which has length $2\pi/nN_1$; so that the maximum slope of $h_{m,1}$ is almost $n$.

Suppose that $k \geq 2$ and that $h_{m,k-1}$ has been defined so that it is a piecewise linear map from $[0,2\pi]$ onto $[0,2\pi]$, and so that the discontinuities of its derivative occur at the points $\frac{2\pi((j-1)n+m)}{nN_{k-1}}$ and $\frac{2\pi(jn+m-1)}{nN_{k-1}}$ for $0 \leq j \leq N_{k-1}$ that lie in $(0,2\pi)$. Let $F_{m,k} = \{x(m,k,j): j = 0,1,\cdots,N_k - 1\}$ be a finite set chosen so that $x(m,k,pN_{k-1}) = x(m,k-1,p)$ for $0 \leq p < N_k/N_{k-1}$; $\alpha(F_{m,k}) = 1$; and $|x(m,k,j) - h(\frac{2\pi j}{N_k})| < d_{k-1}10^{-k}$ for each $j$. Now let us write $x_j$ instead of $x(m,k,j)$. Let $U_{m,k}$ be a finite set of integers such that for every function $f$ on $F_{m,k}$ with $|f| = 1$, there is an integer $p \in U_{m,k}$ and a real number $\theta$ such that $|e^{i(px_j+\theta)} - f(x_j)| < 10^{-k}$ for every $j$. There is a number $d_{m,k} > 0$ such that

(1) $$p \in U_{m,k}, \ 0 \leq j < N_k, \ |x - x_j| < d_{m,k} \Rightarrow |e^{ipx} - e^{ipx_j}| < 10^{-k}.$$

The function $h_{m,k}$ will be strictly increasing and will map $[0,2\pi]$ onto itself. First we specify that

$$h(\frac{2\pi j}{N_k}) = x_j \quad \text{for } 0 \leq j < N_k; \quad h_{m,k}(2\pi) = 2\pi.$$

Then we require that $h_{m,k}$ increase linearly by the small amount $d_{m,k}/2$ on each of the $N_k + 1$ intervals

(2) $$I_j = [\frac{2\pi((j-1)n+m)}{nN_k}, \frac{2\pi(jn+m-1)}{nN_k}] \quad (0 \leq j \leq N_k).$$

This defines $h_{m,k}$ on some points that lie outside $[0,2\pi]$, so we restrict it now to $\bigcup_{j=0}^{N_k}([0,2\pi] \cap I_j)$. Finally we complete the definition by making $h_{m,k}$ linear on each of the $N_k$ intervals that make up the complement of $\bigcup_{j=0}^{N_k} I_j$ in $[0,2\pi]$, each of which has length $1/nN_k$. Thus $h_{m,k}$ agrees with $h_{m,k-1}$ on the multiples of $\frac{2\pi}{N_{k-1}}$; and on each interval on which the old function $h_{m,k-1}$ has constant slope, the new function has almost $n$ times as great a slope $1/n$th of the time, and has a much smaller slope $(n-1)/n$ths of the time.

It is easy to see that $\{h_{m,k}:k = 1,2,\cdots\}$ is a Cauchy sequence in the metric of uniform convergence on $[0,2\pi]$. The uniform limit of the sequence is the promised homeomorphism $h_m$. It has the important property that for each $k$,

(3) $$|h_m(t) - x(m,k,j)| < d_{m,k} \quad \text{for} \quad t \in I_j ,$$

where $I_j$ is as defined by (2) and $x(m,k,N_k)$ means $2\pi$. When we behold all the $h_m$'s at once, we see that for each $t \in [0,2\pi]$, and for each $k$, the distance from $h_m(t)$ to the set $F_{m,k}$ is less than $d_{m,k}$ for every value of $m = 1,2,\cdots,n$ except one (at most).

Let $\mu$ be a measure with $\|\mu\|_M = 1$ and with support in the curve $E = \{\underline{h}(t) = (h_1(t),\cdots,h_n(t)):0 \leq t \leq 2\pi\}$. Let $\varepsilon > 0$. Then there exists a continuous function $f$ on $E$ with $|f| = 1$ such that $\int_0^{2\pi} f(\underline{h}(t))d\mu(\underline{h}(t)) > 1 - \varepsilon$. Fix $k$ sufficiently large so that

(4) $$10^{-k} < \varepsilon, \quad \text{and} \quad |f(\underline{h}(t)) - f(\underline{h}(t'))| < \varepsilon \quad \text{whenever} \quad |t - t'| \leq \frac{2\pi}{N_k} .$$

For $1 \leq m \leq n$, let $E_m$ denote the image under $\underline{h}$ of the complement of $\bigcup_{j=0}^{N_k} I_j$ in $[0,2\pi]$, with $I_j$ defined by (2). The sets $E_m$ are pairwise disjoint. For at least one value of $m$, say $q$, the total variation of $\mu$ on $E_q$ will be less than $1/n$. For $x \in T$, let $\pi_q(x)$ denote the point on $E$ whose $q$th coordinate is $x$. There is an integer $p \in U_{q,k}$ and a real number $\theta$ such that $|e^{i(px_j+\theta)} - f(\pi_q(x_j))| < 10^{-k}$ for all $x_j \in F_{q,k}$. But then by (1), (3), and (4),

$$|e^{i(ph_q(t)+\theta)} - f(\pi_q h_q(t))| < 3\varepsilon \quad \text{whenever} \quad \pi_q h_q(t) \notin E_q .$$

On the other hand, this difference may be as great as 2 when $\pi_q h_q(t) \in E_q$. Thus

$$\left| \int_0^{2\pi} e^{i(ph_q(t)+\theta)} d\mu(\underline{h}(t)) \right| > 1 - 4\varepsilon - \frac{2}{n} .$$

The left-hand side of this inequality is the value of $|\hat{\mu}|$ at the element of

$T^n = Z^n$ whose qth coordinate is p and whose other coordinates are zero.

We have proved that for every $\mu \in M(E)$ , $\|\mu\|_{PM} \geq (1 - \frac{2}{n})\|\mu\|_M$. The theorem is established.

Remark 1. Evidently the homeomorphisms $h_m$ are not Lipschitz-one. However, we can require them to belong to any larger Lipschitz class we choose. Let b be a continuous subadditive function defined on $[0,\infty)$ such that $b(t)t^{-1} \to \infty$ as $t \to 0$. Let $\lambda_b$ denote the class of functions f on $[0,2\pi]$ such that the modulus of continuity of f , defined by

$$\omega_f(\eta) = \sup\left\{|f(t) - f(t')| : |t - t'| \leq \eta\right\},$$

satisfies the condition: $\omega_f(\eta) = o(b(\eta))$ as $\eta \to 0$ . Then $\lambda_b$ is a metric space with the distance from f to g given by

$$d(f,g) = \sup_{0 \leq \eta \leq \pi} \frac{\omega_{f-g}(\eta)}{b(\eta)} + |f(0) - g(0)| \ .$$

In the proof of the theorem, at the kth step, $h_{m,k}$ may be made as close as desired to $h_{m,k-1}$ in the sense of this metric by taking $N_k$ sufficiently much larger than $N_{k-1}$ ; and thus $\{h_{m,k}\}_{k=1}^{\infty}$ will be a Cauchy sequence in $\lambda_b$ .

Remark 2. In the proof of Theorem I , $U_{m,k}$ corresponds to a finite set of characters on the group T whose restrictions to $F_{m,k}$ are $(10^{-k})$-dense in the unimodular functions on that set. We wish now to consider an alternative procedure in that part of the proof. First of all, at the kth step, we may choose the sets $F_{m,k}$ so that $\alpha(F_k) = 1$ , where $F_k = \bigcup_{m=1}^{n} F_{m,k}$ . Then we may let $U_k$ be a finite set of functions $g \in A(T)$ , with $\|g\|_{A(T)} \leq 1$ , whose restrictions to $F_k$ are $(10^{-k})$-dense in the unit ball of $C(F_k)$ ; that is, $U_k$ may be chosen so that for every function f bounded by one on $F_k$ , there is a g in $U_k$ such that $|g(x) - f(x)| < 10^{-k}$ for $x \in F_k$ . Then we may find $d_k > 0$ such that

$$g \in U_k\ ,\ x \in F_k\ ,\ |x - y| < d_k \Rightarrow |g(y) - g(x)| < 10^{-k}\ .$$

Now the last paragraph of the proof may be replaced as follows.

Let $f \in C(E)$, $\|f\|_{C(E)} \leq 1$, $\varepsilon > 0$. Fix $k$ sufficiently large so that $10^{-k} < \varepsilon$, and $|f(\underline{h}(t)) - f(\underline{h}(t'))| < \varepsilon$ whenever $|t - t'| < 2\pi/N_k$. Then there is a function $g \in U_k$ such that $|g(x) - f(\pi_m(x))| < \varepsilon$ if $x \in F_{m,k}$ and $1 \leq m \leq n$. Thus for each $t \in [0,2\pi]$, $|g(h_m(t)) - f(\underline{h}(t))| < 3\varepsilon$ for $n - 1$ out of the n possible values of $m$; for the other value, the difference may be as great as 2. Let $\tau_m$ map a point of $T^n$ onto its mth coordinate. Let $G = \frac{1}{n}\sum_{m=1}^{n} g \circ \tau_m$. Then $G \in A(T^n)$, $\|G\|_{A(T^n)} \leq 1$, and $\|G - f\|_{C(E)} < \frac{3\varepsilon + 2}{n}$. It follows by an iterative procedure that there is a function $g^* \in A(T)$ with $\|g^*\|_{A(T)} \leq \frac{n}{n-3\varepsilon-2}$, such that if $G^* = \frac{1}{n}\sum_{m=1}^{n} g^* \circ \tau_m$, then $\|G^*\|_{A(T^n)} \leq \frac{n}{n-3\varepsilon-2}$ and $G^* = f$ on $E$. This proves again that $\alpha(E) \leq \frac{n}{n-2}$; and the result now has the following form, which extends the theorems of Kahane ([10, Section VII.9]).

Theorem II. For $n \geq 3$, there exist homeomorphisms $h_m$ of $T$ onto $T$ for $m = 1,\cdots,n$, such that for every $f \in C(T)$ of norm one and every $\varepsilon > 0$, there is a function $F$ with $\|F\|_{A(T)} \leq \frac{1+\varepsilon}{n-2}$ such that $f$ is the sum of the $n$ rearrangements of $F: f = \sum_{m=1}^{n} F \circ h_m$. In particular, every $f \in C(T)$ may be written

$$f(t) = \sum_{k=-\infty}^{\infty} \sum_{m=1}^{n} c_{k,m} e^{ikh_m(t)} ,$$

where $\sum |c_{k,m}| \leq \frac{n+\varepsilon}{n-2} \|f\|_{C(T)}$.

Remark 3. We do not know whether the constants $\frac{n}{n-2}$ in Theorem I are the best possible. It is easy to prove that if $\alpha(E) = 1$ and $E \subseteq T^n$ for some finite $n$, then $E$ must be totally disconnected. The proof is essentially the same as for Theorem 5.2.9 in Rudin's book [15], but we include it here for the sake of completeness. If $F$ is a finite subset of $E$, then $\alpha(F) = 1$, so that $A(F)$ is isometrically isomorphic to $A(F^*)$, where $F^*$ is an independent finite set of the same cardinality as $F$. The isomorphism from $A(F^*)$ onto $A(F)$ must be given by the rule $f \to f \circ \phi$ where $\phi$ maps $F$ onto $F^*$. It is well known [2] that $\phi = \theta + \chi$, where $\chi$ is a character on $T^n_d$ and hence preserves all arithmetic relations, and $\theta$ is a constant.

Claim. Given an integer $p > 0$, each element of $T^n$ can be written in at most one way as the sum of $p$ elements of $F$. This property clearly holds for $F^*$, and if we have two sums of $p$ points of $F$ that are equal: $\sum_{j=1}^{p} x_j = \sum_{j=1}^{p} x'_j$, then $\sum \chi(x_j) = \sum \chi(x'_j)$, so that $\sum(\theta + \chi(x_j))$ and $\sum(\theta + \chi(x'_j))$ are two sums of p points of $F^*$ that are equal ; hence these two sets of summands are the same, and hence also $\{x_j\} = \{x'_j\}$. The claim is proved. Now if $E$ were not totally disconnected, there would be $n + 1$ pairwise disjoint connected compact subsets $X_1, \cdots, X_{n+1}$ of $E$ ; then $X = X_1 \times \cdots \times X_{n+1}$ would be an $(n+1)$-dimensional space. But the mapping $(x_1, \cdots, x_{n+1}) \to \sum_{j=1}^{n+1} x_j$ is a homeomorphism from $X$ into the n-dimensional space $T^n$. Therefore $E$ must be totally disconnected if $\alpha(E) = 1$.

On the other hand, one might conjecture that for every $n \geq 3$ and $\varepsilon > 0$, there is a continuous curve $E$ such that $\alpha(E) \leq 1 + \varepsilon$. We would guess that this is not the case. The corresponding statement in the discrete group $Z^n$ is true, however; for $n \geq 2$ and $\varepsilon > 0$ there is a "curve" $E$ in $Z^n$ (a set whose every coordinate projection is surjective) with $\alpha(E) < 1 + \varepsilon$. This is not difficult to show.

Remark 4. We have said nothing about the case $n = 2$. There is a version of Theorem I for this case, as Kahane has shown ([10, p. 101];[9]), but we can offer no improvement in the bound on the Helson constant obtainable by his methods. As Kahane observes, there can be no Theorem II in the case $n = 2$.

Remark 5. The higher-dimensional version of Theorem I states that for $n > 2k$, there is a k-dimensional manifold $E \subset T^n$ such that $\alpha(E) \leq \frac{n}{n-2k}$.

## REFERENCES

[1] J. W. S. Cassels, An Introduction to Diophantine Approximation, Cambridge University Press, 1957.

[2] K. de Leeuw and Y. Katznelson, On certain homomorphisms of quotients of group algebras, Israel J. Math. 2(1964), 120-126.

[3] R. Doss, Elementary proof of a theorem of Helson, to appear.

[4] S. W. Drury, Sur les ensembles de Sidon, C.R.A.S., Paris 271A(1970), 161-162.

[5] H. Helson, Fourier transforms on perfect sets, Studia Math. 14(1954), 209-213.

[6] C. S. Herz, Fourier transforms related to convex sets, Ann. Math. 75(1962), 81-92.

[7] C. S. Herz, Drury's Lemma and Helson sets, to appear in Studia Math.

[8] J.-P. Kahane, Sur les réarrangements des suites de coefficients de Fourier-Lebesgue, C.R.A.S. Paris, 265A(1967), 310-312.

[9] J.-P. Kahane, Sur les réarrangements de fonctions de la classe A , Studia Math. 31(1968), 287-293.

[10] J.-P. Kahane, Séries de Fourier Absolument Convergentes, Springer-Verlag, 1970.

[11] J.-P. Kahane and R. Salem, Ensembles Parfaits et Séries Trigonométriques, Hermann, Paris, 1963.

[12] T. W. Körner, Some results on Kronecker, Dirichlet and Helson sets II, to appear in J. d'Analyse.

[13] D. Rider, Gap series on groups and spheres, Canad. J. Math. 18(1966), 389-397.

[14] W. Rudin, Fourier-Stieltjes transforms of measures on independent sets, Bull. Am. Math. Soc. 66(1960), 199-202.

[15] W. Rudin, Fourier Analysis on Groups, Interscience Publishers, New York, 1962.

[16] J. D. Stegeman, Studies in Fourier and Tensor Algebras, Pressa Trajectina, Utrecht, 1971.

[17] N. Th. Varopoulos, Sets of multiplicity in locally compact abelian groups, Ann.Inst. Fourier, Grenoble 16(1966), 123-158.

[18] N. Th. Varopoulos, Sidon sets in $R^n$ , Math. Scand. 27(1970), 39-49.

# RECENT ADVANCES IN SPECTRAL SYNTHESIS

by

Yves Meyer
Faculté des Sciences, Orsay, France

Introduction. A function of the time is denoted $\varphi(t)$ ("function" means here complex valued continuous function). Observations are made over a long interval $[0,L_1]$. The first feeling is that $\varphi(t)$ is periodic. Let $T_1$ be this period and $\nu_1 = T_1^{-1}$ be the corresponding frequency. For the sake of simplicity, it will be assumed that $\varphi(t) = a(0) + a(1)\cos 2\pi\nu_1 t + a(2)\sin 2\pi\nu_1 t$. But more careful observations on greater intervals show that $a(0)$, $a(1)$ and $a(2)$ are actually three functions of the time whose period $T_2$ could not be detected on the interval $[0,L_1]$ : $T_2$ is very great compared to $L_1$ or to $T_1$. For the sake of simplicity, it will be assumed that $a(j) = a(j,0) + a(j,1)\cos 2\pi\nu_2 t + a(j,2)\sin 2\pi\nu_2 t$ $(0 \leqslant j \leqslant 2)$ and we have $\nu_2 = \frac{1}{T_2}$.

Now $\varphi(t)$ becomes an almost periodic function with nine coefficients ; the frequencies of $\varphi$ are $\varepsilon_1\nu_1 + \varepsilon_2\nu_2$ $(\varepsilon_1,\varepsilon_2 = -1,\ 0$ or $1)$.

Better observations show that each of these nine coefficients of $\varphi$ has itself "secular perturbations" whose period $T_3$ is very great compared to $T_2$, and so on.

If $\varphi$ is assumed to have a nice behavior at infinity, for instance if $\varphi$ is bounded, a mathematical model can be built to explain this "projective sequence" of approximations of $\varphi$, to compute the corresponding error terms and to find the global behavior of $\varphi$.

Let $E$ be the set of all infinite sums $\sum_1^\infty \varepsilon_k\nu_k$, $\varepsilon_k = -1,\ 0$ or $1$, when the

sequence $T_k$, $k \geq 1$, of periods of "secular perturbations" has a sufficiently rapid growth to have $\sum_1^\infty \frac{1}{T_k} < +\infty$. If $\sum_1^\infty (\frac{\vartheta_{k+1}}{\vartheta_k})^2 < +\infty$, then this mysterious function $\varphi$ is any complex valued bounded continuous function on the line whose spectrum lies in E. For each $k \geq 1$, we define a finite set $F_k$ of $3^k$ sums $\varepsilon_1 \vartheta_1 + \ldots + \varepsilon_k \vartheta_k$, $\varepsilon_k = -1, 0,$ or $1$. Then, for each $k \geq 1$, $\varphi(t)$ is identically $\sum_{\lambda \in F_k} \varphi_\lambda(t) \exp 2\pi i \lambda t$, where the spectrum of each $\varphi_\lambda$ lies in $[0, \bar{\vartheta}_{k+1}]$, $\bar{\vartheta}_{k+1} = \vartheta_{k+1} + \vartheta_{k+2} + \ldots$ We put $\varphi_k(t_o, t) = \sum_{\lambda \in F_k} \varphi_\lambda(t_o) \exp 2\pi i \lambda t$ and we get $|\varphi(t) - \varphi_k(t_o, t)| \leq C \frac{|t - t_o|}{T_{k+1}} \sup_{\mathbb{R}} |\varphi|$.
Hence if $L_k = \sqrt{T_k T_{k+1}}$ and $t_o = 0$, $|\varphi(t) - \varphi_k(0, t)| \leq C \sqrt{\frac{T_k}{T_{k+1}}} \sup_{\mathbb{R}} |\varphi|$ on $[0, L_k]$, and we could not distinguish between $\varphi(t)$ and $\varphi_k(0, t) = \varphi_k(t)$.

Our functions $\varphi$ have excellent local approximations by almost periodic functions. However the global behavior of $\varphi$ can be quite different from the expected one of an almost periodic function, as it will be shown later (theorem III of § 3).

In § 1 and § 2 these local approximations of $\varphi$ will be specified in two important cases. In § 3 the main theorem about global behaviors of $\varphi$ is stated. The proof of theorem III is given in § 4 and 5. The concept of Sidon sequences in Banach spaces plays a fundamental role in the proof and leads to fascinating problems (§ 6).

1. Pisot numbers and spectral synthesis. Let $\theta > 2$ be a real number and $E_\theta$ be the compact set of all sums $\sum_{k \geq 1} \varepsilon_k \theta^{-k}$, $\varepsilon_k = 0$ or $1$. A Pisot number $\theta$ is an algebraic integer (let $n$ be the degree of $\theta$) whose all conjugates $\theta_2, \ldots, \theta_n$ (with the exception of $\theta$) have an absolute value less than one.

*Theorem I. Let $\theta > 2$ be a Pisot number. For each continuous bounded function $\varphi : \mathbb{R} \longrightarrow \mathbb{C}$ whose spectrum lies in $E_\theta$, there is a sequence $(\varphi_k)_{k \geq 1}$ of trigonometric sums whose frequencies lie in $E_\theta$ and such that*

a) $\sup_{\mathbb{R}} |\varphi_k| \leqslant \sup_{\mathbb{R}} |\varphi|$ _for each_ $k \geqslant 1$

b) $\varphi_k(t) \longrightarrow \varphi(t)$ _uniformly on each compact subset of_ $\mathbb{R}$.

Roughly speaking, theorem I asserts that, if $\theta$ is a Pisot number, $E_\theta$ is a set of synthesis in a strong sense. It is not known if $E_\theta$ can be a set of synthesis when $\theta$ is not a Pisot number.

_2. Ultrathin symmetric sets and spectral synthesis._ Let $(t_k)_{k\geqslant 1}$ be a sequence of positive real numbers such that, for each $k \geqslant 1$,

$$t_k > \bar{t}_{k+1} = t_{k+1} + t_{k+2} + \dots$$

and let $E$ be the corresponding compact set of all real numbers $t = \sum_1^\infty \varepsilon_k t_k$, $\varepsilon_k = 0$ or $1$ ; $E$ is called a symmetric set of real numbers. If $\sum_{k\geqslant 1} \left(\frac{t_{k+1}}{t_k}\right)^2 < +\infty$, _$E$ is called an ultrathin symmetric set of real numbers_. Let $F_k$, $k \geqslant 1$, be the finite set of the $2^k$ sums $\sum_1^k \varepsilon_j t_j$, $\varepsilon_j = 0$ or $1$.

Each bounded continuous function $\varphi : \mathbb{R} \longrightarrow \mathbb{C}$ whose spectrum lies in $E$ can be written as a perturbed trigonometric sum

$$\varphi(t) = \sum_{\lambda \in F_k} \varphi_\lambda(t) \exp 2\pi i \lambda t \; ; \tag{2.1}$$

$k \geqslant 1$ is an arbitrary integer and $\varphi_\lambda(t)$ are not constant coefficients but functions of the time with slow variations : the spectrum of each $\varphi_\lambda$ lies in $[0 , \bar{t}_{k+1}]$.

In a neighborhood of $t_o$ it is tempting to replace the perturbed trigonometric sum $\varphi(t)$ by the corresponding pure trigonometric sum $\varphi_k(t_o,t)$ defined by

$$\varphi_k(t_o,t) = \sum_{\lambda \in F_k} \varphi_\lambda(t_o) \exp 2\pi i \lambda t. \tag{2.2}$$

The quality of this approximation is specified by the following theorem.

_Theorem_ II (R. Schneider). _For each ultrathin symmetric set_ $E$ _of real numbers_, *there exists a constant* $C$ *such that all bounded continuous functions* $\varphi : \mathbb{R} \longrightarrow \mathbb{C}$ *whose spectrum lies in* $E$ *can be approximated by the following process*

i) *for all real* $t$ *and* $t_o$ *and each* $k \geqslant 1$ $|\varphi(t)-\varphi_k(t_o,t)| \leqslant Ct_{k+1}|t-t_o| \sup_{\mathbb{R}}|\varphi|$

ii) *for each* $t_o$ *and each* $k \geqslant 1$, $\sup_{t\in\mathbb{R}}|\varphi_k(t_o,t)| \leqslant C \sup_{\mathbb{R}}|\varphi|$.

Theorem II is proved in [4] or in [2] ch. VIII. As a corollary, each ultrathin symmetric set admits spectral synthesis.

3. *Behavior at infinity*. In § 1 and § 2 we have obtained for a class of compact sets $E$ of real numbers, excellent *local approximations* of any bounded continuous $\varphi : \mathbb{R} \longrightarrow \mathbb{C}$ whose spectrum lies in $E$ by trigonometric sums. It will be shown that the *global behavior* of $\varphi$ may be quite different from the expected one of an almost periodic function.

*Theorem* III. *Let* $E$ *be a symmetric set of real numbers and let* $(I_k)_{k \geqslant 1}$ *be any sequence of intervals of real numbers whose lengths* $|I_k|$ *tend to infinity with* $k$.
*A subsequence* $(J_k)_{k \geqslant 1}$ *of these intervals can be choosen in such a way that, for each bounded sequence* $(b_k)_{k \geqslant 1}$ *of complex numbers, a bounded continuous function* $\varphi : \mathbb{R} \rightarrow \mathbb{C}$ *can be found with the following property*:

a) *the spectrum of* $\varphi$ *lies in* $E$

b) *for each* $k \geqslant 1$, $\frac{1}{|J_k|}\int_{J_k} \varphi(t)dt = b_k$.

Since each symmetric set contains an ultrathin symmetric set, we may restrict our attention to this last case.

4. *Proof of theorem III*

4.1. The proof depends on the concept of Sidon sequences in normed vector spaces over the complex field.

Definition 1. Let A be a normed vector space and for any $X \in A$ let $\|X\|$ be the norm of X. A sequence $X_k$, $k \geq 1$, of elements of A is a Sidon sequence if $\|\sum_{k\geq 1} a_k X_k\|$ and $\sum_{k\geq 1} |a_k|$ are equivalent norms on the vector space generated by the $X_k$, $k \geq 1$ ($a_k \in \mathbf{C}$ and $a_k = 0$ for all k with the exception of a finite number of them).

An alternative definition is the following : $(X_k)_{k\geq 1}$ is a Sidon sequence if there exist two positive constants c and d such that $\|X_k\| \leq d$ for every k and $\|\sum_1^n a_k X_k\| \geq c \sum_1^n |a_k|$ for each integer $n \geq 1$.

4.2. Properties of Sidon sequences. First of all, the stability of Sidon sequences will be studied.

Proposition 1. Let A be a normed vector space, $(X_k)_{k\geq 1}$ and $(Y_k)_{k\geq 1}$ be two sequences of elements of A, c, d and $\varepsilon$ be three positive real numbers such that

(4.2.1) $0 < \varepsilon < c$

(4.2.2) $\|Y_k - X_k\| \leq \varepsilon$ for each $k \geq 1$

(4.2.3) $\|X_k\| \leq d$ for each $k \geq 1$

(4.2.4) $\|\sum_1^n a_k X_k\| \geq c \sum_1^n |a_k|$ for each $n \geq 1$ and complex numbers $a_k$, $1 \leq k \leq n$.

Then $(Y_k)_{k\geq 1}$ is also a Sidon sequence.

Proof. We have $\|Y_k\| \leq d+\varepsilon$ for each $k \geq 1$ while $\|\sum_1^n a_k Y_k\| \geq \|\sum_1^n a_k X_k\| - \varepsilon \sum_1^n |a_k| \geq (c - \varepsilon)\sum_1^n |a_k|$.

Corollary. Let A be a normed vector space and let $(X_k)_{k\geq 1}$ and $(Y_k)_{k\geq 1}$ be two sequences of elements of A such that $\|Y_k - X_k\| \longrightarrow 0$ $(k \longrightarrow +\infty)$. If $(X_k)_{k\geq 1}$ is a Sidon sequence, there exists an integer n such that $(Y_k)_{k\geq n}$ is a Sidon sequence.

*Proposition 2. Let* A *be a normed vector space and* $(X_k)_{k\geq 1}$ *be a bounded sequence of elements of* A. *Then the three following properties of the sequence* $(X_k)_{k\geq 1}$ *are equivalent:*

(4.2.5) $(X_k)_{k\geq 1}$ *is a Sidon sequence.*

(4.2.6) *there exists a constant* C *such that for any bounded sequence* $(b_k)_{k\geq 1}$ *of complex numbers, an element* Y *can be found in the dual space* $A^*$ *of* A *such that* $\|Y\| \leq C \sup_{k\geq 1} |b_k|$ *and* $\langle X_k , Y\rangle = b_k$ $(k \geq 1)$.

(4.2.7) *for any bounded sequence* $(b_k)_{k\geq 1}$ *of complex numbers, an element* Y *can be found in the dual space* $A^*$ *of* A *such that* $\langle X_k , Y\rangle = b_k$ $(k\geq 1)$.

The obvious proof is left to the reader.

More interesting is the following

*Proposition 3. Let* A *be a normed vector space and* $(X_k)_{k\geq 1}$ *be a Sidon sequence of elements of* A. *For any* $X_o$ *in* A, *an integer* m *can be found such that* $(X_o + X_k)_{k\geq m}$ *be a Sidon sequence.*

Proof. Let D be the closed vector space spanned by the $X_k$, $k\geq 1$. Elements X of D are series $\sum_{k\geq 1} a_k X_k$ for which $\sum_{k\geq 1} |a_k| < +\infty$. Two cases may occur. In the first case $X_o$ does not belong to D. Then a **Z** may be found in the dual space $A^*$ of A such that $\langle X_o , Z\rangle = 1$ while $\langle X , Z\rangle = 0$ for all $X\in D$. Let $(b_k)_{k\geq 1}$ be any bounded sequence of complex numbers and $Y\in A^*$ be such that $\langle X_k , Y\rangle = b_k$ for any $k\geq 1$. Put $b_o = \langle X_o , Y\rangle$ and let Y' be $Y - b_o Z$. Then $\langle X_o + X_k , Y'\rangle = b_k$ for each $k\geq 1$ and $(X_o + X_k)_{k\geq 1}$ is a Sidon sequence.

In the second case $X_o = \sum_1^\infty c_k X_k$ where $\sum_1^\infty |c_k| < +\infty$. Let the positive constant c be defined by (4.2.4) and m be an integer defined by $\sum_{k\geq m} |c_k| \leq \frac{1}{2}$. Then

$(X_o + X_k)_{k \geqslant m}$ is a Sidon sequence. In fact, $\sum_m^n a_k(X_o + X_k) = \sum_{k\geqslant 1} b_k X_k$, where $b_k = a_k + (\sum_m^n a_j)c_k$ if $k \geqslant m$. Since $(X_k)_{k\geqslant 1}$ is a Sidon sequence,

$$\|\sum_{k\geqslant 1} b_k X_k\| \geqslant c \sum_{k\geqslant m} |b_k| \geqslant c \sum_{k\geqslant m} |a_k| - c\,|\sum_m^n a_j| \sum_{k\geqslant m} |c_k| \geqslant \frac{c}{2} \sum_{k\geqslant m} |a_k| .$$

**4.3.** Sidon sequences in quotient algebras of a group algebra. Let $E$ be an ultrathin symmetric compact set of real numbers. Let $A(\mathbf{R})$ be the Banach algebra of all Fourier transforms $\hat{f}$ of all $f \in L^1(\mathbf{R})$ ; the product is the pointwise multiplication of functions and the norm is $\|\hat{f}\| = \|f\|_1$. Let $I(E)$ be the ideal of all functions in $A(\mathbf{R})$ vanishing on $E$ and let $A = A(E)$ be the quotient algebra $A(\mathbf{R})/I(E)$ equipped with the quotient norm. Elements of $A$ may be viewed as restrictions to $E$ of functions of $A(\mathbf{R})$.

Each ultrathin symmetric set is a set of spectral synthesis (theorem II). Hence the dual space $PM(E)$ of $A(E)$ consists of all distributions $S$ carried by $E$ whose Fourier transform $\hat{S}$ is bounded on the real line. The norm of $S$ in $PM(E)$ is $\|\hat{S}\|_\infty = \sup_{\mathbf{R}} |\hat{S}|$.

Let $\Omega$ be the compact space $\{0,1\}^{\mathbf{N}}$ of all sequences $\omega = (\varepsilon_k)_{k\geqslant 1}$ of $0$ or $1$ and $H : \Omega \longrightarrow E$ be the canonical homeomorphism defined by $H(\omega) = \sum_{k\geqslant 1} \varepsilon_k t_k$, $\varepsilon_k = 0$ or $1$.

Proposition 4. *Let* $1 < p_1 < q_1 < p_2 < q_2 < \cdots$ *be a double increasing sequence of natural integers. Assume that* $p_{k+1} - q_k \geqslant 3$ *for each* $k \geqslant 1$. *Let* $(f_k)_{k\geqslant 1}$ *be a sequence of elements of* $A(E)$ *with the following properties* :

(4.3.1) $f_k o H$ *is carried by* $\varepsilon_1 = \varepsilon_2 = \ldots = \varepsilon_{p_k} = 0$

(4.3.2) $f_k o H$ *depends only on those* $\varepsilon_j$ *for which* $p_k < j \leqslant q_k$

(4.3.3) *there exist two positive constants* $c$ *and* $d$ *such that* $0 < c \leqslant \|f_k\| \leqslant d$ *for each* $k \geq 1$.

*Then* $(f_k)_{k \geq 1}$ *is a Sidon sequence in* $A(E)$.

Before giving the proof of proposition 4, some corollaries must be stated.

*Corollary* 1. *Let* $(p_k)_{k \geq 1}$ *be a sequence of integers tending to infinity and* $(f_k)_{k \geq 1}$ *be a sequence of elements of* $A(E)$ *fulfilling* (4.3.1) *and* (4.3.3). *Then* $(f_k)_{k \geq 1}$ *contains a Sidon sequence*.

*Proof of corollary* 1. Let $E_k$ be the compact set of all sums $\sum_{j \geq k} \varepsilon_j t_j$, $\varepsilon_j = 0$ or $1$, and let $I_k : E \longrightarrow \{0,1\}$ be defined by $I_k = 1$ on $E_k$ and $I_k = 0$ elsewhere. Then $\|I_k\|_{A(E)} \leq C$ where $C$ depends only on $E$. Since locally constant functions in $A(E_k)$ are dense in $A(E_k)$, a function $g_k$ in $A(\mathbb{R})$, locally constant on $E_k$, may be found such that

$$(4.3.4) \qquad \|f_k - g_k\|_{A(E_k)} \leq (kC)^{-1}.$$

Multiplying, if necessary, $g_k$ by $I_k$ we obtain a new $g_k : E \longrightarrow \mathbb{C}$ vanishing on the complement of $E_k$ in $E$ such that

$$(4.3.5) \qquad \|f_k - g_k\|_{A(E)} \leq \frac{1}{k}.$$

Since $g_k : E \longrightarrow \mathbb{C}$ is locally constant, an integer $q_k \geq 1$ may be defined by the following property : $g_k \circ H$ depends only on those $\varepsilon_j$ for which $j \leq q_k$.

Replacing, if necessary, the sequence of $g_k$, $k \geq 1$, by a subsequence, it may be assumed that $p_{k+1} - q_k \geq 3$. Then proposition 4 and the corollary of proposition 1 give the result.

*Corollary* 2. *Let* $f_k : E \longrightarrow \mathbb{C}$ *be a sequence of elements of* $A(E)$ *such that* $\|f_k\|_{A(E)} = 1$ *for each* $k \geq 1$ *and such that* $\|f_k\|_{A(K)} \longrightarrow 0$ $(k \longrightarrow +\infty)$ *for each compact subset* $K$ *of* $E$ *not containing* $0$. *Then* $(f_k)_{k \geq 1}$ *contains a Sidon sequence*.

Proof. A sequence $(g_k)_{k\geq 1}$ can be found in $A(E)$ such that $\|f_k-g_k\|_{A(E)} \longrightarrow 0$ and such that, on each compact subset $K$ of $E$ not containing $0$, $g_k = 0$ for $k \geq k_o(K)$. Corollary 1 of proposition 4 and the corollary of proposition 1 give the result.

A more explicit version of corollary 2 is the following

Corollary 3. Let $f_k : E \longrightarrow \mathbf{C}$ be defined by $f_k(x) = e^{-i\lambda_k x} \dfrac{\sin\mu_k x}{\mu_k x}$. If $\mu_k \longrightarrow +\infty$, then $(f_k)_{k\geq 1}$ contains a Sidon sequence.

Using proposition 2, we get theorem III.

## 5. Proof of proposition 4

5.1. Sketch of the proof. Norms in $A(E)$ will be denoted $\| \ \|$. We have to show that a positive constant $c$ exists with the following property : for any $n \geq 1$ and complex numbers $a_1, \ldots, a_n$,

$$(5.1.1) \qquad \|\sum_1^n a_k f_k\| \geq c \sum_1^n |a_k| \, \|f_k\|.$$

We write $a_k = e^{-i\alpha_k}|a_k|$ $(1 \leq k \leq n \ ; \ 0 \leq \alpha_k < 2\pi)$ and the idea is to construct an element $S$ of the dual space $PM(E)$ of $A(E)$ such that

$$(5.1.2) \qquad \langle S \, , \, f_k \rangle = e^{i\alpha_k}\|f_k\| \quad \text{for each } k \geq 1 \text{ and}$$

$$(5.1.3) \qquad \|\hat{S}\|_\infty \leq C \quad \text{(the constant } C \text{ depends on } E \text{ but not on } n \text{ and } a_1, \ldots, a_n).$$

This implies $\langle S \, , \sum_1^n a_k f_k \rangle = \sum_1^n |a_k| \|f_k\|$ and $\|\sum_1^n a_k f_k\| \geq C^{-1} \sum_1^n |a_k| \|f_k\|$. We get (5.1.1) with $c = C^{-1}$.

5.2. The construction of $S$. We obtain $S$ as a sum of $n$ pieces ; $S = 4(\mu_1' + \ldots + \mu_n')$ where $\mu_1', \ldots, \mu_n'$ are finitely supported complex measures. Each $\mu_k'$ is defined with the help of a measure $\mu_k$ and we shall begin with specifying $\mu_k$.

Let $Q_k$, $k \geq 1$, be the finite set of all sums

$\sum_{p_k<j\leqslant q_k} \varepsilon_j t_j + t_{q_k+1}$ for which $\varepsilon_j = 0$ or 1. Using theorem II, it may easily be checked that $\mu_k$ can be choosen with the two following properties

(5.2.1) $\langle \mu_k , f_k \rangle = \|f_k\|$

(5.2.2) $\|\hat{\mu}_k\|_\infty \leqslant C$.

To get (5.1.2), we need two extra conditions : $\langle \mu_\ell , f_k \rangle = 0$ if $\ell \neq k$ and $\|\mu_1 +\dots+ \mu_n\| \leqslant C$ (the norm is taken in the dual space of $A(E)$). To obtain these properties it is necessary to change $\mu_k$ into $\mu'_k$ defined by formula (5.2.3).

Let $\rho_k$ be the measure $\dfrac{\delta(x) + \delta(x-t_k)}{2}$ ($\delta(x)$ is the unit mass put at $x$) and $\sigma_k$ be the measure $\dfrac{\delta(x) - \delta(x-t_k)}{2}$.

We put

(5.2.3)
$$\begin{aligned}
\mu'_n &= e^{i\alpha_n} \mu_n * \sigma_{p_n} * \sigma_{p_n-1} \\
\mu'_{n-1} &= e^{i\alpha_{n-1}} \mu_{n-1} * \rho_{p_n} * \rho_{p_n-1} * \sigma_{p_{n-1}} * \sigma_{p_{n-1}-1} \\
&\dots\dots\dots\dots\dots\dots\dots\dots \\
\mu'_2 &= e^{i\alpha_2} \mu_2 * \rho_{p_n} * \rho_{p_n-1} * \dots * \rho_{p_3} * \rho_{p_3-1} * \sigma_{p_2} * \sigma_{p_2-1} \\
\mu'_1 &= e^{i\alpha_1} \mu_1 * \rho_{p_n} * \rho_{p_n-1} * \dots * \rho_{p_3} * \rho_{p_3-1} * \rho_{p_2} * \rho_{p_2-1} * \sigma_{p_1} * \sigma_{p_1-1}.
\end{aligned}$$

and

(5.2.4) $$S = 4(\mu'_1 +\dots+ \mu'_n).$$

5.3. <u>Computation of</u> $\langle S, f_k \rangle$. We shall prove that

(5.3.1) $$\langle S , f_k \rangle = e^{i\alpha_k} \|f_k\|.$$

To get (5.3.1) we need the three identities

(5.3.2) if $\ell < k$, $\langle \mu'_\ell , f_k \rangle = 0$

(5.3.3) $\langle \mu'_k , f_k \rangle = \dfrac{e^{i\alpha_k}}{4} \|f_k\|$

(5.3.4) if $\ell > k$, $\langle \mu'_\ell , f_k \rangle = 0$.

Proof of (5.3.2). $\mu'_\ell$ is carried by the compact subset $G_\ell$ of $E$ defined by $\varepsilon_{q_\ell+1} = 1$ while $f_k = 0$ on $G_\ell$ by (4.3.1).

The proofs of (5.3.1) and (5.3.2) depend on the following lemma.

Lemma 1. Let $T_1$ and $T_2$ be two finite sets of real numbers and let $\pi : T_1 \times T_2 \longrightarrow T_1 + T_2$ be defined by $\pi(x_1, x_2) = x_1 + x_2$. Assume that $\pi$ is (1,1). Let $g_1 : T_1 \longrightarrow \mathbb{C}$, $g_2 : T_2 \longrightarrow \mathbb{C}$ be two functions and $\nu_1$, $\nu_2$ be two measures carried respectively by $T_1$ and $T_2$. Define $h : T_1 + T_2 \longrightarrow \mathbb{C}$ by $h(x_1 + x_2) = g_1(x_1) g_2(x_2)$. Then $\langle h, \nu_1 * \nu_2 \rangle = \langle g_1, \nu_1 \rangle \langle g_2, \nu_2 \rangle$.

Proof of lemma 1. Obvious by definition of the convolution product.

Proof of (5.3.4). In our case $T_1$ is the set of all sums $\sum_{1 \leq j \leq q_k} \varepsilon_j t_j$ ($\varepsilon_j = 0$ or 1), $g_1 = f_k$, $\nu_1$ is the unit mass put in 0, $T_2$ is the set of all sums $\sum_{q_k < j \leq n} \varepsilon_j t_j$ ($\varepsilon_j = 0$ or 1), $g_2 = 1$, $\nu_2 = \mu'_\ell$. Then $\langle f_k, \mu'_\ell \rangle = f_k(0) \langle 1, \mu'_\ell \rangle = 0$.

Proof of (5.3.3). The argument is quite similar : $T_1$ is the set of all sums $\sum_{p_k < j \leq n} \varepsilon_j t_j$, $g_1 = f_k$, $\nu_1 = e^{i\alpha_k} \mu_k * \rho_{p_n} * \dots * \rho_{p_{k+1}-1}$, $T_2$ is the set of all sums $\sum_{1 \leq j \leq p_k} \varepsilon_j t_j$, $g_2$ is the characteristic function of 0 and $\nu_2 = \sigma_{p_k} * \sigma_{p_k - 1}$. Then $\langle f_k, \mu_k \rangle = \langle f_k, \nu_1 \rangle \langle g_2, \nu_2 \rangle$ ; $\langle g_2, \nu_2 \rangle = \frac{1}{4}$ and the computation of $\langle f_k, \nu_1 \rangle$ can be done by a second use of lemma 1. We call $T_1$ the set of all sums $\sum \varepsilon_j t_j$ for which $\varepsilon_j = 0$ or 1 and $p_k + 1 \leq j \leq q_k + 1$ and $T_2$ the set of all sums $\sum \varepsilon_j t_j$ for which $\varepsilon_j = 0$ or 1 and $p_{k+1} - 1 \leq j \leq n$ ; $g_1 = f_k$, $\nu_1 = e^{i\alpha_k} \mu_k$, $g_2 = 1$ and $\nu_2 = \rho_{p_n} * \dots * \rho_{p_{k+1}-1}$. Obviously $\langle 1, \nu_2 \rangle = 1$ and $\langle f_k, \nu_1 \rangle = e^{i\alpha_k} \|f\|_k$ which ends the proof of (5.3.3).

5.4. Computation of $\sup_{\mathbb{R}} |\hat{S}|$. We write $x_k$ instead of $t_k \frac{x}{2}$ ; then

$|\hat{\rho}_k(x)| = |\cos x_k|$ and $|\hat{\sigma}_k(x)| = |\sin x_k|$. If $a_1, \dots, a_{2n}$ is any sequence of real numbers, we have

$$\begin{aligned}
& |\sin a_1 \sin a_2| \\
+ \quad & |\cos a_1 \cos a_2 \sin a_3 \sin a_4| \\
+ \quad & |\cos a_1 \cos a_2 \cos a_3 \cos a_4 \sin a_5 \sin a_6| \\
+ \quad & \dots\dots\dots\dots\dots\dots\dots\dots\dots\dots \\
+ \quad & |\cos a_1 \cos a_2 \dots\dots \cos a_{2n-3} \cos a_{2n-2} \sin a_{2n-1} \sin a_{2n}| \\
+ \quad & |\cos a_1 \cos a_2 \dots\dots \cos a_{2n-3} \cos a_{2n-2} \cos a_{2n-1} \cos a_{2n}| \leqslant 1.
\end{aligned}$$

This inequality can be easily checked by induction on $n$ and by using the obvious inequality $|\sin a \sin b| + |\cos a \cos b| \leqslant 1$.

In our case $|\hat{\mu}'_k| = |\hat{\mu}_k||\cos x_{p_n}||\cos x_{p_n-1}|\dots|\cos x_{p_{k+1}}||\cos x_{p_{k+1}-1}||\sin x_{p_k}|$ $|\sin x_{p_k-1}| \leqslant C|\cos x_{p_n}||\cos x_{p_n-1}|\dots|\sin x_{p_k} \sin x_{p_k-1}|$ and we get $\|\hat{S}\|_\infty \leqslant 4C$.

## 6. Sidon sequences and Sidon spaces

Some results will be given now without proofs (these proofs are straightforward).

6.1. From now on $A$ will be a normed vector space whose elements are denoted by $X$ ; the norm of $X$ is $\|X\|$.

We recall that a sequence $(X_k)_{k \geqslant 1}$ of elements of $A$ is called a Sidon sequence if $\|\sum_{k\geqslant 1} a_k X_k\|$ and $\sum_{k\geqslant 1} |a_k|$ are equivalent norms on the vector space of all finite sums $\sum_{k\geqslant 1} a_k X_k$ $(a_k \in \mathbf{C})$.

For example, if $A$ is the Banach space of all complex valued continuous $2\pi$-periodic functions, the sequence $(\exp i 2^k t)_{k\geqslant 1}$ is a Sidon sequence (Sidon).

<u>Definition 2. A normed vector space is called a Sidon space if any bounded sequenc</u>

$(X_k)_{k\geq 1}$ of elements of A does have one of the following two properties:

a) a subsequence $(X_k)_{k\in\Lambda}$ is a Cauchy sequence ($\Lambda$ is an infinite set of integers)

b) a subsequence $(X_k)_{k\in\Lambda}$ is a Sidon sequence.

6.2. Examples of Sidon spaces.

(6.2.1) Each finite dimensional space is a Sidon space.

(6.2.2) If $(E_k)_{k\geq 1}$ is a sequence of normed vector space, $A = \bigoplus_{k\geq 1} E_k$ is by definition the space of all sequences $X = (X_k)_{k\geq 1}$ such that $X_k \in E_k$ for each $k\geq 1$ and $\|X\| = \sum_{k\geq 1} \|X_k\| < +\infty$ ($\|X_k\|$ is the norm of $X_k$ in $E_k$).

Then if each $E_k$, $k\geq 1$, has a finite dimension, A is a Sidon space.

(6.2.3) $\ell^1$ is a Sidon space.

(6.2.4) By definition $\ell^1 \hat{\otimes}_\varepsilon \ell^1$ is the completion of $\ell^1 \otimes \ell^1$ with respect to the norm induced on $\ell^1 \otimes \ell^1$ by that of $\mathcal{L}(c_o, \ell^1)$ (the Banach space of all continuous linear maps T from $c_o$ into $\ell^1$ with the norm of operators).

Then $\ell^1 \hat{\otimes}_\varepsilon \ell^1$ may also be defined as the space of all matrices $(a_{p,q})_{p\geq 1, q\geq 1}$ such that for a constant $C > 0$ and each bounded sequences $(x_p)_{p\geq 1}$ and $(y_q)_{q\geq 1}$ of complex numbers

$$\left|\sum a_{p,q}\, x_p y_q\right| \leq C \sup_{p\geq 1} |x_p| \sup_{q\geq 1} |y_q|$$

($x_p = 0$ and $y_q = 0$ for large enough p and q).

With these definitions, we can state : $\ell^1 \hat{\otimes}_\varepsilon \ell^1$ is a Sidon space.

(6.2.5) If $r\geq 1$ is a fixed integer, $\ell^1 \hat{\otimes}_\varepsilon \ell^1 \hat{\otimes}_\varepsilon \dots \hat{\otimes}_\varepsilon \ell^1$ (r times) is a Sidon space.

(6.2.6) Let $(n_k)_{k\geq 1}$ be an increasing sequence of positive integers such that

$n_{k+1} \geqslant 3n_k$ $(k \geqslant 1)$ and $n_1 \geqslant 1$. Let **A** be the space of all complex-valued continuous $2\pi$-periodic functions whose spectra lie in the set of all sums $n_p + n_q$, $1 \leqslant q < p$. <u>Then</u> A <u>is a Sidon space</u> (as a Banach space **A** is isomorphic to the subspace of $\ell^1 \hat{\otimes}_\varepsilon \ell^1$ defined by $a_{p,q} = a_{q,p}$ for all $p \geqslant 1$ and $q \geqslant 1$ and $a_{p,p} = 0$ for all $p \geqslant 1$).

(6.2.7) <u>Let</u> A <u>be a Sidon space and let</u> $A^*$ <u>be the dual space of</u> A. <u>If a sequence</u> $(X_k)_{k\geqslant 1}$ <u>of elements of</u> A <u>tends to</u> 0 <u>in the topology</u> $\sigma(A,A^*)$ (i.e. <u>if for each</u> $Y \in A^*$, $\langle X_k, Y\rangle \longrightarrow 0$) <u>then</u> $\|X_k\| \longrightarrow 0$ $(k \longrightarrow +\infty)$.

(6.2.8) Let K be a compact space, C(K) be the space of all continuous $g : K \longrightarrow \mathbf{C}$ with sup-norm and **A** be a closed subspace of C(K), always with sup-norm. <u>If</u> A <u>is a Sidon space, the following property is true : each sequence</u> $(g_k)_{k\geqslant 1}$ <u>in</u> A <u>such that</u> $g_k(x) \longrightarrow 0$ <u>for all</u> $x \in K$ <u>and</u> $\sup_K |g_k(x)| \leqslant 1$ <u>tends uniformly to</u> 0 <u>on</u> K.

This is the case when $K = [0,2\pi]$ and A is the space of all continuous $g : K \longrightarrow \mathbf{C}$ whose spectra lie in the set of all sums $3^p + 3^q$, $0 < q < p$.

(6.2.9) <u>The dual space of any subspace</u> H <u>of</u> $c_0$ <u>is a Sidon space</u>.

(6.2.10) Let T be a distribution defined on the real line with the two following properties : the support K of T is compact and the Fourier transform $\hat{T}$ of T vanishes at infinity. <u>Then, the restriction algebra</u> A(K) <u>is a Sidon space</u>.

### 6.3. <u>Negative results</u>

If K is any infinite compact set, the Banach space C(K) of all continuous functions $f : K \longrightarrow \mathbf{C}$ is not a Sidon space.

$\ell^p$, $p \neq 1$ is not a Sidon space.

$L^p[0,1]$, $1 \leqslant p < +\infty$ is never a Sidon space (the vector space spanned by any infinite subsequence of $(\exp i 2^k t)_{k \geq 1}$ is isomorphic to $\ell^2$).

<u>Conjectures</u>. If E and F are two Sidon spaces, their projective tensor product $E \hat{\otimes}_\pi F$ is also a Sidon space.

If K is any symmetric set of real numbers, the restriction algebra A(K) is a Sidon space.

## References


[1] MEYER, Y. Les nombres de Pisot et la synthèse harmonique. Ann. Sc. E.N.S., 4ème série, t. 3, fasc. 2 (1970)

[2] MEYER, Y. Algebraic numbers and harmonic analysis. North-Holland (to appear)

[3] SCHNEIDER, R. Some theorems in Fourier analysis on symmetric sets. Pacific J. Math., 31 (1969), 175-196

[4] Thin sets in harmonic analysis, edited by L. Å. Lindahl and F. Poulsen. Lecture Notes in Pure and Applied Mathematics. Marcel Dekker (1971)

# HARMONIC ANALYSIS ON HOMOGENEOUS VECTOR BUNDLES

by

K. Okamoto
Hiroshima University

## §1. Introduction

Let $G$ be a non-compact real form of a simply connected complex semisimple Lie group $G_c$, and let $K$ be a maximal compact subgroup of $G$. We suppose that $G/K$ is hermitian symmetric. For an irreducible unitary representation $\tau_\Lambda$ of $K$ with highest weight $\Lambda$, we denote by $E_\Lambda$ the holomorphic vector bundle on $G/K$ associated to the contragredient representation. Then we have the Dolbeault complex;

$$\cdots \overset{\overline{\partial}}{\to} C^{p,q}(E_\Lambda) \overset{\overline{\partial}}{\to} C^{p,q+1}(E_\Lambda) \overset{\overline{\partial}}{\to} \cdots$$

where $C^{p,q}(E_\Lambda)$ denotes the space of $E_\Lambda$-valued $C^\infty$-differential forms of type $(p,q)$. Denoting by $\theta$ the formal adjoint operator of $\overline{\partial}$, we define the laplacian $\Delta$ by $\Delta = \overline{\partial}\theta + \theta\overline{\partial}$. Let $L_2^{p,q}(E_\Lambda)$ be the space of square-integrable forms of type $(p,q)$ and let $\tilde{\Delta}$ be the weak extension of $\Delta$. Then one can show that $\tilde{\Delta}$ is self-adjoint.

Our concern, here, is to study the spectral resolution of the self-adjoint operator $\tilde{\Delta}$.

Using a formula of the laplacian $\Delta$ which we gave in [18], the problem is essentially reduced to the Plancherel formula for $G$ which was obtained by Harish-Chandra. On the contrary to the compact case, all eigenspaces are of infinite dimension. Therefore, to express the "components" of the resolution explicitly, one has to make a deeper investigation of the structure of eigenspaces. Let $H_2^{p,q}(E_\Lambda)$ denote the eigenspace corresponding to the zero eigenvalue. The elements of

$H_2^{p,q}(E_\Lambda)$ are called square-integrable harmonic forms. If $\Lambda$ is "sufficiently" regular, we know that $H_2^{o,q}(E_\Lambda)$ gives an irreducible unitary representation of discrete series (See [15]). In §4, we shall show that, in case $p = q = 0$, the zero is isolated in the spectrum of $\tilde{\Delta}$ if $\Lambda + \rho$ is regular.

The above-mentioned problem about the explicit realization of the "components" leads us to a notion that is analogous to the "Dirichlet Problem". Let us explain this in the following. Using the Borel imbedding, we regard G/K as an open submanifold of its compact dual. Then the Plancherel theorem shows that each "component" is realized as a section of a certain vector bundle over a "boundary component" of G/K (we regard G/K as a boundary component of G/K itself). Strictly speaking, these "boundary components" are certain fibre bundles over the "topological" boundary components with respect to the Borel imbedding. To express the "component" more explicitly one must transform the section into an element of the original space $L_2^{p,q}(E_\Lambda)$ (i.e. a section of a vector bundle over G/K). This mapping is essentially given by Harish-Chandra's "Eisenstein integral". Using the Harish-Chandra imbedding, we realize G/K as a bounded domain which is of course symmetric. Then under the suitable condition it happens that the "Eisenstein integral" transforms sections of vector bundles on the boundary into harmonic forms in the domain. For lack of a suitable name, let us for a moment call this the "Poisson transform". When one speaks of an intertwining operator it is usually meant between the same series of representations. The "Poisson transform", however, gives us intertwining operators between different series, namely discrete series and (non unitary) other series. The "Poisson transform" of the class one case (i.e. associated to the trivial representation of K) were studied by Hua and Look for classical domains. However, there exists no spherical (i.e. class one) unitary representation of discrete series. Moreover, in class one case there exists only one type of series of representations, so that the above-mentioned property of the "Poisson transform", giving the intertwining operators between different series of representations, appears only if one considers non-trivial vector bundles. When one tries to define an intertwining operator between different series, the first obstruction comes from the fact that representations of different series are realized on spaces of sections of different vector bundles. The method to avoid such obstruction which we gave in [8], was the following. Applying the Borel-Weil theorem to both vector bundles, we realize both spaces as sections of line bundles. Since both total spaces of the associated principal bundles can be identified with $G$, both spaces are realized as certain subspaces of the space of complex valued functions on $G$. Then the intertwining operator is defined by taking the limiting process of functions of one space to another. Anyway, if we start with vector bundles associated to representations on the "same" vector space $V$, then both spaces of sections can be realized as certain subspaces of $V$-valued functions on $G$. This enables us to avoid the above mentioned obstruction. In this case, it is necessary to consider induced representations from "reducible" representations. Let $\overline{T}^*_o(G/K)$ be the anti-holomorphic cotangent space at the origin of $G/K$ and let $\sigma_-$ be the representation of $K$ on $\overline{T}^*_o(G/K)$ which is induced from the linear isotropy representation. Then $C^{o,q}(E_\Lambda)$ is identified with the space of the induced representation from $\tau_\Lambda \otimes \Lambda^q \sigma_-$. The representation $\tau_\Lambda \otimes \Lambda^q \sigma_-$ is, of course, highly reducible. Therefore, to realize an "irreducible" representation on $H_2^{o,q}(E_\Lambda)$, it was necessary for us to consider the induced representation from the "reducible" one of $K$. There is another reason why we consider induced representations from "reducible" ones of subgroups (See §3). When one deals with non-spherical representations such as discrete series, there appears a more crucial difficulty. This is roughly explained as follows. If we consider a hermitian symmetric space of "compact" type, then we have the

Hodge-Kodaira decomposition:

$$C^{p,q}(E_\Lambda) = H^{p,q}(E_\Lambda) \oplus \bar\partial C^{p,q-1}(E_\Lambda) \oplus \theta C^{p,q+1}(E_\Lambda),$$

where $H^{p,q}(E_\Lambda)$ denotes the space of harmonic forms. However, in our case where $G/K$ is non-compact, this decomposition does not hold anymore. The method of the theory of harmonic integral leads us to the square-integrable harmonic forms and we get the decomposition theorem in $L_2$-category:

$$L_2^{p,q}(E_\Lambda) = H_2^{p,q}(E_\Lambda) \oplus B_{\bar\partial}^{p,q}(E_\Lambda) \oplus B_\theta^{p,q}(E_\Lambda),$$

where $B_{\bar\partial}^{p,q}(E_\Lambda)$ and $B_\theta^{p,q}(E_\Lambda)$ denote the "closure" of the images of $\bar\partial$ and $\theta$, respectively. Here, of course we must modify the definition of $\bar\partial$ and $\theta$ so that we would necessarily have unbounded operators between Hilbert spaces. The essential question will be to find "nice" topological vector subspaces of $L_2^{p,q}(E_\Lambda)$ (which should be compatible with the decomposition) such that $\bar\partial$ and $\theta$ are continuous operators between them. Remark that for any non-zero $C^\infty$-differential form with compact support its component in $H_2^{p,q}(E_\Lambda)$ is never with compact support. A completely satisfactory answer to this question is given by Harish - Chandra's Schwartz space. Moreover, this space is compatible with the spectral resolution in the sense that every eigenspace is contained in this space. The notion of the Schwartz space is indispensable in our following discussion.

In case that $p = q = 0$ and that $\Lambda = 0$, our Poisson integral coincides with the classical Poisson integral (See [17]). However, if one considers the case where $\Lambda \neq 0$, he notices that the above Poisson integral generalizes also the Cauchy formula (cf. [1]). To explain this situation we shall give an example at the end of this note.

§2. Notation and Preliminaries

We keep the notation in §1. If $E$ is a vector bundle we denote by $\mathcal{E}(E)$ (resp. $\mathcal{D}(E)$) the space of smooth sections (resp. smooth sections with compact support) equipped with the usual topology. We write $\mathcal{E}(E)^*$ (resp. $\mathcal{D}(E)^*$) for the dual space of $\mathcal{E}(E)$ (resp. $\mathcal{D}(E)$). Let $E^*$ denote the dual bundle of $E$. If $\tau$ is a unitary representation of $K$ on a finite dimensional vector space $V_\tau$, we denote by $E_\tau$ the vector bundle on $G/K$ associated to $\tau$. Let $dx$ be the Haar measure of $G$. If $f \in \mathcal{D}(E_\tau^*)$ and $g \in \mathcal{E}(E_\tau)$, by integrating $\int_G \langle f(x), g(x)\rangle dx$, we obtain the canonical pairing $\mathcal{D}(E_\tau^*) \times \mathcal{E}(E_\tau) \to \mathcal{C}$ (where $\mathcal{C}$ denotes the set of all complex numbers) so that we have the natural inclusion map $\mathcal{E}(E_\tau) \hookrightarrow \mathcal{D}(E_\tau^*)^*$. We put $\mathcal{D}'(E_\tau) = \mathcal{D}(E_\tau^*)^*$. Similarly, we have $\mathcal{D}(E_\tau) \hookrightarrow \mathcal{E}(E_\tau^*)^* = \mathcal{E}'(E_\tau)$. If $P$ is a parabolic subgroup of $G$, we write $P = MAN$ for the Langlands decomposition of $P$. Let $\mathfrak{a}$ (resp. $\mathfrak{n}$) be the Lie algebra of $A$ (resp. $N$). Let $\mathfrak{a}^*$ denote the dual space of $\mathfrak{a}$, and $\mathfrak{a}_c^*$ the complexification of $\mathfrak{a}^*$. We define $\rho \in \mathfrak{a}^*$ by $\rho(H) = \frac{1}{2}$ trace $\mathrm{ad}(H)|\mathfrak{n}$ for $H \in \mathfrak{a}$. Put $K_M = K \cap M$. If $\lambda \in \mathfrak{a}_c^*$ we denote by $F_{\tau,\lambda}$ the vector bundle on $G/K_MAN$ associated to the representation of $K_MAN$: $man \to \tau(m)e^{-i\lambda(\log a)}$, where $\log a$ denotes the unique element of $\mathfrak{a}$ such that $a = \exp(\log a)$. Let $L_\rho$ be the line bundle on $G/K_MAN$ associated to the representation of $K_MAN$: $man \to e^{\rho(\log a)}$. If $\phi \in \mathcal{D}(F_{\tau,\lambda}^* \otimes L_\rho)$ and $\psi \in \mathcal{E}(F_{\tau,\lambda} \otimes L_\rho)$, by integrating $\int_{KM} \langle \phi(km), \psi(km)\rangle\, dk\, dm$, where $dk$ and $dm$ denote the Haar measures of $K$ and $M$, we get the (equivariant) pairing $\mathcal{D}(F_{\tau,\lambda}^* \otimes L_\rho) \times \mathcal{E}(F_{\tau,\lambda} \otimes L_\rho) \to \mathcal{C}$ which gives the (equivariant) inclusion map $\mathcal{E}(F_{\tau,\lambda} \otimes L_\rho) \hookrightarrow \mathcal{D}(F_{\tau,\lambda}^* \otimes L_\rho)^*$. We put $\mathcal{D}'(F_{\tau,\lambda} \otimes L_\rho) = \mathcal{D}(F_{\tau,\lambda}^* \otimes L_\rho)^*$. Let $\mathfrak{g}$ (resp. $\mathfrak{k}$) be the Lie algebra of $G$ (resp. $K$) and let $\mathfrak{p}$ be the orthogonal complement of $\mathfrak{k}$ in $\mathfrak{g}$ with respect to the Killing form. If $x \in G$ we write $\kappa(x)$, $\mu(x)$ and $H(x)$ for the unique elements $\kappa(x) \in K$, $\mu(x) \in M \cap \exp\mathfrak{p}$, and $H(x) \in \mathfrak{a}$, such that $x = \kappa(x)\mu(x)\exp H(x)n$, where $n \in N$.

Now suppose $P$ is a minimal parabolic subgroup and let $P = MAN$ be the corresponding Langlands decomposition. Then $M$ is compact and we have an Iwasawa decompostion $G = KAN$. We put $\Xi(x) = \int_K e^{-\rho(H(xk))}dk$. If $x = k\exp X$ ($x \in G$, $k \in K$, $X \in \mathfrak{p}$), we denote by $\sigma(x)$ the norm of $X$ with respect to the Killing form. If $\mathfrak{m}$ is a Lie algebra we write $\mathfrak{m}_c$ for the complexification of $\mathfrak{m}$. Let $\mathcal{G}$ be the universal enveloping algebra of $\mathfrak{g}_c$ and let $\mathcal{Z}$ be the center of $\mathcal{G}$. We regard elements of $\mathcal{G}$ as right-invariant differential operators on $G$. If $f \in \mathcal{E}(E_\tau)$ we put $\nu_{u,r}(f) = \sup_G |uf(x)|_{V_\tau} \Xi(x)^{-1}(1+\sigma(x))^r$ for $u \in \mathcal{G}$ and $r \geqq 0$. The Schwartz space $\mathcal{C}(E_\tau)$ is defined as a locally convex Hausdorff space topologized by means of seminorms $\nu_{u,r}$.

Now the holomorphic vector bundle $E_\Lambda$ introduced in §1 is defined as follows. Since we assumed that $G/K$ is hermitian, $\mathfrak{p}_c$ is a direct sum of two abelian subalgebras $\mathfrak{p}_+$ and $\mathfrak{p}_-$ of $\mathfrak{g}_c$ such that $\mathfrak{p}_+ = \overline{\mathfrak{p}}_-$ and $[\mathfrak{k}_c, \mathfrak{p}_\pm] \subset \mathfrak{p}_\pm$. We denote by $K_c$(resp. $P_+$, $P_-$) the complex analytic subgroup of $G_c$ corresponding to $\mathfrak{k}_c$ (resp. $\mathfrak{p}_+$,$\mathfrak{p}_-$). Then $K_cP_+$ is a parabolic subgroup of $G_c$ ($P_+$ is normal in $K_cP_+$). Let $\tau_\Lambda$ be an irreducible unitary representation of $K$ with highest weight $\Lambda$. Then $\tau_\Lambda$ is uniquely extended to a holomorphic representation of $K_cP_+$ which is trivial on $P_+$. Let $\tilde{E}_\Lambda$ denote the holomorphic vector bundle on $G_c/K_cP_+$ associated to the contragredient $\tau^*_\Lambda$. We notice that $G \cap K_cP_+ = K$. Then $G/K$ can be canonically identified with the open $G$-orbit of the origin. We denote by $E_\Lambda$ the restriction of $\tilde{E}_\Lambda$ to the open submanifold $G/K$ of $G_c/K_cP_+$. Put $W = GK_cP_+$. If $w \in W$ we write $z(w)$(resp. $\gamma(w)$, $\zeta(w)$) for the unique element of $\mathfrak{p}_-$ (resp. $K_c$, $P_+$) such that $w = \exp z(w)\gamma(w)\zeta(w)$. Let $\Omega$ denote the set of elements of the form $z(x)$ ($x \in G$). Then the map of $\Omega \times K_c \times P_+ : (z,\gamma,\zeta) \to \exp z\gamma\zeta$ to $W$ is a complex analytic isomorphism. Clearly one has $z(xk) = z(x)$ ($x \in G$, $k \in K$). The Harish-Chandra imbedding says that the map of $G/K : xK \to z(x)$ to $\Omega$ is a complex analytic isomorphism.

## §3. The Fourier transform

In this section we describe some profound results of Harish-Chandra [4], in a form suitable for our formulation. Let $P$ be a parabolic subgroup of $G$ and let $P = MAN$ be the corresponding Langlands decomposition. Let $(\hat{M})_d$ be the discrete series for $M$. We assume that $P$ is cuspidal so that $(\hat{M})_d$ is not empty. Fix a finite dimensional unitary representation $\tau$ of $K$ and let $E_\tau$ be the vector bundle on $G/K$ associated to $\tau$. Let $\lambda \in \mathfrak{a}_c^*$ and consider the vector bundle $F_{\tau,\lambda}$ on $G/K_M AN$ which was defined in §2. Using the functions $\sigma_M$ and $\Xi_M$ on $M$ (which is defined in the same way as in §2) for $\phi \in \mathcal{E}(F_{\tau,\lambda} \otimes L_\rho)$ we put

$$\tilde{\nu}_{u,r}(\phi) = \sup |u\phi(km)|_{V_\tau} \Xi_M(m)^{-1}(1+\sigma_M(m))^r \quad \text{for } u \in \mathcal{G} \text{ and } r \geqq 0.$$

By means of these seminorms we define the Schwartz space $\mathcal{C}(F_{\tau,\lambda} \otimes L_\rho)$. For any $\omega \in (\hat{M})_d$ let $d(\omega)$ (resp. $\theta_\omega$) be the formal degree (resp. the character) of $\omega$. We define the projection operator (to the discrete part) by

$$\phi^\circ(x) = \sum_{\omega \in (\hat{M})_d} d(\omega) \int_M \overline{\theta}_\omega(m)\phi(xm)dm \quad (\phi \in \mathcal{C}(F_{\tau,\lambda} \otimes L_\rho)).$$

We denote by $\mathcal{C}(F_{\tau,\lambda} \otimes L_\rho)^\circ$ the image of this map. For any $f \in \mathcal{C}(E_\tau)$ we also put

$$f^\circ(x) = \sum_{\omega \in (\hat{M})_d} d(\omega) \int_M \overline{\theta}_\omega(m)f(xm)dm.$$

We notice that in both formulas the sum is actually finite because $\phi$ and $f$ transform according to $\tau|K_M$ under the right action by $K_M$.

Now we define the Fourier transform (with respect to P)

$$\mathcal{F}_\lambda : \mathcal{D}(E_\tau) \to \mathcal{D}(F_{\tau,\lambda} \otimes L_\rho)^o$$

by

$$\mathcal{F}_\lambda f(x) = \int_{AN} f^o(xan) e^{(-i\lambda+\rho)(\log a)} da dn.$$

We remark that if $\lambda$ is real (i.e. $\lambda \in \mathfrak{a}^*$) $\mathcal{F}_\lambda$ is extended to:

$$\mathcal{F}_\lambda : \mathcal{C}(E_\tau) \to \mathcal{C}(F_{\tau,\lambda} \otimes L_\rho)^o.$$

We also define a mapping

$$\mathcal{P}_\lambda : \mathcal{E}(F_{\tau,\lambda} \otimes L_\rho) \to \mathcal{E}(E_\tau)$$

by

$$\mathcal{P}_\lambda \phi(x) = \int_K \tau(k)\phi(xk) dk.$$

The map $\mathcal{P}_\lambda$ restricted to $\mathcal{C}(F_{\tau,\lambda} \otimes L_\rho)^o$ is unconditionally more important than $\mathcal{P}_\lambda$ itself. We define an inner product ( , ) on $\mathcal{C}(E_\tau)$ by

$$(f, g) = \int_G (f(x), g(x))_{V_\tau} dx.$$

Let $L_2(E_\tau)$ be the completion. Then it is clear that the canonical action of G on $L_2(E_\tau)$ defines a unitary representation which we denote by $\pi_\tau$. On the other hand we define an inner product $( , )_\lambda$ on $\mathcal{C}(F_{\tau,\lambda} \otimes L_\rho)^o$ by

$$(\phi, \psi)_\lambda = \int_{KM} (\phi(km), \psi(km))_{V_\tau} dk\, dm.$$

Denoting by $L_2(F_{\tau,\lambda} \otimes L_\rho)^o$ the completion we obtain a bounded representation $\pi_{\tau,\lambda}$ of G on $L_2(F_{\tau,\lambda} \otimes L_\rho)^o$. If $\lambda$ is real (i.e. $\lambda \in \mathfrak{a}^*$) then it is clear that $\pi_{\tau,\lambda}$ is unitary.

Lemma 3.1. Both $F_\lambda$ and $P_\lambda$ are intertwining operators. Moreover, if $\lambda$ is real we have

$$(F_\lambda f, \phi)_\lambda = (f, P_\lambda \phi) \quad \text{for} \quad f \in C(E_\tau) \text{ and } \phi \in C(F_{\tau,\lambda} \otimes L_\rho)^o.$$

Let $\mu$ be the function on $(\hat{M})_d \times \mathfrak{a}^*$ which was defined in ([4], Th.11). For each $\lambda \in \mathfrak{a}^*$ we define a linear endomorphism $\mu(\lambda)$ on $C(F_{\tau,\lambda} \otimes L_\rho)^o$ by

$$\mu(\lambda)\phi(x) = \sum_{\omega \in (\hat{M})_d} d(\omega)\mu(\omega,\lambda) \int_M \overline{\theta}_\omega(m)\phi(xm)dm.$$

We denote by $\mathfrak{m}$ the Lie algebra of M. If $f \in C(E_\tau)$ we put

$$f_P(x) = \int_{\mathfrak{a}^*} P_\lambda \mu(\lambda) F_\lambda f(x) d\lambda,$$

where $d\lambda$ denotes the Euclidean measure with respect to the Killing form. Let $\mathfrak{h}_i$ $(1 \leqq i \leqq r)$ be a set of representatives of conjugacy classes of Cartan subalgebras which are stable under the Cartan involution. For each i $(1 \leqq i \leqq r)$ fix a cuspidal parabolic subgroup $P_i$ such that the split component of $P_i$ coincides with the vector part of $\mathfrak{h}_i$. Then Harish-Chandra's Plancherel formula gives the following

Lemma 3.2.

$$f = \sum_{1 \leqq i \leqq r} f_{P_i} \quad (f \in C(E_\tau)).$$

Remark. Let P = MAN be a minimal parabolic subgroup such that M is contained in K and let $W(G/A)$ be the Weyl group of A in G. Suppose $\lambda \in \mathfrak{a}^*$ such that $s\lambda \neq \lambda$ for all $e \neq s \in W(G/A)$. For any

$\omega \in (\hat{M})_d$ let $\pi_{\omega,\lambda}$ denote the unitary representation of $G$ induced from the irreducible unitary representation of $MAN : man \to \omega(m)e^{-i\lambda(\log a)}$. Then Bruhat's criterion shows that $\pi_{\omega,\lambda}$ is irreducible. It follows from Frobenius reciprocity law that the multiplicity of $\pi_{\omega,\lambda}$ in $\pi_\tau$ coincides with the multiplicity of $\omega$ in $\tau$. Hence even if $\tau$ is irreducible $\pi_\tau$ contains irreducible components with various multiplicities in general. However, if one starts with the representation $\tau|M$ (which is, in general, reducible) the above Plancherel formula shows that the representation $\pi_{\tau,\lambda}$ (which we constructed in our above formulation) occurs in $\pi_\tau$ exactly once. (cf. [16])

## §4. The spectral resolution of the laplacian

Let $T$ be a maximal torus in $K$ and let $h$ be the Lie algebra of $T$. Since we assumed that $G/K$ is hermitian and that $G$ has a complexification $G_c$, $T$ is a Cartan subgroup of both $K$ and $G$. Let $g = k + p$ be a Cartan decomposition. Then the real tangent space of $G/K$ at the origin is canonically identified with $p$. Hence, using the Killing form restricted to $p$ (which is clearly invariant under the adjoint action of $K$) we can introduce an invariant riemannian metric on $G/K$. This metric is easily seen to be hermitian with respect to the complex structure on $G/K$ which was introduced in §2. Let $\langle\ ,\ \rangle$ denote the bilinear form on $h_c^*$ which is defined by means of the Killing form. Choose an ordering on the roots of $(g_c, h_c)$ compatible with the complex structure on $G/K$ and denote by $\delta$ half the sum of the positive roots. Let $\Lambda$ be an integral form on $h$ which is dominant with respect to $k$. Since we assumed that $G_c$ is simply connected there exists an irreducible unitary representation $\tau_\Lambda$ of $K$ with highest weight $\Lambda$ on a space $V_\Lambda$. Let $E_\Lambda$ be the holomorphic vector bundle which was constructed in §2. We denote by $\theta$ the formal adjoint operator of $\bar{\partial}$ with respect to the above defined hermitian metric and the metrics along fibres induced from the inner product on $V_\Lambda$. Let $C^{o,q}(E_\Lambda)$ denote the space of $E_\Lambda$-valued $C^\infty$-differential forms of type (0, q). Then we have the following lemma (See [18]).

Lemma 4.1.

$$2(\bar{\partial}\theta + \theta\bar{\partial}) = \langle\Lambda + 2\delta, \Lambda\rangle - C,$$

where $C$ denotes the Casimir operator.

Let $Ad_-^q$ denote the representation of $K$ on $\Lambda^q p_-$ induced by the adjoint action of $K$ on $p_-$. Then one can show that $C^{o,q}(E_\Lambda)$ is isomorphic to the space of smooth sections of the vector bundle on $G/K$ associated to the representation $\tau_\Lambda^* \otimes Ad_-^q$ of $K$. We apply the results of the previous section to the case where $\tau = \tau_\Lambda^* \otimes Ad_-^q$. In this case if $P = P_i$ $(1 \leq i \leq r)$ we write $M_i$, $C_i(F_{\tau,\lambda}\otimes L_\rho)^o$, $F_\lambda^i$, $P_\lambda^i$, $a_i$, and $\mu_i(\omega, \lambda)$ for $M$, $C(F_{\tau,\lambda}\otimes L_\rho)^o$, $F_\lambda$, $P_\lambda$, $a$, and $\mu(\omega, \lambda)$ which were defined in §3. We put $C^{o,q}(E_\Lambda) = C(E_\tau)$. If $\phi \in C_i(F_{\tau,\lambda}\otimes L_\rho)$ we define $e_\omega^i\phi(x) = \int_M \overline{\theta}_\omega(m)\phi(xm)dm$ $(\omega \in (\hat{M})_d)$. Then one can show that there exists a $\chi_{\omega,\lambda}^i \in Hom(Z, C)$ such that $ze_\omega^i\phi = \chi_{\omega,\lambda}^i(z)e_{\bar{\omega}}^i\phi$ for all $z \in Z$. Since we assumed that $P_i$ is cuspidal $m_i$ has a compact Cartan subalgebra which we denote by $h_i$. For any $\omega \in (\hat{M}_i)_d$ we denote by $\nu_\omega$ the unique element of $(\sqrt{-1}h_i)^*$ dominant with respect to $K_M$ which corresponds to $\omega$ in ([3] Th.16). Using Lemma 3.2. and Lemma 4.1. we can prove the following

Theorem 4.2. For any $f \in C^{o,q}(E_\Lambda)$, we have

$$(\bar{\partial}\theta + \theta\bar{\partial})f(x) = \sum_{i=1}^{r} \sum_{\omega\in(\hat{M}_i)_d} d(\omega)\int_{a_i^*} c^i(\omega,\lambda)P_\lambda^i e_\omega^i F_\lambda^i f(x)\mu_i(\omega,\lambda)d\lambda$$

$$c^i(\omega,\lambda) = \frac{1}{2}\{|\Lambda+\rho|^2 + |\lambda|^2 - |\nu_\omega|^2\}.$$

Making use of the method in [10] one can prove the following theorem.

Theorem 4.3. *Assume that* $\Lambda + \rho$ *is regular and that* $p = q = 0$; *then the zero is isolated in the spectrum of the laplacian* $\tilde{\Delta}$.

We fix again a cuspidal parabolic subgroup $P = MAN$ and we use the notation in §2 and §3. We can assume that $\mathfrak{a}$ is stable under the Cartan involution. Let $\tau = \tau_\Lambda^* \otimes Ad_-^q$ as above. Fix any $\lambda \in \mathfrak{a}_c^*$. We denote by $\mathcal{D}'(F_{\tau,\lambda} \otimes L_\rho)^\circ$ the dual space of $\mathcal{D}(F_{\tau^*,\lambda} \otimes L_\rho)^\circ$ (equipped with the relative topology). For any $T \in \mathcal{D}'(F_{\tau,\lambda} \otimes L_\rho)^\circ$ we define $P_\lambda T \in \mathcal{D}'(E_\tau)$ by $P_\lambda T(f) = T(F_\lambda^* f)$ $(f \in \mathcal{D}(E_\tau^*))$, where $F_\lambda^* = F_{-\lambda}$.

Proposition 4.4. $P_\lambda$ *is given by an integral operator with a smooth kernel* $K_\lambda$ *such that*

$$K_\lambda(x,y) = \sum_{\omega \in (\hat{M})_d} d(\omega) \int_{K_M} \bar{\theta}_\omega(k\mu(x^{-1}y))e^{-(i\lambda+\rho)(H(x^{-1}y))}(\tau_\Lambda^* \otimes Ad_-^q)(\kappa(x^{-1}y)k^{-1})dk.$$

(*The sum is actually finite.*)

Let $\mathfrak{h}_-$ be a compact Cartan subalgebra of $\mathfrak{m}$ which may be assumed to be contained in $\mathfrak{h}$. Then $\mathfrak{a} + \mathfrak{h}_-$ is a Cartan subalgebra of $\mathfrak{g}$ so that there exists an element $u \in G_c$ such that $Ad(u)\mathfrak{h} = \sqrt{-1}\mathfrak{a} + \mathfrak{h}_-$. Suppose that $(\Lambda + \delta)(H) = (i_\lambda + \nu_\omega)(Ad(u)H)$ for all $H \in \mathfrak{h}$. Then we have $\langle \Lambda + \delta, \Lambda + \delta \rangle = \langle i\lambda + \nu_\omega, i\lambda + \nu_\omega \rangle$. This implies that $\chi_{\omega,\lambda}(C) = \langle \Lambda + 2\delta, \Lambda \rangle$. We recall that the total space of the principal bundle associated to $E_\Lambda$ is $W = \exp\Omega K_c P_+$. If $x = \exp z(x)\gamma(x)\zeta(x)$ we put

$$\tilde{K}_\lambda(x, y) = (\tau_\Lambda^* \otimes Ad_-^q)(\gamma(x))K_\lambda(x, y).$$

Then we have $\tilde{K}_\lambda(xk, y) = \tilde{K}_\lambda(x, y)$ $(k \in K, x, y \in G)$. If $z = z(x)$ $(x \in G)$ we write $\tilde{K}_\lambda(z, y)$ for $\tilde{K}_\lambda(x, y)$. In the same way we realize $C^{o,q}(E_\Lambda)$ as the space of $V^*_\Lambda \otimes \Lambda^q p_-$-valued $C^\infty$-functions on $\Omega$. We put

$$H^{o,q}(E_\Lambda) = \{f \in C^{o,q}(E_\Lambda) \mid (\bar\partial\theta + \theta\bar\partial)f = 0\}.$$

Let $\mathcal{D}'(F_{\tau,\lambda} \otimes L_\rho)^o_\Lambda$ denote the set of all $T \in \mathcal{D}'(F_{\tau,\lambda} \otimes L_\rho)^o$ such that for any $\phi \in \mathcal{D}(F_{\tau,\lambda}^* \otimes L_\rho)^o$ we have $T(e_\omega \phi) = 0$ whenever $\langle \Lambda + \delta, \Lambda + \delta \rangle \neq \langle i\lambda + \nu_\omega, i\lambda + \nu_\omega \rangle$. Then we have the following theorem.

Theorem 4.5. *$P_\lambda$ maps $\mathcal{D}'(F_{\tau,\lambda} \otimes L_\rho)^o$ into $H^{o,q}(E_\Lambda)$. Moreover, $P_\lambda$ is given by an integral operator with a smooth kernel* $\tilde{K}_\lambda(z, y)$ $(z \in \Omega, y \in G)$.

Remark. In case $\Lambda = 0$ and $q = 0$ it can be shown that $C(F_{\tau,\lambda} \otimes L_\rho)^o = \{0\}$ unless $P$ is minimal parabolic. On the other hand for a minimal parabolic one has $C(F_{\tau,\lambda} \otimes L_\rho)^o = C(F_{\tau,\lambda} \otimes L_\rho) = \mathcal{D}(F_{\tau,\lambda} \otimes L_\rho) = \mathcal{E}(F_{\tau,\lambda} \otimes L_\rho)$. In the case that $\Lambda = 0$, that $q = 0$, and that $P$ is minimal parabolic, the map $P_\lambda$ in Proposition 4.4. was studied by Helgason [5]. He obtained many interesting results in [5], only a little of which we have generalized to our operator $P_\lambda$ (in general case) (See [17]).

Finally we give an example.

Notation

$G = SU(1, 1) \simeq SL(2, \mathbf{R})$.

$$u_\theta = \begin{pmatrix} e^{i\theta} & 0 \\ 0 & e^{-i\theta} \end{pmatrix}, \quad a_t = \begin{pmatrix} \text{ch}t & \text{sh}t \\ \text{sh}t & \text{ch}t \end{pmatrix}, \quad n_\xi = \begin{pmatrix} 1+i\xi & -i\xi \\ i\xi & 1-i\xi \end{pmatrix}$$

$K = \{u_\theta;\ \theta \in \mathbf{R}\}$, $A = \{a_t;\ t \in \mathbf{R}\}$, $N = \{n_\xi;\ \xi \in \mathbf{R}\}$.

$$M = \left\{ \begin{pmatrix} 1 & 0 \\ 0 & 1 \end{pmatrix}, \begin{pmatrix} -1 & 0 \\ 0 & -1 \end{pmatrix} \right\}, \quad P = MAN.$$

$$G_c = SL(2, C).$$

$$K_c = \left\{ \begin{bmatrix} \gamma & 0 \\ 0 & \gamma^{-1} \end{bmatrix} ; \gamma \epsilon C^* \right\}, \quad P_+ = \left\{ \begin{bmatrix} 1 & 0 \\ \zeta & 1 \end{bmatrix} ; \zeta \epsilon C \right\}.$$

$$u = \begin{bmatrix} \alpha & \alpha \\ -\alpha & \alpha \end{bmatrix} \quad \left( \alpha = \frac{1}{\sqrt{2}} \right) \quad G \cap u K_c P_+ u^{-1} = MAN.$$

$$a = \left\{ \begin{bmatrix} 0 & t \\ t & 0 \end{bmatrix} ; t \epsilon R \right\}, \quad h = \left\{ \begin{bmatrix} i\theta & 0 \\ 0 & -i\theta \end{bmatrix} ; \theta \epsilon R \right\}$$

$$\rho \begin{bmatrix} 0 & t \\ t & 0 \end{bmatrix} = t, \quad \delta \begin{bmatrix} i\theta & 0 \\ 0 & -i\theta \end{bmatrix} = -i\theta.$$

$$\rho \left[ Ad(u) \begin{bmatrix} i\theta & 0 \\ 0 & -i\theta \end{bmatrix} \right] = -i\theta = \delta \begin{bmatrix} i\theta & 0 \\ 0 & -i\theta \end{bmatrix}.$$

$$\Omega = \left\{ \begin{bmatrix} 0 & z \\ 0 & 0 \end{bmatrix} ; |z| < 1 \right\}, \quad P_- = \left\{ \begin{bmatrix} 1 & z \\ 0 & 1 \end{bmatrix} ; z \epsilon C \right\}.$$

For any integer $n$ we put $\Lambda = -n\delta$. Then $<\Lambda + \delta, \Lambda + \delta> = \frac{1}{8}(n-1)^2$. The laplacian $\Delta$ is given by

$$-\frac{1}{2}(1-|z|^2)^2 \frac{\partial^2}{\partial z \partial \bar{z}} + \frac{n}{2}\bar{z}(1-|z|^2)\frac{\partial}{\partial \bar{z}}.$$

For any complex number $s$ we put $\lambda = s\rho$. Then $<i\lambda, i\lambda> = -\frac{s^2}{8}$ so that $<\Lambda + \delta, \Lambda + \delta> = <i\lambda, i\lambda>$ if and only if $s = \pm i(n-1)$. The kernel function $\overset{\sim}{K}_\lambda$ is given by

$$\overset{\sim}{K}_\lambda \left( \begin{bmatrix} 0 & z \\ 0 & 0 \end{bmatrix}, u_\phi \right) = \left( \frac{e^{i\phi}}{e^{i2\phi} - z} \right)^n \left( \frac{1-|z|^2}{|e^{i2\phi} - z|^2} \right)^{\frac{is+1-n}{2}}$$

and satisfies the differential equation:

$$\Delta \overset{\sim}{K}_\lambda \left( \begin{bmatrix} 0 & z \\ 0 & 0 \end{bmatrix}, u_\phi \right) = \frac{1}{8} \left\{ (n-1)^2 + s^2 \right\} \overset{\sim}{K}_\lambda \left( \begin{bmatrix} 0 & z \\ 0 & 0 \end{bmatrix}, u_\phi \right).$$

We denote by $\overset{\sim}{P}_\lambda$ the integral operator with the kernel $\overset{\sim}{K}_\lambda$. For any integer $k$ we define the function $\psi_k$ on $K$ by $\psi_k(u_\phi) = e^{ik\phi}$.

(I) $i\lambda = (n-1)\rho$ (i.e. $s = -i(n-1)$).

1) $n = 0$.

$(\mathfrak{P}_\lambda \psi_k)(z) = 0$ for all $k \neq 0$.

2) $n \geqq 1$.

$$(\mathfrak{P}_\lambda \psi_k)(z) = \begin{cases} \binom{\frac{n+k-2}{2}}{n-1} z^{\frac{k-n}{2}} & (k \geqq n \text{ and } k \equiv n \pmod 2), \\ 0 & (\text{otherwise}). \end{cases}$$

(In this case $\mathfrak{P}_\lambda$ gives us the discrete series ($n>1$) and a limit of the discrete series ($n=1$).)

3) $n \leqq -1$.

$$(\mathfrak{P}_\lambda \psi_k)(z) = \begin{cases} \binom{|n|}{\frac{|n|+k}{2}} (-z)^{\frac{|n|+k}{2}} & (|k| \leqq |n| \text{ and } k \equiv n \pmod 2), \\ 0 & (\text{otherwise}). \end{cases}$$

(In this case $\mathfrak{P}_\lambda$ gives us finite dimensional representations.)

(II) $i\lambda = -(n-1)\rho$ (i.e. $s = i(n-1)$).

1) $n = 0$.

$$(\mathfrak{P}_\lambda \psi_k)(z) = \begin{cases} z^{-\frac{k}{2}} & (k \leqq 0 \text{ and } k \equiv 0 \pmod 2), \\ \bar{z}^{\frac{k}{2}} & (k > 0 \text{ and } k \equiv 0 \pmod 2), \\ 0 & (k \not\equiv 0 \pmod 2). \end{cases}$$

2) $n = 1$.

$$(\mathfrak{P}_\lambda \psi_k)(z) = \begin{cases} z^{\frac{k-1}{2}} & (k \geqq 1 \text{ and } k \equiv 1 \pmod 2), \\ 0 & (\text{otherwise}). \end{cases}$$

(In this case $\mathfrak{P}_\lambda$ gives us a limit of the discrete series.)

3) $n > 1$.

$$(\mathcal{P}_\lambda \psi_k)(z) = \begin{cases} z^{\frac{k-n}{2}} & (k \geq n \text{ and } k \equiv n \pmod 2), \\ \text{(the formula given below)} & (|k| \leq n-1 \text{ and } k \equiv n \pmod 2), \\ 0 & \text{(otherwise)}. \end{cases}$$

$$\frac{1}{(1-|z|^2)^{n-1}} \sum_{0 \leq \ell \leq \frac{n+k}{2}-1} \binom{n-1}{\ell} z^{\frac{n+k}{2}-\ell-1} (-\bar{z})^{n-1-\ell}$$

4) $n < 0$.

$$(\mathcal{P}_\lambda \psi_k)(z) = \begin{cases} \frac{1}{(1-|z|^2)^{n-1}} \frac{1}{|n|!} \frac{\partial^{|n|}}{z^{|n|}} \left( \frac{z^{-\frac{n+k}{2}}}{1-|z|^2} \right) & (k \leq -n \text{ and } k \equiv n \pmod 2), \\ \frac{1}{(1-|z|^2)^{n-1}} \frac{\bar{z}^{\frac{n+k}{2}}}{|n|!} \frac{\partial^{|n|}}{\partial z^{|n|}} \left( \frac{1}{1-|z|^2} \right) & (k > -n \text{ and } k \equiv n \pmod 2), \\ 0 & (k \not\equiv n \pmod 2). \end{cases}$$

<u>References</u>


[1] S. Bochner, Group invariance of Cauchy's formula in several variables, Ann. of Math., 45(1944).

[2] H. Furstenberg, A Poisson formula for semisimple Lie groups, Ann. of Math., 77(1963).

[3] Harish-Chandra, Discrete series for semisimple Lie groups II, Acta Math., 116(1966).

[4] ———, On the theory of the Eisenstein integral, Lecture notes in Math., Springer (1972).

[5] S. Helgason, A duality for symmetric spaces with applications to group representations, Advances in Math., vol.5, no.1, (1970).

[6] R. Hermann, Geometric aspect of Potential theory in bounded symmetric domains I, II, III, Math. Annalen, 148(1962), 151(1963), 153(1964).

[7] L. K. Hua, Harmonic analysis of functions of several complex variables in the classical domains, A. M. S. (Translations of mathematical monographs, vol. 6)(1963).

[8] A. W. Knapp and K. Okamoto, Limits of holomorphic discrete series, Jour. of Functional Analysis, 9(1972).

[9] A. Koranyi, The Poisson integral for generalized half-planes and bounded symmetric domains , Ann. of Math., 82(1965).

[10] B. Kostant, Lie algebra cohomology and the generalized Borel-Weil theorem, Ann. of Math., 74(1961).

[11] R. A. Kunze and E. M. Stein, Uniformly bounded representations, II. Analytic continuation of the principal series of representations of the n×n complex unimodular group, Amer. J. Math., 83(1961).

[12] D. B. Lowdenslager, Potential theory in bounded symmetric homogeneous complex domains, Ann. of Math., 67(1958).

[13] J. Mitchell, Potential theory in the geometry of matrices, Trans. Amer. Math. Soc., 79(1955).

[14] C. C. Moore, Compactifications of symmetric spaces II, The Cartan domains, Amer. J. Math., 86(1964).

[15] M. S. Narasimhan and K. Okamoto, An anlogue of the Borel-Weil-Bott theorem for hermitian symmetric pairs of non-compact type, Ann. of Math., 91(1970).

[16] K. Okamoto, On induced representations, Osaka J. Math., 4(1967).

[17] K. Okamoto, The Poisson integral for vector bundle valued harmonic forms on bounded symmetric domains, to appear.

[18] ——— and H. Ozeki, On square-integrable $\bar{\partial}$-cohomology spaces attached to hermitian symmetric space, Osaka J. Math., 4(1967).

[19] P. J. Sally, Analytic continuation of the irreducible unitary representations of the universal covering group of SL(2, $\mathbf{R}$), Memoirs of the A. M. S., No.69(1967).

[20] I. Satake, On representations and compactifications of symmetric Riemannian spaces, Ann. of Math., 71(1960).

[21] E. M. Stein, G. Weiss and M. Weiss, $H^p$ classes of holomorphic functions in tube domains, Proc. Nat. Acad. Sci. U. S. A., 52(1964).

[22] M. Takeuchi, On orbits in a compact hermitian symmetric space, Amer. J. of Math., vol. 90. no. 3 (1968).

[23] J. A. Wolf and A. Koranyi, Generalized Cayley transformations of bounded symmetric domains, Amer. J. of Math., 87(1965).

# ACTION OF ALGEBRAIC GROUPS OF AUTOMORPHISMS ON THE DUAL OF A CLASS OF TYPE I GROUPS

by

L.Pukanszky

University of Pennsylvania

Let G be a real linear algebraic group operating on the finite dimensional real vector space V . If $G_0$ is the connected component of the neutral element in G , by virtue of a known theorem of C.Chevalley ( cf. [2] , p.316, and [4] , p.183 bottom ) the orbit space $V/G_0$ is countably separated .The purpose of the present talk is to outline the proof of the following result ,of which the statement just quoted is a special case . The details will be published elsewhere .

Theorem. *Suppose that* $\mathfrak{g}$ *and* $\mathfrak{h}$ *are algebraic Lie algebras of endomorphisms of* V , *such that* $\mathfrak{h}$ *is an ideal of* $\mathfrak{g}$ . *Let* G *be a connected and simply connected Lie group belonging to* $\mathfrak{g}$ *,and* H *the analytic subgroup of* G *determined by* $\mathfrak{h}$ . *Then the natural action of* G *on the dual* $\hat{H}$ *of* H *is countably separated* .

An assertion of the indicated type plays an important role in J.Dixmier's proof of the semifiniteness of the left

---

This research was partially supported by a grant from the National Science Foundation .

ring of an arbitrary connected Lie group ( cf. [5] and Remark 3 below ) .[1)]

Remark 1 . We recall that if $\mathfrak{g}$ and G are as above , then G is of type I ( cf. [5] ,2.1 ,Proposition , p.425 ) .

Remark 2 .One can show that our theorem is equivalent to the following . Let $\mathfrak{g}$ be a Lie algebra over the reals , $\mathfrak{h}$ an ideal in $\mathfrak{g}$ such that the image of $\mathfrak{g}$ and $\mathfrak{h}$ resp. in the adjoint representation of $\mathfrak{g}$ is an algebraic Lie algebra of endomorphisms of the underlying space of $\mathfrak{g}$ . Then , upon forming G and H as above , $\hat{H}/G$ is countably separated .

Remark 3 . Let us observe that in [5] only the following special case is needed . Assume that $\mathfrak{g}_1$ is a Lie algebra of endomorphisms of V , such that the greatest ideal which is nilpotent ( =nilradical ) consists of nilpotent endomorphisms. Then take for $\mathfrak{g}$ the algebraic closure of $\mathfrak{g}_1$ and set $\mathfrak{h} = \mathfrak{n} + [\mathfrak{g}_1,\mathfrak{g}_1]$.

To start we recall ( cf. [6] , Theorem 1 , p.124 ) that if G is a locally compact group acting as transformation group on the locally quasicompact and almost Hausdorff space M ,G and M both being assumed to satisfy the second axiom of countability , then , in particular , the following conditions are equivalent :

---

1) One of the motivations of the present research is to complete this proof . Our theorem implies easily 4.6 Lemme in [5] .The demonstration given for this loc.cit. , however , is incomplete since it is not shown , that the maps $\Psi$ and $\Phi$ ( cf. p.433 ) are bijective , as claimed .

1) M/G is countably separated , 2) For any m in M , Gm is locally closed ( cf. (iii) and (i) loc. cit. ) . If G and H are as in our Theorem , then (G,H) satisfies the conditions formulated above for (G,M) , and thus it is enough to show that for any fixed $\lambda \in \hat{H}$ the orbit G is locally closed in H . We shall proceed by mathematical induction , assuming the validity of our statement for pairs (G',H') as above with dim(G') + dim(H') < dim(G) + dim(H) . The reduction procedure employed in the sequel is partly inspired by that of [3] ( cf. p. 326 ) .

Let $\mathfrak{m}$ be the greatest ideal of nilpotency of the identical representation of $\mathfrak{g}$. In the following we shall distinguish notions relative to $\mathfrak{h}$, analogous to ones already defined for $\mathfrak{g}$, by an index zero . Thus we have $\mathfrak{m}_o = \mathfrak{m} \cap \mathfrak{h}$. We can assume that $\dim(\mathfrak{m}_o) > 0$ . In fact otherwise , since $\mathfrak{m}_o \supset [\mathfrak{h}, \mathfrak{r}_o]$ ($\mathfrak{r}$ = radical of $\mathfrak{g}$ ) , $\mathfrak{h}$ is reductive , and then the desired conclusion follows by an easy application of Chevalley's theorem . We shall distinguish two cases: A) There is a $\mathfrak{g}$ ideal $\mathfrak{a}$ in $\mathfrak{m}_o$ , such that either $\dim(\mathfrak{a}) > 1$ , or $\dim(\mathfrak{a}) = 1$ and $\mathfrak{a}$ is not central in $\mathfrak{g}$; B) Condition A) can not be fulfilled .

Ad A) Let us put $A = \exp_H(\mathfrak{a})$ for the analytic subgroup , determined by $\mathfrak{a}$, of H .By virtue of Chevalley's result the spaces $\hat{A}/H$ and $\hat{A}/G$ are countably separated . Hence , in particular , $\lambda$ in $\hat{H}$ ( as above ) restricts on A to a transitive quasi orbit $H\omega$ ( $\omega \in A$ ) , and $\Omega = G\omega \subset \hat{A}$ is locally closed . There are two possibilities : AI) $\dim \Omega > 0$ , AII) $\dim \Omega = 0$ .

Ad AI) . The following two lemmas contain more than needed for the given context . Suppose , that H and N are closed , invariant and type I subgroups of the separable locally compact group G , such that $G \supset H \supset N$ . Let $\lambda$ be a fixed element of $\hat{H}$ . We assume that , on N , $\lambda$ restricts to a transitive quasi orbit , $H\omega$ ( $\omega \in \hat{N}$ ) say . We put $\Omega = G\omega$ ($\subset \hat{N}$ ) and write $\hat{H}_\Omega$ for the set of all elements $\eta$ in $\hat{H}$ , for which the measure class , corresponding to $\eta|N$ on $\hat{N}$ , is carried by $\Omega$ . Let us observe , incidentally , that $\Omega$ is Borel in $\hat{N}$ , and if $\Omega$ is locally closed then so is $\hat{H}_\Omega$ ( for the latter cf. [5] , 4.2.Lemma ) .Let $S = H_\omega$ ( $G_\omega$ ) be the stabilizer of $\omega$ in H ( G resp. ) . With these assumptions and notations we have

Lemma 1 .$\hat{H}_\Omega/G$ <u>is countably separated if and only if so is</u> $\hat{S}_\omega/G_\omega$ .

Lemma 2 .<u>Assume in addition, that all</u> $\lambda$ <u>in</u> $\hat{H}$ <u>restrict on</u> N <u>to transitive quasi orbits and that</u> $\hat{N}/G$ <u>is countably separated</u> . <u>Then</u> $\hat{H}/G$ <u>is countably separated if and only if so is</u> $\hat{S}_\omega/G_\omega$ <u>for all</u> $\omega$ <u>in</u> $\hat{N}$ .

Let now G and H be as in our Theorem and A ($\subset$ H ) and $\omega \in \hat{A}$ as above . Then , $\hat{H}_\Omega$ being locally closed in $\hat{H}$ , since $\lambda \in \hat{H}_\Omega$ , it is enough to establish that $\hat{H}_\Omega/G$ is countably separated . For this , by virtue of Lemma 1 , it suffices to show that so is $\hat{S}_\omega/G_\omega$ .

Lemma 3 . <u>For any</u> $\omega$ <u>in</u> $\hat{A}$ , $S = H_\omega$ <u>is of type</u> I .

If $\dim(\Omega) > 0$ , then $\dim(G_\omega) < \dim(G)$ .Using the assumption of our inductive procedure and Lemma 2 , one can show that

$\hat{S}/G_\omega$ is countably separated , which implies the same for $\hat{S}_\omega/G_\omega$

Ad AII) Let us put J for the connected component of the kernel of $\omega \in \hat{A}$ . We have $\dim(J) > 0$ . Let j be the Lie algebra of J ; j is an ideal in $\mathfrak{g}$. Recalling that $j \subset \mathfrak{a} \subset \mathfrak{m}_o$ , we conclude that j is algebraic . One can show the existence of a rational representation $\rho$ of $\mathfrak{g}$( cf. for this and the following [2] , pp.47-48, and [3] , Lemma 11 , p.325 ), the kernel of which coincides with j .This being so, the desired conclusion follows by applying the assumption of our induction to ( $\rho(\mathfrak{g})$ , $\rho(\mathfrak{g})$ ) in place of ($\mathfrak{g}$ , $\mathfrak{h}$) .

Ad B) Given a Lie algebra $\mathfrak{g}$ ( or group G ) we denote by $\mathfrak{g}^\natural$ ( $G^\natural$ resp. ) its center . Our assumption implies that $\dim(\mathfrak{m}_o) = 1$ and that $\mathfrak{m}_o \subset \mathfrak{g}^\natural$ . In addition one can show , proceeding as in [3] Lemma 10 ( p.325 ) , that $\mathfrak{m}_o$ is a Heisenberg algebra . Let us put $M_o = \exp_H(\mathfrak{m}_o)$ . There is a character $\chi$ of $M_o^\natural$ such that $\lambda | M_o^\natural$ is a multiple of $\chi$ . We can assume that $\chi$ is not identically one , since otherwise we could prove our point by using the reasoning of AII). It is known that there is a uniquely determined element $\omega$ of $\hat{M}_o$ which , on $M_o^\natural$ , restricts to a multiple of $\chi$ . Since $\chi$ is G invariant , so is $\omega$, and hence G operates on the closed subset $\hat{S}_\omega$ , containing $\lambda$ , of $\hat{H}$ ( observe , that now $S = H_\omega = H$ ) .

Lemma 4 . Let $\mathfrak{g}$ be a finite dimensional Lie algebra over the the reals , and $\mathfrak{h}$ an ideal of $\mathfrak{g}$ . There is an ideal $\mathfrak{g}_1$ of $\mathfrak{g}$, such that $\mathfrak{g} = \mathfrak{g}_1 + \mathfrak{h}$ and $[\mathfrak{g}_1, \mathfrak{h}] \subset [\mathfrak{r}, \mathfrak{g}] \cap \mathfrak{h}$, where

$\mathfrak{r}$ is the radical of $\mathfrak{g}_1$.

Applying this to $\mathfrak{g}$ and $\mathfrak{h}$ as in our Theorem , we obtain an ideal $\mathfrak{g}_1$ of $\mathfrak{g}$, such that $\mathfrak{g} = \mathfrak{g}_1 + \mathfrak{h}$ and $[\mathfrak{g}_1, \mathfrak{h}] \subset [\mathfrak{g}, \mathfrak{r}] \cap \mathfrak{h} \subset \mathfrak{m}_o$. Let us put $G_1 = \exp_G(\mathfrak{g}_1)$ ; we have $G = G_1 H$ . Since evidently $G\lambda = G_1\lambda$ , to complete our proof it suffices to show that $\hat{S}_\omega/G_1$ is countably separated . Let us put $K = H/M_o$ ; K is reductive and simply connected .Furthermore , if r and h are arbitrary elements of $G_1$ and H resp. , we have $rhr^{-1} \subset hM_o$ . For the notions and notations used in the following cf. [1] , Ch.I ,Section 4 . We denote by $\omega'$ a projective $\iota(\alpha)^{-1}$ extension of $\omega$ to H ( $\alpha \in Z^2(K,T)$ ) .

Lemma 5 . There is a continuous homomorphism $\tau$ of $G_1$ into the group of characters of K , such that

$$a\,\omega' \equiv \iota(\tau(a))\,\omega'$$

for all a in $G_1$ in the sense of unitary equivalence .

We put $E = (K , \alpha)^\wedge$ and recall that E is now standard .

Lemma 6 . The map $\phi(\eta) = \omega' \otimes \iota(\eta)$ ( $\eta \in E$ ) defines a Borel isomorphism between E and $\hat{S}_\omega$ .

We observe now that if $a \in G_1$ , then $a\phi(\eta) = \phi(\tau(a)\eta)$ . From here the final conclusion is obtained by showing that the action of $G_1$ on E , corresponding to multiplication by $\tau$, is countably separated .

## References

[1] L.Auslander and C.C.Moore , Unitary representations of solvable Lie groups , Memoirs of the Am.Math.Soc. nr. 62 ( 1966 ) .

[2] C.Chevalley , Théorie des groupes de Lie , vol. 3, Act.Sci.Ind. n° 1226 , Paris ,Hermann ,1955 .

[3] J.Dixmier , Sur les répresentations unitaires des groupes de de Lie algébriques , Ann. Inst. Fourier , vol. 7 ( 1957 ) pp. 315 - 328 .

[4] J.Dixmier , Répresentations induites holomorphes des groupes résolubles algebriques , Bull.Soc. Math. France ,vol.94 (1966 ) pp. 181 - 206 .

[5] J.Dixmier , Sur la répresentation régulière d'un groupe localement compact connexe , Ann.Sci. de l'Ec.Norm. Sup., 4th series ,vol.2 ( 1969 ) pp. 423 - 436 .

[6] J.Glimm , Locally compact transfomation groups , Trans. Am. Math. Soc., vol. 101 ( 1961) pp. 124 - 138 .

# $L^1$- ALGEBRAS AND SEGAL ALGEBRAS

by

Hans Reiter
University of Vienna

Let $G$ be a locally compact group. A subset $S^1(G)$ of the Banach algebra $L^1(G)$ is said to be a Segal algebra if it satisfies the following conditions:

(i) $S^1(G)$ is dense in $L^1(G)$;

(ii) $S^1(G)$ is a Banach space, with norm $\|\cdot\|_S$, and

$$\|f\|_1 \leq \|f\|_S \quad \text{for all} \quad f \in S^1(G);$$

(iii) $f \in S^1(G) \Rightarrow L_y f \in S^1(G)$ for all $y \in G$, where

$$L_y f(x) = f(y^{-1}x), \quad x \in G,$$

and the mapping $y \to L_y f$ of $G$ into $S^1(G)$ is continuous;

(iv) $f \in S^1(G) \Rightarrow \|L_y f\|_S = \|f\|_S$ for all $y \in G$.

It readily follows from the conditions above that $S^1(G)$ is a left ideal of the algebra $L^1(G)$ (with convolution as multiplication) and that

$$\|h * f\|_S \leq \|h\|_1 \cdot \|f\|_S$$

holds for all $f \in S^1(G)$ and all $h \in L^1(G)$; in particular it follows that $S^1(G)$ is, in fact, a Banach algebra.

Examples:

1. The continuous functions in $L^1(G)$ that vanish at infinity form a Segal algebra, with norm $\|f\|_S = \|f\|_1 + \|f\|_\infty$.

2. $L^1(G) \cap L^p(G)$, $1 < p < \infty$, is a Segal algebra, with norm $\|f\|_S = \|f\|_1 + \|f\|_p$.

3. Let $G = \mathbb{R}$, the real numbers; denote by $\mathbb{Z}$ the integers. The continuous functions on $\mathbb{R}$ such that the norm

$$\|f\|_S = \sup_{y\in\mathbb{R}} \sum_{n\in\mathbb{Z}} \max_{0\leq x\leq 1} |f(y+x+n)|$$

is finite, form a Segal algebra.

4. Let $G = \mathbb{U}$, the unit circle, and consider the functions holomorphic in the open annulus $1/2 \leq |z| \leq 2$ and continuous on the boundary. The restrictions of these functions to $\mathbb{U}$ form a Segal algebra on $\mathbb{U}$, with norm

$$\|f\|_S = \max|f(z)| \ (1/2 \leq |z| \leq 2).$$

Let us put for any function $f$ on $G$,

$$R_y f(x) = f(xy^{-1})\Delta_G(y^{-1}), \quad y \in G,$$

where $\Delta_G$ denotes the Haar modulus of $G$. A Segal algebra is said to be symmetric if, in addition to the conditions (iii) and (iv) above, it also satisfies the analogous conditions for $R_y$. Examples 1 and 2 above are symmetric if and only if $G$ is unimodular.

Now consider any Banach algebra $B$, with norm $\|\cdot\|$. We say that $B$ has approximate left units if, given any $f \in B$ and $\varepsilon > 0$, there is a $u \in B$ such that

$$\|u\cdot f - f\| < \varepsilon.$$

We say that $B$ has multiple approximate left units if, given any finite set $F \subset B$ and $\varepsilon > 0$, there is a $u \in B$ such that

$$\|u\cdot f - f\| < \varepsilon \quad \text{for all} \quad f \in F.$$

A Segal algebra $S^1(G)$ possesses multiple approximate left units of $L^1$-norm 1; but, as was observed by J. T. Burnham, these cannot be bounded in the Segal norm if $S^1(G)$ is symmetric unless $S^1(G)$ coincides with $L^1(G)$.

We now proceed to discuss some results.

I. Consider a closed subgroup $H$ of $G$ (which may coincide with $G$). We define $D_H S^1(G)$ as the closed linear subspace of $S^1(G)$ spanned by the set

$$\{L_y f - f \mid f \in S^1(G),\ \ y \in H\}.$$

It is readily seen that $D_H S^1(G)$ is a right ideal of $S^1(G)$; in particular it is a Banach algebra (with the Segal norm). If $H$ is a normal subgroup of $G$ and $S^1(G)$ is symmetric, then it can be shown that

$$D_H S^1(G) = J^1(G,H) \cap S^1(G),$$

where $J^1(G,H)$ is the kernel of the canonical mapping $T_H$ of $L^1(G)$ onto $L^1(G/H)$, as discussed in H. Reiter, Classical harmonic analysis and locally compact groups (Oxford University Press, 1968), Chap. 3, §§4.2 - 4.5 and §5.3. We shall now give some applications of the property $P_1$ (discussed loc. cit., Chap. 8, §3) in this context.

$I_1$. If the closed subgroup $H$ has the property $P_1$, then $D_H S^1(G)$ has multiple approximate left units, of $L^1$-norm $\leq 2$.

$I_2$. If $H$ is normal and $S^1(G)$ is symmetric, then $H$ has the property $P_1$ if and only if $D_H S^1(G)$ has approximate left units bounded in $L^1$-norm; moreover, this statement remains true if we replace 'approximate left units' by 'multiple approximate left units'.

$I_3$. It follows that, if $D_H S^1(G)$ has approximate left units bounded in $L^1$-norm, then it has multiple approximate left units bounded in $L^1$-norm. It is an open question whether this result can be proved directly (i.e., without recourse to the property $P_1$).

II. Let G be a compact group.

$II_1$. Every closed two-sided ideal of a symmetric Segal algebra $S^1(G)$ has approximate (left) units of the form

$$(1) \qquad u = \sum_n c_n \cdot \chi_{D_n}, \quad c_n > 0,$$

where $\chi_{D_n}$ denotes the character of an irreducible unitary (finite-dimensional) representation of G and $\sum_n$ is a finite sum. We note that u lies in the centre of $L^1(G)$ and is positive-definite.

$II_2$. A closed two-sided ideal I of $L^1(G)$ has bounded approximate left units if and only if I is of the form

$$I = \mu * L^1(G),$$

where $\mu$ is a central, idempotent measure on G; moreover, we may take these approximate (left) units of the form (1) and of $L^1$-norm not larger than the norm of the measure $\mu$. We may again replace here 'approximate left units' by 'multiple approximate left units'.

III. Let G be a locally compact, but not compact, abelian group.

$III_1$. Not every closed ideal of $L^1(G)$ has approximate units. This is contained in the well-known work of Malliavin.

$III_2$. (a) It is known that there is a bijective correspondence $I_S \leftrightarrow I$ between the closed ideals $I_S$ of $S^1(G)$ and the closed ideals I of $L^1(G)$: I is the closure of $I_S$ in $L^1(G)$, $I_S$ is the intersection of I with $S^1(G)$ (cf. the book cited above, Chap. 6, §2.4). (b) The following new result is proved: $I_S$ has approximate units if and only if I has approximate units. J. T. Burnham has extended (a) to general locally compact groups (and even further); in the proof of (b) the abelian group structure enters decisively at one particular point and it is not known whether a corresponding extension holds.

$III_3$. A closed ideal I of $L^1(G)$ has bounded approximate units if and only if I is of the form

$$I = \bigcap_n \chi_n \cdot T_{H_n}^{-1}(\mu_n * L^1(G/H_n)),$$

where, for each n, $H_n$ is a closed subgroup of G, $\mu_n$ is a bounded, idempotent measure on $G/H_n$, $\chi_n$ is a character of G, and $T_{H_n}^{-1}$ is the inverse of the canonical mapping $T_{H_n} : L^1(G) \to L^1(G/H_n)$ (as introduced above); $\bigcap_n$ is a finite intersection. Moreover, we can say here "multiple approximate units' instead of 'approximate units'; also, the (multiple) approximate units may be chosen so as to have a positive Fourier transform, with compact support.

IV. Various extensions of the preceding results to soluble groups are possible; likewise to groups that contain abelian normal subgroups with compact quotient groups.

The proofs of these results, and of some others, and references, are given in the author's Springer Lecture Notes (Lecture Notes in Mathematics, Nr. 231, Springer-Verlag, Berlin-Heidelberg-New York, 1971).

# NORMS OF CHARACTERS AND CENTRAL $\Lambda_p$ SETS FOR U(n)

by

Daniel Rider
University of Wisconsin

The study of lacunary or thin sets in the dual object of a compact group began with the circle group (cf. [8]). Most of the results obtained for the circle were easily extended, with only minor modifications, to compact abelian groups [9; section 5.7]. More recently, in [3] definitions of Sidon and $\Lambda_p$ sets in the dual object of any compact group were given which permitted generalizations of many of the results on abelian groups. One difference, however, is the following. It is very easy to see that an infinite compact abelian group has infinite sets which are Sidon and $\Lambda_p$ for every p. On the other hand SU(2) has no infinite $\Lambda_2$ sets [6] and a priori no infinite Sidon sets.

In his thesis Parker [5] defined the notions of central Sidon and central $\Lambda_p$ sets. These differ from Sidon and $\Lambda_p$ sets in several ways. Perhaps the most important is that they deal with spaces of central functions which are not invariant under translations of the group. This greatly limits their useful applications. Another difference is that a central Sidon set need not be central $\Lambda_p$ for any p [5], [7]. It was also shown in [7] that the union of two central Sidon sets need not be central Sidon, as opposed to the remarkable theorem of Drury [2] for abelian groups.

The problem of the existence of central lacunary sets is much more tractable than that of lacunary sets. In [7] it is shown that a compact connected group has an infinite central Sidon set exactly when it is not a semi-simple Lie

---

This research was supported in part by NSF Grant GP-24182. The author is a fellow of the Alfred P. Sloan Foundation.

group. In this paper it is shown for which p the groups U(n) and SU(n) have infinite central $\Lambda_p$ sets. For example each such group has a set which is central $\Lambda_p$ for all $p < 3$ but is not central $\Lambda_3$. It is not known if there is for any compact group, even the circle, a set which is $\Lambda_2$ but not $\Lambda_3$.

We first find estimates for the norms of characters on U(n) and use them to determine the existence of central $\Lambda_p$ sets. The same things are then done for SU(n). In the final section we consider some problems for general compact groups. All the definitions used may be found in section 37 of [4].

1. Norms of characters on U(n). U(n) is the group of unitary $n \times n$ matrices $(n > 1)$. The maximal torus, $T_n$, of U(n) consists of the diagonal matrices and the Weyl integration formula states

(1) $$\int_{U(n)} f(g)\, dg = \frac{1}{n!} \int_{T_n} f(t)\, |D_n(t)|^2\, dt$$

whenever $f \in$ center of $L_1(U(n))$. For $t \in T_n$ having $z_1, \ldots, z_n$ as eigenvalues,

(2) $$D_n(t) = \det(z_i^{n-j}) = \prod_{i<j} (z_i - z_j) .$$

It is not difficult to see that

(3) $$\int_{T_n} |D_n(t)|^{-\alpha} < \infty \quad \text{if } \alpha < \frac{2}{n} .$$

The irreducible characters of U(n) are obtained in the following way [10; p. 198]: to each decreasing set of m integers, $m = (m_1, \ldots, m_n)$, corresponds an irreducible character $\chi_m$. If $g \in U(n)$ then g is conjugate to some $t \in T_n$; if t has distinct eigenvalues, $z_i$, then

(4) $$\chi_m(g) = \frac{\det(z_i^{m_j})}{D_n(t)} .$$

THEOREM 1. *For* $p < 2 + \frac{2}{n}$ *there is a constant* $B(p, n)$ *such that*

$$\|\chi_m\|_p \leq B(p, n)$$

*for all characters* $\chi_m$ *on* $U(n)$.

The theorem follows immediately from (1), (3) and (4). The following example shows it is the best possible.

EXAMPLE 2. *There is a sequence of characters* $\chi_m$ *on* $U(n)$ *with*

$$\|\chi_m\|_{2+2/n} \to \infty .$$

Let $m = ((n-1)N, (n-2)N, \ldots, N, 0)$ so that by (2)

$$D_n(t)\,\chi_m(t) = \prod_{i<j} (z_i^N - z_j^N) .$$

If $p = 2 + \frac{2}{n}$ then (1) gives

$$\|\chi_m\|_p^p = \frac{1}{n!} \int_T \prod |z_i^N - z_j^N|^p \, |D_n(t)|^{-2/n} \, dt . \tag{5}$$

Some standard calculations give that (5) increases as $\log N$.

Let $dm$ be the degree of the character $\chi_m$. If a set of characters has bounded degrees then it also has bounded sup norms. On the other hand we have the following:

THEOREM 3. *On* $U(n)$, $\|\chi_m\|_3 \to \infty$ *as* $dm \to \infty$.

The theorem is easily verified if $n = 2$. Assume it is true for $n-1$. For $m = (m_1, \ldots, m_n)$ let $m' = (m_2, \ldots, m_n)$ and $m'' = (m_1, \ldots, m_{n-1})$ represent characters on $U(n-1)$. For $t \in T_n$ with eigenvalues $z_1, \ldots, z_n$, let $t' \in T_{n-1}$ have eigenvalues $(z_2, \ldots, z_n)$. Then by (2) and (4),

(6) $$\prod_{2}^{n} (z_i - z_j)\, \chi_m^3(t)\, D_n^2(t) = z_1^{3m_1}\, \chi_{m'}^3(t')\, D_{n-1}^2(t') + Q,$$

where $Q$ is a polynomial with powers of $z_1$ less than $3m_1$. From (6) it follows that

$$2^{n-1} \int_{T_n} |\chi_m^3 D_n^2| \geq \int_{T_{n-1}} |\chi_{m'}^3 D_{n-1}^2| ,$$

so that by (1)

(7) $$2^{n-1} n \|\chi_m\|_3^3 \geq \|\chi_{m'}\|_3^3 .$$

In the same way (7) can be obtained with $m''$ in place of $m'$. Now as $dm \to \infty$ so does $dm' + dm''$, and the theorem follows by induction.

Theorem 3 is best possible in the following sense.

EXAMPLE 4. <u>There is a sequence of characters</u> $\chi_m$ <u>on</u> $U(n)$ <u>with</u> $dm \to \infty$ <u>and, for all</u> $p < 3$,

$$\|\chi_m\|_p \leq B(p) < \infty .$$

Let $m = (N, n-2, \ldots, 1, 0)$. Then $dm \to \infty$ with $N$. Now

(8) $$\chi_m(t) = \sum_{r=1}^{n} z_r^N \prod_{i \neq r} (z_r - z_i)^{-1} .$$

It follows that, if $p < 3$,

$$\|\chi_m\|_p^p \leq C \int_{T_n} \prod_{i=2}^{n} |z_1 - z_i|^{2-p} \leq B(p) < \infty .$$

2. <u>Central</u> $\Lambda_p$ <u>sets for</u> $U(n)$. The simple results of the previous section enable us to classify the types of central $\Lambda_p$ sets that exist for $U(n)$.

THEOREM 5. Suppose $E$ is a subset of the dual object of $U(n)$.

a) If $E$ is a central $\Lambda_3$ set then $dm(m \in E)$ is bounded.

b) There is an infinite set $E$ with $dm \to \infty$ that is a central $\Lambda_p$ set for all $p < 3$.

c) There is an infinite set $E$ which contains no infinite central $\Lambda_{2+2/n}$ set.

d) Every infinite set $E$ contains an infinite subset which is central $\Lambda_p$ for all $p < 2 + 2/n$.

a) and c) follow immediately from Theorem 3 and Example 2 respectively since, at the very least, a central $\Lambda_p$ set, $p > 2$, has $\|\chi_m\|_p$ bounded.

Proof of b). Let $E$ be the set having characters $\chi_m$ as in Example 4 with $\{N\}$ a $\Lambda_3$ set of integers. Let $f = \Sigma a_m x_m$ be an $E$-polynomial. Then by (8)

$$|f(t)| \leq \sum_{r=1}^{n} \left|\sum_m a_m z_r^N \prod_{i \neq r} (z_r - z_i)^{-1}\right| . \tag{9}$$

It follows from (1) and (9) that

$$\|f\|_p^p \leq \frac{n^p}{n!} \int_{T_n} \left|\sum_m a_m z_1^N \prod_{i\neq 1} (z_1 - z_i)^{-1}\right|^p |D_n|^2 \tag{10}$$

$$\leq C(n, p) \int_{T_n} \left|\sum_m a_m z_1^N\right|^p \prod_{i\neq 1} |z_1 - z_i|^{2-p} .$$

Since $\{N\}$ is a $\Lambda_3$ set of integers and $p < 3$, (10) is then bounded by $B(n, p)(\Sigma |a_m|^2)^{p/2}$ and this shows that $E$ is central $\Lambda_p$.

Proof of d). Let $E$ be an infinite set. If $dm(m \in E)$ is bounded then $E$ contains an infinite Sidon set which is an infinite $\Lambda_p$ set for all $p$. If $dm$ is unbounded then one can pick an infinite subset, $F$, such that either

$\{m_1\}$ or $\{m_n\}$ is a Sidon set of integers. Suppose the former. Let $f = \Sigma a_m x_m$ be an F-polynomial. Now

$$f(t) = \Sigma a_m \sum_{r=1}^{n} z_r^{m_1} \theta_{m,r} D_n^{-1}(t),$$

where $\theta_{m,r}$ is a polynomial that doesn't depend on $z_r$ and

(11) $$|\theta_{m,r}| \leq (n-1)!$$

As in the proof of b),

$$\|f\|_p^p \leq \frac{n^p}{n!} \int_{T_n} |\sum_m a_m z_1^{m_1} \theta_{m,1}|^p \, |D_n|^{2-p} .$$

If $p < 2 + 2/n$ pick $t > 1$ so small that $(p-2)t < 2/n$. Letting $s$ be conjugate to $t$, Hölders inequality gives

(12) $$\|f\|_p^p \leq \frac{n^p}{n!} \left( \int_{T_n} |\sum_m a_m z_1^{m_1} \theta_{m,1}|^{sp} \right)^{1/s} \left( \int_{T_n} |D_n|^{(2-p)t} \right)^{1/t} .$$

Since $\{m_1\}$ is a Sidon set of integers it is a $\Lambda_{ps}$ set as well. Thus for some constants $B$ and $B'$ it follows from (12), (3) and (11) that

$$\|f\|_p^{ps} \leq B \int_{T_{n-1}} (\Sigma |a_m \theta_{m,1}|^2)^{ps/2}$$

and

$$\|f\|_p \leq B'(\Sigma |a_m|^2)^{1/2} = B' \|f\|_2 .$$

This shows F is central $\Lambda_p$.

3. <u>Central</u> $\Lambda_p$ <u>sets for</u> SU(n). SU(n) is the subgroup of U(n) of matrices of determinant 1. Since U(n) = SU(n)·center of U(n), each irreducible character on U(n) is still irreducible when restricted to SU(n). Two characters $\chi_m$ and $\chi_q$ are the same on SU(n) when $m_i - m_n = q_i - q_n$ for all i. Since $\|\chi_m\|_p$ is the same for SU(n) as for U(n), Theorems 1 and 3 and examples 2 and 4 also hold for SU(n).

Theorem 5 also holds for SU(n). The proofs of parts b) and d) are slightly different but won't be given. Since SU(n) has only finitely many characters of any given degree, part a) can be started as: a central $\Lambda_3$ set for SU(n) is finite.

4. <u>Central</u> $\Lambda_p$ <u>sets for compact groups</u>. In some sense the previous sections have only used the fact that U(n) and SU(n) are compact connected Lie groups and thus have maximal tori. It seems reasonable to conjecture that Theorem 5 holds for any such group with an appropriate n. We can say the following at least.

<u>THEOREM</u> 6. <u>An infinite compact connected Lie group</u>, G, <u>has an infinite central</u> $\Lambda_2$ <u>set</u>.

The Weyl integration formula [1; p. 142] states that

$$\int_G f = \frac{1}{w} \int_T f(t)\ |D(t)|^2$$

for all central functions on G where w is the order of the Weyl group, T is a maximal torus of G and D is a polynomial on T with integral coefficients. Now a sequence of characters $E = \{\chi_n\}$ on G and a Sidon set of characters $\gamma_n$ on T can be chosen so that $\varphi_n = \chi_n\ |D|^2\ | T$, which is a polynomial on T with integral coefficients, has $\hat{\varphi}_n(\gamma_n) \neq 0$ and $\hat{\varphi}_n(\gamma_m) = 0$ if $m \neq n$.

Suppose $f = \sum_E a_n \chi_n \in L_{3/2}(G)$. Then

$$|\Sigma\, a_n \chi_n|^{3/2}\ |D|^3 \leq\ |D|_\infty |\Sigma\, a_n \chi_n|^{3/2}\ |D|^2 \in L_1(T),$$

and hence $F = \Sigma\, a_n \varphi_n \in L_{3/2}(T)$. Now since $\hat{F}(\gamma_n) = a_n \hat{\varphi}_n(\gamma_n)$ and $\gamma_n$ is a Sidon set for $T$, it follows [8; Theorem 5.4] that $\Sigma\, |a_n|^2 \leq \Sigma\, |\hat{F}(\gamma_n)|^2 < \infty$. Thus $f \in L_2(G)$ so that $E$ is a central $\Lambda_2$ set.

<u>COROLLARY</u> 7. <u>An infinite compact connected group</u> $G$ <u>has an infinite central</u> $\Lambda_2$ <u>set</u>.

This follows from Theorem 6 since $G$ has a compact connected Lie group as a factor group.

It does not seem to be known if "connected" is necessary.

## References


[1] J. F. Adams, Lectures on Lie Groups, Benjamin, New York (1969).

[2] S. Drury, Sur les ensembles de Sidon, C. R. Acad. Sci. Paris 271 (1970), 162-163.

[3] A. Figá-Talamanca and D. Rider, A theorem of Littlewood and lacunary series for compact groups, Pacific J. Math. 16 (1966), 505-514.

[4] E. Hewitt and K. A. Ross, Abstract Harmonic Analysis II, Springer-Verlag, New York (1970).

[5] W. A. Parker, Central Sidon sets and central $\Lambda_p$ sets, Doctoral dissertation, University of Oregon (1970).

[6] J. F. Price, Non ci sono insiemi infiniti di tipo $\Lambda(p)$ per $SU(2)$, to appear Boll. Un. Mat. It.

[7] D. Rider, Central lacunary sets, to appear.

[8] W. Rudin, Trigonometric series with gaps, J. Math. and Mech. 9 (1960), 203-227.

[9] _______, Fourier Analysis on Groups, Interscience, New York (1962).

[10] H. Weyl, The Classical Groups, Princeton Univ. Press, Princeton (1946).

# SOME NEW CHARACTERIZATIONS OF SIDON SETS

by

Kenneth A. Ross
University of Oregon

This note is an announcement of results obtained jointly with Robert E. Edwards and Edwin Hewitt; the details will appear in [1].

Let $G$ be a compact Abelian group with character group $X$. For any subset $P$ of $X$ and any subset $E$ of $L^1(G)$, we write $E_P$ for the set of all $f$ in $E$ such that $\hat{f}(\chi) = 0$ for $\chi \in X \setminus P$. The space of functions with absolutely convergent Fourier series will be denoted by $A(G)$.

Recall that a subset $P$ of $X$ is said to be a *Sidon set* if and only if $C_P(G) \subset A(G)$. This characterization of Sidonicity remains intact if $C(G)$ is replaced by any one of a large class of smaller function spaces on $G$.

*Theorem*. *A subset* $P$ *of* $X$ *is a Sidon set if and only if* $E_P \subset A(G)$, *where* $E$ *is any of the following linear subspaces of* $C(G)$:

(a) $E = A^p(G)$ *for* $1 < p < 2$, *where*

$$A^p(G) = \{f \in C(G) : \hat{f} \in \ell^p(X)\}.$$

(b) $E = A(G;w)$ *where* $w$ *is in* $c_0(X)$ *and*

$$A(G;w) = \{f \in C(G) : \hat{f}w \in \ell^1(X)\}.$$

(c) $E = A^{1+}(G) = \{f \in C(G) : \hat{f} \in \ell^p(X)$ *for all* $p > 1\}$.

(d) *for* $G = T$, *let* $E = U(T)$, *the space of all* $f$ *in* $C(T)$ *for which the sequence of symmetric partial sums of its Fourier series converges uniformly.*

The spaces $E$ of this theorem all have the following properties for any subset $P$ of $X$:

---

This research supported by NSF Grant GP-19274 and NSF Grant GP-20226.

(i) every P-spectral trigonometric polynomial belongs to E;

(ii) there is a topology on $E_P$ making $E_P$ into a barrelled locally convex topological vector space;

(iii) for each $\chi$ in P, $f \to \hat{f}(\chi)$ is a continuous function on $E_P$;

(iv) if L is a continuous linear functional on $E_P$, then there are a number $\alpha$ in (0,1), a measure $\mu$ in M(G), a function $\psi$ in $\ell^\infty(X)$, and a finite subset $\Phi$ of P, such that $|\psi(\chi)| \leqq \alpha$ for all $\chi \in P \setminus \Phi$ and

$$L(f) = \mu^* f(e) + \sum_{\chi \in P} \psi(\chi)\hat{f}(\chi)$$

for all P-spectral trigonometric polynomials f on G.

Our general theorem asserts that the inclusion $E_P \subset A(G)$ is equivalent to the Sidonicity of P for any space E satisfying (i) - (iv). In particular, it applies to a fairly general class of spaces generalizing U(T), i.e., spaces of functions on G whose Fourier series converge uniformly in some sense.

In many cases the inclusions $E_P \subset A(G)$ are equivalent to certain inequalities. As a simple example, we observe that a subset P of X is a Sidon set if and only if there exist p in $(1,\infty]$, w in $c_o(X)$, and a positive constant k such that

$$\|\hat{f}\|_1 \leqq k\left\{\|f\|_u + \|\hat{f}\|_p + \|\hat{f}w\|_1\right\}$$

for all P-spectral trigonometric polynomials f on G.

The foregoing results generalize in a straightforward manner to the setting of compact non-Abelian groups.

## Reference

[1] R. E. Edwards, E. Hewitt and K. A. Ross, Lacunarity for compact groups, II, to appear in Pacific J. Math.

# THE FOURIER TRANSFORM OF INVARIANT DISTRIBUTIONS

by

Paul J. Sally, Jr. and Garth Warner*

The Institute for Advanced Study and University of Chicago
University of Washington

§0 - Introduction. Let $G$ be a locally compact, separable unimodular group and let $dx$ denote a Haar measure on $G$. If $\pi$ is an irreducible unitary representation of $G$ on a Hilbert space and $f \in L^1(G)$, then the (operator-valued) Fourier transform of $f$ at $\pi$ is

$$\pi(f) = \int_G f(x)\pi(x)dx. \tag{0.1}$$

Denote by $E(G)$ the set of equivalence classes of irreducible unitary representations of $G$. If $\pi(f)$ is of trace class, then $\Theta_\pi(f) = \operatorname{tr} \pi(f)$ depends only on the equivalence class $\omega$ of $\pi$ and we shall write $\Theta_\omega(f)$ for $\Theta_\pi(f)$. For $y \in G$, set ${}^y f(x) = f(yxy^{-1})$. Then $\pi(f)$ is of trace class if and only if $\pi({}^y f)$ is of trace class for all $y$ in $G$, and, in this case, we have $\Theta_\omega(f) = \Theta_\omega({}^y f)$.

We assume that there exists a topological vector space $V$ of continuous functions on $G$ which is suitable for harmonic analysis in the sense of [7(i)], p. 530. A distribution on $G$ is a continuous linear functional $\Lambda$ on $V$, and we say that $\Lambda$ is an invariant distribution if $\Lambda({}^y f) = \Lambda(f)$ for $f \in V$, $y \in G$. For each $f \in V$, the (scalar-valued) Fourier transform $\hat{f}(\omega) = \Theta_\omega(f)$ exists and, if $\Lambda$ is an invariant distribution on $G$, then the Fourier transform of $\Lambda$ is a linear functional $\hat{\Lambda}$ on $E(G)$ satisfying

$$\hat{\Lambda}(\hat{f}) = \Lambda(f), \quad f \in V. \tag{0.2}$$

We do not attempt to discuss here either the precise nature of the space of functions $\{\hat{f} : f \in V\}$ on $E(G)$ or the continuity properties of $\hat{\Lambda}$. Moreover, in order

*Research of both authors supported by the National Science Foundation.

to compute the Fourier transforms of the invariant distributions that we consider, it is necessary to define $\hat{f}$ on a set which may be larger than $E(G)$.

Our interest centers on invariant distributions which arise as follows. Fix an element $\gamma \in G$ and let $G_\gamma$ be the centralizer of $\gamma$ in $G$. Assume that $G_\gamma$ is unimodular and let $d\dot{x}$ be an invariant measure on $G/G_\gamma$. Set

$$\Lambda_\gamma(f) = \int_{G/G_\gamma} f(x\gamma x^{-1})d\dot{x}, \quad f \in V. \tag{0.3}$$

Assuming that $\Lambda_\gamma$ is a distribution on $G$ (obviously invariant), we consider the problem of computing $\hat{\Lambda}_\gamma$. In the particular case when $\gamma = 1$, the identity in $G$, $\Lambda_\gamma(f) = \delta(f) = f(1)$ and $\hat{\delta}$ is the Plancherel measure for $G$.

Now, we turn to the situations in which something concrete can be stated. For $G = SL(2, k)$, where $k$ is a non-archimedean local field of odd residual characteristic, $\hat{\Lambda}_\gamma$ is computed in [10(b)] for $\gamma$ a regular element in a compact Cartan subgroup of $G$ and also for $\gamma$ a unipotent element of $G$. If $G$ is a connected semisimple Lie group of real rank one which has discrete series, then the formula for $\hat{\Lambda}_\gamma$, $\gamma$ a regular element in a compact Cartan subgroup, is announced in [11(a)]. In the latter case, it is necessary to consider the Fourier transform not only on $E(G)$ but on a collection $I(G)$ of invariant eigendistributions on $G$. Along with the characters $\Theta_\omega(f)$, $\omega \in E(G)$, $I(G)$ also contains a collection $I^S(G)$ of invariant eigendistributions introduced by Harish-Chandra ([7(f)], [7(g)]) which, as yet, have no character theoretic interpretation except in special cases. These special cases include the representations introduced by Knapp-Okamoto in [6] and by Hirai in [8]. Even though the invariant eigendistributions in $I^S(G)$ are not related to characters of the discrete series for $G$, the members of $I^S(G)$ appear as discrete summands in the formula for $\hat{\Lambda}_\gamma$. It is a central problem to identify explicitly the character theoretic nature of the distributions in $I^S(G)$.

The Plancherel formula for the groups mentioned above ([7(g)], [9], [10(a)] can be obtained from $\hat{\Lambda}_\gamma$ by using a suitable limit formula. This limit formula is given by Harish-Chandra [7(b)], [7(d)] in the real case and by Shalika [12] in the p-adic case.

In §1 of this note, the problem of computing $\hat{\Lambda}_\gamma$ for compact groups is solved by two different approaches. The first approach is applicable to arbitrary compact groups and provides motivation for the techniques employed in [10(b)] for $G = SL(2, k)$. The second approach deals with compact connected semisimple Lie groups and illustrates the method used for semisimple Lie groups of real rank one in [11(b)]. Of course, compact groups have only discrete series. For non-compact groups, we first analyze the discrete series $E_2(G)$ as in the case of compact groups. We are left with a remainder term which must be expressed in terms of the principal series.

In §2, we give a detailed outline of the derivation of $\hat{\Lambda}_\gamma$ for $G = SU(n, 1)$. This presents most of the facets of the general rank one situation. Then, for $G = SU(1, 1)$ ($\cong SL(2, \mathbb{R})$), we show how the Plancherel formula can be obtained from $\hat{\Lambda}_\gamma$. We note that, for $SL(2, \mathbb{R})$, the formula for $\hat{\Lambda}_\gamma$ is given in [2], [3] and [7(a)].

Throughout, the cardinal number of a set $W$ is denoted by $[W]$. For any subset $B$ of a semisimple group, we denote by $B'$ the set of regular elements in $B$. For the purposes of this paper, a regular element may be considered simply as a matrix with all eigenvalues distinct. We also write $B^G = \{xbx^{-1} : x \in G, b \in B\}$. The general facts that we use for compact groups in §1 may be found in [1], [5], [13], [14]. For §2, we draw on [13] along with the papers of Harish-Chandra cited in the References.

## §1 - The Fourier Transform on Compact Groups

(a) The General Case. Let $G$ be a compact group and $\gamma$ a fixed element of $G$. If $\omega \in E_2(G) = E(G)$, we denote the degree of $\omega$ by $d_\omega$ and the character of $\omega$ by $\Theta_\omega$. For any $x \in G$, we have the functional equation for characters,

$$\Theta_\omega(\gamma)\overline{\Theta_\omega(x)} = d_\omega \int_G \Theta_\omega(u\gamma u^{-1}x^{-1})du, \tag{1.1}$$

where, as usual, $\int_G du = 1$.

Furthermore, for suitably nice functions $f$ on $G$, we have

$$f(x) = \sum_{\omega \in E(G)} d_\omega (f * \Theta_\omega)(x), \tag{1.2}$$

where $*$ denotes convolution on $G$ and the series converges absolutely and uniformly to $f$. For simplicity, we consider the invariant linear functional $\Lambda_\gamma(f) = \int_G f(u\gamma u^{-1})du$. This differs from $\Lambda_\gamma$ in (0.3) only by a constant factor.

From (1.1), we have, for $f$ as above,

$$\sum_{\omega \in E(G)} \Theta_\omega(\gamma)\int_G f(x)\overline{\Theta_\omega(x)}dx \tag{1.3}$$

$$= \sum_{\omega \in E(G)} d_\omega \int\int_G f(x)\Theta_\omega(u\gamma u^{-1}x^{-1})du\, dx.$$

Then, the last sum can be written as

$$\int_G \sum_{\omega \in E(G)} d_\omega \int_G f(x)\Theta_\omega(u\gamma u^{-1}x^{-1})dx\, du$$

$$= \int_G \sum_{\omega \in E(G)} d_\omega (f * \Theta_\omega)(u\gamma u^{-1})du$$

$$= \int_G f(u\gamma u^{-1})du$$ by (1.2). Thus,

$$\Lambda_\gamma(f) = \sum_{\omega \in E(G)} \hat{f}(\omega)\overline{\Theta_\omega(\gamma)} = \hat{\Lambda}_\gamma(\hat{f}). \tag{1.4}$$

The Plancherel formula for $G$, follows immediately from (1.4) by taking

$\gamma = 1$, that is

$$\delta(f) = f(1) = \sum_{\omega \in E(G)} \hat{f}(\omega)\overline{\Theta_\omega(1)} = \sum_{\omega \in E(G)} d_\omega \hat{f}(\omega). \tag{1.5}$$

(b) Compact Lie Groups. Let $G$ be a compact, connected semisimple Lie group and let $T$ be a maximal torus in $G$. We write $\underline{g}$ and $\underline{t}$ for the Lie algebras of $G$ and $T$ respectively and $\underline{g}_{\mathbb{C}}$, $\underline{t}_{\mathbb{C}}$ for the complexifications of $\underline{g}$ and $\underline{t}$. Let $P$ denote the set of positive roots of $(\underline{g}_{\mathbb{C}}, \underline{t}_{\mathbb{C}})$ with respect to some ordering and, as usual, set $\rho = \frac{1}{2}\sum_{\alpha \in P} \alpha$. The character group $\hat{T}$ of $T$ may be identified with a lattice $L_T$ in $\sqrt{-1}\,\underline{t}^*$, $\underline{t}^*$ the dual of $\underline{t}$. For $\tau \in L_T$, the corresponding character $\xi_\tau$ on $T$ is defined by

$$\xi_\tau(\exp H) = e^{\tau(H)}, \quad H \in \underline{t}. \tag{1.6}$$

Now let $W = W(\underline{g}_{\mathbb{C}}, \underline{t}_{\mathbb{C}}) = N(T)/T$, the Weyl group of $G$. $W$ acts on $\underline{t}$ and $T$, and on $L_T$ and $\hat{T}$ by duality. Thus,

$$\xi_{w\tau}(\exp H) = e^{w\tau(H)} = e^{\tau(w^{-1}H)}, \quad w \in W. \tag{1.7}$$

An element $\tau \in L_T$ is called regular if $w\tau \neq \tau$ for all $w \neq 1$ in $W$. Otherwise $\tau$ is called singular. A character $\xi_\tau \in \hat{T}$ is also called regular or singular accordingly. We assume that $G$ is acceptable ([7(e)]) so that $\rho \in L_T$. Set

$$\Delta(t) = \xi_\rho(t) \prod_{\alpha \in P} (1-\xi_\alpha(t^{-1})) = \prod_{\alpha \in P} (e^{\alpha(H)/2} - e^{-\alpha(H)/2}), \tag{1.8}$$

where $t = \exp(H)$, $H \in \underline{t}$.

Denote the set of regular elements in $L_T$ by $L_T'$. The irreducible unitary representations of $G$ are indexed by the orbits of $W$ in $L_T'$.

For any $\tau \in L_T$, we define

$$\Theta_\tau(t) = \Delta(t)^{-1} \sum_{w \in W} \det(w)\xi_{w\tau}(t), \quad t \in T'. \tag{1.9}$$

Recall that T' denotes the set of regular elements in T. If $\tau$ is singular, then $\Theta_\tau \equiv 0$ on T' and we simply extend $\Theta_\tau$ to be identically zero on G. If $\tau$ is regular, then it follows from Weyl's character formula that the character of the irreducible unitary representation corresponding to $\tau$ is

(1.10) $$\Theta_{\omega(\tau)}(t) = \varepsilon(\tau)\Theta_\tau(t), \quad t \in T',$$

where $\varepsilon(\tau) = \mathrm{sgn}(\prod_{\alpha \in P}(\tau, \alpha))$. (The classical Weyl character formula in terms of the highest weight $\Lambda$ results upon taking $\tau = w_0(\Lambda+\rho)$ for some $w_0 \in W$. Then $\varepsilon(\tau) = \det(w_0)$.)

Normalize the Haar measures on G and T so that $\int_G dx = \int_T dt = 1$. We assume that the invariant measure $d\dot{x}$ on G/T is normalized in such a way that

(1.11) $$\int_G f(x)dx = \int_{G/T}\int_T f(xt)dt\, d\dot{x}.$$

For $f \in C^\infty(G)$, define

(1.12) $$F_f(t) = \Delta(t)\int_{G/T} f(xtx^{-1})d\dot{x}, \quad t \in T.$$

Then, from Weyl's integral formula, we have

(1.13) $$\int_G f(x)dx = [W]^{-1}\int_T \overline{\Delta(t)}F_f(t)dt.$$

Observe that $F_f \in C^\infty(T)$ and

(1.14) $$F_f(wt) = \det(w)F_f(t), \quad w \in W.$$

Let t be a fixed element of T'. Our goal is to determine the Fourier transform of the invariant distribution $f \mapsto F_f(t)$, $f \in C^\infty(G)$. For any $\tau \in L_T$, we define

(1.15) $$\check{f}(\tau) = \int_G f(x)\Theta_\tau(x)dx.$$

For $\tau$ singular, $\check{f}(\tau) = 0$, and, for $\tau$ regular $\varepsilon(\tau)\check{f}(\tau)$ is just the scalar-valued Fourier transform at the representation indexed by $\tau$. Of course, $\check{f}(w\tau) = \det(w)\check{f}(\tau)$ for $w \in W$. The point here is that we wish to employ the classical inversion theorems for multiple Fourier series and, hence, we must include the singular $\tau$ in our analysis to obtain the full character group of T.

Set $r = [P]$. Then, from (1.9) and (1.13),

$$\check{f}(\tau) = [W]^{-1}(-1)^r \sum_{w \in W} \det(w) \int_T \xi_{w\tau}(t) F_f(t)dt.$$

Employing (1.7) and (1.14), we obtain

$$\check{f}(\tau) = (-1)^r \int_T \xi_\tau(t) F_f(t) dt = (-1)^r \hat{F}_f(\tau),$$

where $\hat{F}_f(\tau)$ is the Fourier coefficient of $F_f$ at $\xi_\tau$.

Since $F_f \in C^\infty(T)$, we have

$$F_f(t) = \sum_{\tau \in L_T} \hat{F}_f(\tau)\overline{\xi_\tau(t)}, \quad t \in T,$$

where the series converges uniformly and absolutely. Thus,

$$\begin{aligned} F_f(t) &= (-1)^r \sum_{\tau \in L'_T} f^\vee(\tau)\overline{\xi_\tau(t)} \\ &= (-1)^r [W]^{-1} \sum_{w \in W} \sum_{\tau \in L'_T} f^\vee(w\tau)\overline{\xi_{w\tau}(t)} \\ &= (-1)^r [W]^{-1} \sum_{\tau \in L'_T} f^\vee(\tau) \sum_{w \in W} \det(w)\overline{\xi_{w\tau}(t)} \\ &= \Delta(t)[W]^{-1} \sum_{\tau \in L'_T} f^\vee(\tau)\overline{\Theta_\tau(t)}. \end{aligned} \tag{1.16}$$

Theorem. Let $f \in C^\infty(G)$ and, for $\tau \in L'_T$, set $\hat{f}(\tau) = \varepsilon(\tau) f^\vee(\tau)$. For $t \in T'$, define an invariant linear functional on $C^\infty(G)$ by $\Lambda_t(f) = \int_{G/T} f(xtx^{-1})dt$. Then $\Lambda_t$ is an invariant distribution on G and

$$(1.17) \qquad \hat{\Lambda}_t(\hat{f}) = [W]^{-1} \sum_{\tau \in L_T'} \hat{f}(\tau) \overline{\Theta_{\omega(\tau)}(t)}$$

Observe that, if the equivalence of representations under the action of W is taken into consideration, then (1.17) yields (1.4). The Plancherel formula for G can be obtained as in (1.5) or from (1.16) by using the fact that there exists a differential operator $\mathfrak{D}$ on G and a constant c independent of f such that $\lim_{t \to 1} \mathfrak{D} F_f(t) = c f(1)$ (see [13], Chapter III). It is the latter technique which can be used to derive the Plancherel formula for non-compact semisimple Lie groups of real rank one. Finally, we note that for non-compact semisimple Lie groups there are "singular characters" analogous to those defined by (1.9) for $\tau$ singular. In the non-compact case, the "singular" $\Theta_\tau$ need not vanish and they appear discretely in the formula for $F_f$ in the same fashion as the "regular" $\Theta_\tau$ in (1.16). This will be illustrated in §2.

## §2 - The Fourier Transform on SU(n, 1)

Let $G = SU(n,1)$, the group of automorphisms of $\mathbb{C}^{n+1}$ which preserve the hermitian quadratic form $|z_1|^2 + |z_2|^2 + \ldots + |z_n|^2 - |z_{n+1}|^2$ and have determinant one. A maximal compact subgroup of $G$ is given by $K = \left\{\begin{pmatrix} u & 0 \\ 0 & \gamma \end{pmatrix}\right\}$, where $u \in U(n)$, the unitary group of rank n, $\gamma \in \mathbb{T} = \{z \in \mathbb{C} : |z| = 1\}$, and $\gamma \det(u) = 1$. The Lie algebra of $K$ is $\underline{k} = \left\{\begin{pmatrix} X & 0 \\ 0 & Y \end{pmatrix}\right\}$ with $X$ an n by n skew hermitian matrix and $Y$ a complex number such that $tr(X) + Y = 0$. If $\underline{p} = \left\{\begin{pmatrix} 0 & Z \\ {}^t\bar{Z} & 0 \end{pmatrix}\right\}$ with $Z$ an n-dimensional column vector, then $\underline{g} = \underline{k} + \underline{p}$ is a Cartan decomposition of $\underline{g}$, the Lie algebra of $G$. The subgroup $T$ of diagonal matrices in $K$ is a maximal torus in $K$ and a compact Cartan subgroup of $G$. $T$ is isomorphic to $\mathbb{T}^n$.

If $H^* = \begin{pmatrix} 0 & \cdots & 1 \\ \vdots & 0 & \vdots \\ 1 & \cdots & 0 \end{pmatrix} \in \underline{g}$, then $\mathbb{R}H^*$ is a maximal abelian subalgebra of $\underline{p}$.

Set $A_{\underline{p}} = \{\exp(\mathbb{R}H^*)\} = \left\{\begin{pmatrix} \cosh t & 0 & \sinh t \\ 0 & I_{n-1} & 0 \\ \sinh t & 0 & \cosh t \end{pmatrix} : t \in \mathbb{R}\right\}$,

and let $M$ be the centralizer of $A_{\underline{p}}$ in $K$. Then $M = \left\{\begin{pmatrix} e^{\sqrt{-1}\theta} & 0 & 0 \\ 0 & u & 0 \\ 0 & 0 & e^{\sqrt{-1}\theta} \end{pmatrix}\right\}$ where $u \in U(n-1)$ and $e^{2\sqrt{-1}\theta}\det(u) = 1$. Let $A_K$ be the subgroup of diagonal matrices in $M$. The Lie algebras of $M$ and $A_K$ are written $\underline{m}$ and $\underline{a}_K$ respectively. $A = A_K A_{\underline{p}}$ is a Cartan subgroup of $G$ and the pair $(T, A)$ is a complete set (up to conjugacy) of Cartan subgroups of $G$.

The complexification of $\underline{g}$ is $\underline{g}_{\mathbb{C}} = sl(n+1, \mathbb{C})$. If $\underline{t}$ is the Lie algebra of $T$, then the complexification $\underline{t}_{\mathbb{C}}$ of $\underline{t}$ is the Cartan subalgebra of diagonal matrices in $\underline{g}_{\mathbb{C}}$. A matrix in $\underline{t}_{\mathbb{C}}$ with diagonal entries $\lambda_1, \lambda_2, \ldots, \lambda_{n+1}$ will be written $(\lambda_1, \lambda_2, \ldots, \lambda_{n+1})$. We can write

$\underline{t} = \{(\sqrt{-1}\varphi_1, \ldots, \sqrt{-1}\varphi_{n+1}) : \sum_{j=1}^{n+1} \varphi_j = 0\}$. Define $e_j(\lambda_1, \ldots, \lambda_{n+1}) = \lambda_{n+2-j}$, $j = 1, 2, \ldots, n+1$. The set of roots of the pair $(\underline{g}_{\mathbb{C}}, \underline{t}_{\mathbb{C}})$ is $\{(e_i - e_j) : i \neq j\}$, and we choose an ordering so that the positive roots are

$$P_T = \{e_i - e_j : 1 \leq i < j \leq n\}. \tag{2.1}$$

We have

<u>compact roots</u>: $e_i - e_j$, $i, j > 1$;

<u>non-compact roots</u>: $\pm(e_1 - e_j)$, $2 \leq j \leq n+1$;

<u>simple roots</u>: $\alpha_j = e_j - e_{j+1}$, $1 \leq j \leq n$.

As usual, we define $\rho = \frac{1}{2} \sum_{\alpha \in P_T} \alpha$.

The Weyl group $W_{\underline{g}} = W(\underline{g}_{\mathbb{C}}, \underline{t}_{\mathbb{C}})$ of the pair $(\underline{g}_{\mathbb{C}}, \underline{t}_{\mathbb{C}})$ is $S_{n+1}$, the symmetric group on $(n+1)$ letters, and $W_{\underline{g}}$ acts on $\underline{t}_{\mathbb{C}}$ in the obvious way $w : (\lambda_1, \ldots, \lambda_{n+1}) \mapsto (\lambda_{w(1)}, \ldots, \lambda_{w(n+1)})$. The Weyl group $W(G, T) = N(T)/T$ is just the Weyl group of K, that is, $S_n$, and $W(G, T)$ may be regarded as the subgroup of $W_{\underline{g}}$ which fixes $\lambda_{n+1}$.

The unitary dual $\hat{T}$ of T can be identified with a lattice $L_T$ in $\sqrt{-1}\underline{t}^*$, $\underline{t}^*$ the dual of $\underline{t}$, and, for $\tau \in L_T$, we have, as in (1.6), the corresponding character $\xi_\tau$ on T given by

$$\xi_\tau(\exp H) = e^{\tau(H)}, \quad H \in \underline{t}. \tag{2.2}$$

For $w \in W_{\underline{g}}$, we have, as in (1.7),

$$w\tau(H) = \tau(w^{-1}H), \quad \xi_{w\tau}(\exp H) = e^{\tau(w^{-1}H)}, \quad H \in \underline{t}, \ \tau \in L_T. \tag{2.3}$$

We say that $\tau \in L_T$ is <u>regular</u> if $w\tau \neq \tau$ for all $w \neq 1$ in $W_{\underline{g}}$; otherwise $\tau$ is <u>singular</u>. The set of regular $\tau$ will be denoted by $L'_T$ and the set of singular $\tau$ by $L^s_T$. The character $\xi_\tau$ is called regular or singular accordingly. Note here that the singular characters $\xi_\tau$ include not only those which are singular in the sense of §1(b), that is, fixed by a non-trivial element of $W(G,T)$, but also those $\xi_\tau$ which are fixed by an element of $W_{\underline{g}} \setminus W(G,T)$. It is these latter characters which give rise to "singular" invariant eigendistributions $\Theta_\tau$ which do not vanish identically on $T'$.

The character group of $T$ is isomorphic to $\mathbb{Z}^n$ and we can identify the lattice $L_T$ with $\mathbb{Z}^n$. If $r = (r_1, \ldots, r_n) \in \mathbb{Z}^n$, we set

$$(2.4) \qquad r(\sqrt{-1}\varphi_1, \ldots, \sqrt{-1}\varphi_n, \sqrt{-1}\varphi_{n+1}) = (\sqrt{-1}) \sum_{j=1}^{n} r_j \varphi_j .$$

It is also possible to identify $\hat{T}$ with $m = (m_1, \ldots, m_{n+1}) \in \mathbb{Z}^{n+1}$ satisfying $0 \leq \sum_{j=1}^{n+1} m_j < n+1$. In this case, we have

$$(2.5) \qquad m(\sqrt{-1}\varphi_1, \ldots, \sqrt{-1}\varphi_{n+1}) = (\sqrt{-1}) \sum_{j=1}^{n+1} m_j \varphi_j .$$

In either case, it is a simple matter to see from (2.2) and (2.3) that

$$(2.6) \qquad L_T' = \{(r_1, \ldots, r_n) \in \mathbb{Z}^n : r_i \neq r_j \text{ for } i \neq j \text{ and } r_j \neq 0,\ j = 1, \ldots, n\}$$

$$= \{(m_1, \ldots, m_{n+1}) \in \mathbb{Z}^{n+1} : 0 \leq \sum_{j=1}^{n+1} m_j < n+1 \text{ and } m_i \neq m_j \text{ for } i \neq j\}.$$

Thus, the singular characters which interest us correspond to those elements $m = (m_1, \ldots, m_{n+1}) \in L_T^s$ which satisfy $\prod_{1 \leq i < j \leq n} (m_j - m_i) \neq 0$ and $m_{n+1} = m_j$ for some $j$, $1 \leq j \leq n$. If $\prod_{1 \leq i < j \leq n} (m_j - m_i) = 0$, then $\Theta_\tau$ defined below vanishes identically on $T'$.

Now define

$$(2.7) \qquad \Delta_T(\exp H) = \prod_{\alpha \in P_T} (e^{\alpha(H)/2} - e^{-\alpha(H)/2}), \quad H \in \underline{t}.$$

Then, if $t = (e^{\sqrt{-1}\varphi_1}, \ldots, e^{\sqrt{-1}\varphi_{n+1}}) \in T$,

$$(2.8)\qquad \Delta_T(t) = \xi_\rho(t) \prod_{\alpha \in P_T} (1-\xi_\alpha(t)^{-1}) = \prod_{1 \le i < j \le n+1} (e^{\sqrt{-1}\varphi_j} - e^{\sqrt{-1}\varphi_i}).$$

According to a basic result of Harish-Chandra ([7(f)], [7(g)]), to each $\tau \in L_T$, we can associate a tempered invariant eigendistribution $\Theta_\tau$ characterized by certain properties. We have

$$(2.9)\qquad \Theta_\tau(t) = \Delta_T(t)^{-1} \sum_{w \in W(G,T)} \det(w)\xi_{w\tau}(t),\ t \in T'.$$

Next, we want the value of $\Theta_\tau$ on $A'$. For $\tau \in L_T'$, this value is prescribed by the fact that $\Theta_\tau$ is essentially a discrete series character (see below). For $\tau \in L_T^s$, $\Theta_\tau$ could be extended in a number of ways. Set

$$(2.10)\qquad \begin{cases} A_{\underline{p}}^+ = \{\exp(tH^*) : t > 0\},\ A_{\underline{p}}^- = \{\exp(tH^*) : t < 0\}; \\ A^+ = A_K A_{\underline{p}}^+ \cap A',\ A^- = A_K A_{\underline{p}}^- \cap A'. \end{cases}$$

To any $\tau$ in $L_T$, we associate the constants

$$(2.11)\qquad c^{\pm}(\tau) = \begin{cases} \mp 1 & \text{if } \tau(-1,0,\ldots,0,1) > 0, \\ \pm 1 & \text{if } \tau(-1,0,\ldots,0,1) < 0, \\ 0 & \text{if } \tau(-1,0,\ldots,0,1) = 0. \end{cases}$$

If $h \in A'$, we write $h = h_K h_t$ where $h_K = (e^{\sqrt{-1}\theta}, e^{\sqrt{-1}\varphi_2}, \ldots, e^{\sqrt{-1}\varphi_n}, e^{\sqrt{-1}\theta}) \in A_K'$ and $h_t = \exp(tH^*)$, $t \in \mathbb{R}$, $t \neq 0$.

The eigenvalues of $h$ are

$$\lambda_1 = e^{-t+\sqrt{-1}\theta},\ \lambda_{n+1} = e^{t+\sqrt{-1}\theta},\ \lambda_j = e^{\sqrt{-1}\varphi_j},\ j = 2,\ldots,n.$$

We set

$$(2.12)\qquad \Delta_A(h) = \prod_{1 \le i < j \le n+1} (\lambda_j - \lambda_i).$$

Then

$$(2.13)\quad \Theta_\tau(h) = \Delta_A(h)^{-1} \sum_{w \in W(G,T)} \det(w)\xi_{w\tau}(h_K)c^{\pm}(w\tau)\exp(-|w\tau(-t,0,\ldots,0,t)|),$$

where the sign for $c^{\pm}(w\tau)$ corresponds to $h \in A^{+}$ or $h \in A^{-}$ accordingly. Let us agree to write $\exp(-|w\tau(t)|)$ for $\exp(-|w\tau(-t, 0, \ldots, 0, t)|)$.

If $\tau \in L'_T$, set $\varepsilon(\tau) = \mathrm{sgn}(\prod_{\alpha \in P_T} (\tau, \alpha))$. Then $\Theta_{\omega(\tau)} = (-1)^n \varepsilon(\tau)\Theta_\tau$ is the character of a discrete series representation of $G$ and all discrete series characters arise in this way. Moreover, $\Theta_{\omega(\tau_1)} = \Theta_{\omega(\tau_2)}$ if and only if there is an element $w \in W(G, T)$ such that $w\tau_1 = \tau_2$ ([7(h)]).

For $f \in C_c^\infty(G)$, the invariant integral of $f$ relative to $T$ is

$$F_f^T(t) = \Delta_T(t) \int_{G/T} f(xtx^{-1}) d_{G/T}(\dot{x}), \quad t \in T'. \tag{2.14}$$

Relative to $A$, we have

$$F_f^A(h) = \varepsilon_R^A(h) \Delta_A(h) \int_{G/A} f(xhx^{-1}) d_{G/A}(\dot{x}), \quad h \in A', \tag{2.15}$$

where $\varepsilon_R^A(h) = \pm 1$ for $h \in A^{\pm}$ respectively. The invariant measures $d_{G/T}(\dot{x})$ and $d_{G/A}(\dot{x})$ on $G/T$ and $G/A$ are normalized as follows. Pick a fixed Haar measure $dx$ on $G$ and normalize Haar measure $dt$ on $T$ so that $\int_T dt = 1$. Then $\int_G f(x)dx = \int_{G/T}\int_T f(xt)dt d_{G/T}(\dot{x})$ determines $d_{G/T}(\dot{x})$. On $A$, write $dh = dh_K dt$, where $dh_K$ is a Haar measure on $A_K$ satisfying $\int_{A_K} dh_K = 1$ and $dt$ is Lebesque measure on $A_{\underline{p}}$. Then $\int_G f(x)dx = \int_{G/A}\int_A f(xh)dh\, d_{G/A}(\dot{x})$ determines $d_{G/A}(\dot{x})$. With the above normalizations, we have Weyl's integral formula

$$\begin{aligned} \int_G f(x)dx &= [W(G, T)]^{-1} \int_T \overline{\Delta_T(t)} F_f^T(t) dt \\ &+ [W(G, A)]^{-1} \int_A \overline{\Delta_A(h)} \varepsilon_R^A(h) F_f^A(h) dh, \quad f \in C_c^\infty(G). \end{aligned} \tag{2.16}$$

It is known [7(e)] that $F_f^A$ can be extended to a function in $C_c^\infty(A)$ and that $F_f^T \in C^\infty(T')$ and behaves in a regular way at the singular points of $T$.

The principal series of unitary representations of $G$ are indexed by pairs $(\sigma, \nu)$ where $\sigma$ is an irreducible representation of $M$ and $\nu \in \mathbb{R}$. Here, the real number $\nu$ defines a unitary character of $A_{\underline{p}}$ by the formula

$$(2.17) \qquad \nu(h_t) = e^{\sqrt{-1}\nu t}, \quad h_t = \exp(tH^*).$$

To the representation $\sigma$ of M, we can associate a regular character $\chi_\sigma$ in $\hat{A}_K$ as in §1(b). (There are $[W(M,A_K)] = (n-1)!$ possible choices for $\chi_\sigma$.)

For any $\chi \in \hat{A}_K$ define $\varepsilon(\chi) = 1$ for $\chi$ singular and, for $\chi$ regular, define $\varepsilon(\chi)$ as in (1.10) (here $P = P_I^+$ denotes the positive roots of $(\underline{m}, \underline{a}_K)$ or, equivalently, the positive roots of $(\underline{g}_{\mathbb{C}}, \underline{a}_{\mathbb{C}})$ taking pure imaginary values on $\underline{a}_K$).

Let $r_I = [P_I^+]$ and, for $f \in C_c^\infty(G)$, set

$$(2.18) \qquad T^{(\chi,\nu)}(f) = (-1)^{r_I} \varepsilon(\chi) \int_{A_K} \int_{A_{\underline{p}}} \chi(h_K)\nu(h_t) F_f^A(h_K h_t) dh_K dt.$$

Then $T^{(\chi,\nu)}$ is an invariant eigendistribution on G. For $\chi$ regular, $\chi = \chi_\sigma$ for some $\sigma \in E(M)$, and $T^{(\chi_\sigma,\nu)} = T^{(\sigma,\nu)}$ is a character of the principal series representation corresponding to $(\sigma, \nu)$. In the latter case, $T^{(\sigma,\nu)} = T^{(\sigma',\nu')}$ if and only if $(\sigma,\nu)$ and $(\sigma',\nu')$ lie in the same orbit of the restricted Weyl group ([13], Chapter V). In particular, $T^{(\chi,\nu)}(f) = T^{(\chi,-\nu)}(f)$ since $F_f^A(h_K h_t) = F_f^A(h_K h_{-t})$. For $\chi$ singular, it is easy to see that $T^{(\chi,\nu)}(f) = 0$.

Now, let $t_0$ be a fixed regular element in T, and consider the invariant distribution $f \mapsto F_f^T(t_0)$, $f \in C_c^\infty(G)$. We outline below the principal steps in the derivation of the Fourier transform of this invariant distribution. In this exposition, we do not present the necessary convergence arguments to justify the application of Fubini's theorem in all its various guises. These arguments are presented in full in [11(b)].

<u>Step I.</u> For any $\tau \in L_T$, we have

$$\Theta_\tau(f) = \int_G f(x)\Theta_\tau(x)dx$$

$$= \int_{{}_T G} f(x)\Theta_\tau(x)dx + \int_{{}_A G} f(x)\Theta_\tau(x)dx$$

$$= (-1)^r [W(G,T)]^{-1} \sum_{w \in W(G,T)} \det(w) \int_T \xi_{w\tau}(t) F_f^T(t)\, dt$$

$$+ \int_{{}^A G} f(x)\Theta_\tau(x)\, dx.$$

Since $F_f^T(w\ t) = \det(w) F_f^T(t)$ for $w \in W(G,T)$ and $t \in T'$, we have

(2.19) $$\hat{F}_f^T(\tau) = \int_T \xi_\tau(t) F_f^T(t)\, dt = (-1)^r [\Theta_\tau(f) - \int_{{}^A G} f(x)\Theta_\tau(x)\, dx],$$

where $r = 1/2(\dim G - \operatorname{rank} G) = n(n+1)/2$. From the properties of $F_f^T$ given above, it follows that

(2.20) $$F_f^T(t_0) = \sum_{\tau \in L_T} \hat{F}_f^T(\tau)\overline{\xi_\tau(t_0)}.$$

Here, $\sum_{\tau \in L_T}$ signifies $\lim_{R \to \infty} \sum_{\substack{-R \le r_j \le R \\ j = 1, 2, \dots, n}}$.

From [7(f)], it follows that $\sum_{\tau \in L_T} \Theta_\tau(f)\overline{\xi_\tau(t_0)}$ converges absolutely, so that

(2.21) $$I_f(t_0) = \sum_{\tau \in L_T} \overline{\xi_\tau(t_0)} \int_{{}^A G} f(x)\Theta_\tau(x)\, dx$$

is convergent. We thus have

(2.22) $$F_f^T(t_0) = (-1)^r \sum_{\tau \in L_T} \Theta_\tau(f)\overline{\xi_\tau(t_0)} + (-1)^{r+1} I_f(t_0).$$

<u>Step II.</u> $I_f(t_0) = [W(G,A)]^{-1} \sum_{\tau \in L_T} \overline{\xi_\tau(t_0)} \int_A \overline{\Delta_A(h)}\, \varepsilon_R^A(h) F_f^A(h)\Theta_\tau(h)\, dh$

$$= [W(G,A)]^{-1} \sum_{\tau \in L_T} \overline{\xi_\tau(t_0)} \{ \int_{A^+} \overline{\Delta_A(h)} F_f^A(h)\Theta_\tau(h)\, dh - \int_{A^-} \overline{\Delta_A(h)} F_f^A(h)\Theta_\tau(h)\, dh \}.$$

Now $\overline{\Delta_A(h)} = (-1)^{r+1}\Delta_A(h)$ and, since $W(G,T)$ acts on $L_T$, we have for $h = h_K h_t \in A^+$,

$$\sum_{\tau \in L_T} \overline{\xi_\tau(t_0)}\, \overline{\Delta_A(h)}\, \Theta_\tau(h)$$

$$= (-1)^{r+1} \sum_{\tau \in L_T} \overline{\xi_\tau(t_0)} \sum_{w \in W(G,T)} \det(w) \xi_{w\tau}(h_K) c^+(w\tau) \exp(-|w\tau(t)|)$$

$$= (-1)^{r+1} \sum_{w \in W(G,T)} \det(w) \sum_{\tau \in L_T} \overline{\xi_{w\tau}(t_0)}\, \xi_\tau(h_K) c^+(\tau) \exp(-|\tau(t)|).$$

If we replace $h_K h_t \in A^+$ by $h_K h_{-t} \in A^-$ in these last equations, then $c^+(\tau)$ must be replaced by $c^-(\tau) = -c^+(\tau)$.

Thus, using the fact that $F_f^A(h_K h_t) = F_f^A(h_K h_{-t})$, we get

$$I_f(t_0) = (-1)^{r+1}[W(G,A)]^{-1} \sum_{w \in W(G,T)} \det(w) \sum_{\tau \in L_T} \overline{\xi_{w\tau}(t_0)}\, 2 I_f^+(\tau), \tag{2.23}$$

where

$$I_f^+(\tau) = \int_{A^+} c^+(\tau) \xi_\tau(h_K) \exp(-|\tau(t)|) F_f^A(h_K h_t)\, dh_K\, dt. \tag{2.24}$$

Step III. Identify $L_T$ with a subset of $\mathbb{Z}^{n+1}$ as in (2.5). Define

$$\begin{cases} L_T^* = \{(m_1, \ldots, m_{n+1}) \in L_T : m_1 = m_{n+1}\} \\ L_T^e = \{(m_1, \ldots, m_{n+1}) \in L_T : m_1 + m_{n+1} \text{ is even}\} \\ L_T^o = \{(m_1, \ldots, m_{n+1}) \in L_T : m_1 + m_{n+1} \text{ is odd}\}. \end{cases} \tag{2.25}$$

Then $\tau \in L_T^*$ if and only if $c^+(\tau) = c^-(\tau) = 0$. Let $\tau_1 = (-1, 0, \ldots, 0, 1)$ and $\tau_2 = (0, 0, \ldots, 0, 1)$ in $L_T$. Then $L_T^e$ is generated by $\tau_1$ and $L_T^*$, and $L_T^o = \tau_2 + L_T^e$.

We set

$$T_1 = A_K,\ T_2 = \exp\{\mathbb{R}(\sqrt{-1}, 0, \ldots, 0, -\sqrt{-1})\}, \tag{2.26}$$

and write $dh_K = dt_1$.

Then $T_2$ is isomorphic to $\mathbb{T}$, $T = T_1 T_2$ and $T_1 \cap T_2 = \{1, \gamma\}$ where

$\gamma = \{-1, 1, \ldots, 1, -1\}$. Given $w \in W(G, T)$, we can write

(2.27) $$w^{-1} \cdot t_0 = t_1(w) t_2(w), \quad t_1(w) \in T_1', \quad t_2(w) \in T_2'.$$

This decomposition is not unique, but, if we write

(2.28) $$t_2(w) = (e^{\sqrt{-1}\theta_w}, 1, \ldots, 1, e^{-\sqrt{-1}\theta_w}), \quad 0 < \theta_w < \pi,$$

then the decomposition (2.27) is unique. <u>We shall assume</u> (2.28) <u>for the remainder of our development.</u> Observe that $L_T^*$ may be identified with the character group of $T_1/\{1, \gamma\}$. We have $\xi_{\tau_1} | T_1 \equiv 1$ and $\xi_{n\tau_1}(t_2(w)) = e^{-2\sqrt{-1}n\theta_w}$. Moreover, $\exp(-|\tau(t)|) = 1$ and $\xi_\tau | T_2 \equiv 1$ for $\tau \in L_T^*$.

Before proceeding with an analysis of (2.23), we isolate a lemma which plays a useful role. Define

(2.29) $$F(t_1; h_t; \tau_0) = \xi_{\tau_0}(t_1) F_f^A(t_1 h_t) + \xi_{\tau_0}(\gamma t_1) F_f^A(\gamma t_1 h_t),$$

$$t_1 \in T_1, \quad h_t \in A_{\underline{p}}^+, \quad \tau_0 \in L_T.$$

<u>Lemma 2.30.</u> <u>For</u> $h_t$ <u>and</u> $\tau_0$ <u>fixed,</u> $t_1 \mapsto F(t_1; h_t; \tau_0)$ <u>may be regarded as a function on</u> $T_1/\{1, \gamma\}$ <u>and, for any element</u> $t_1' \in T_1$,

$$\sum_{\tau \in L_T^*} \overline{\xi_\tau(t_1')} \int_{T_1} \xi_\tau(t_1) \xi_{\tau_0}(t_1) F_f^A(t_1 h_t) dt_1 = (\tfrac{1}{2}) F(t_1'; h_t; \tau_0).$$

Now, a simple calculation leads to

$$\sum_{\tau \in L_T^e} \overline{\xi_{w\tau}(t_0) I_f^+(\tau)} =$$

$$(\tfrac{1}{2})\left\{\sum_{n=-\infty}^{-1} \overline{\xi_{n\tau_1}(t_2(w))} \int_0^\infty e^{-|2nt|} F(t_1(w);h_t;0)\,dt - \sum_{n=1}^{\infty} \overline{\xi_{n\tau_1}(t_2(w))} \int_0^\infty e^{-|2nt|} F(t_1(w);h_t;0)\,dt\right\}$$

$$= (\tfrac{1}{2}) \int_0^\infty F(t_1(w);h_t;0) \left[\frac{e^{-2t}(e^{-2\sqrt{-1}\theta_w} - e^{2\sqrt{-1}\theta_w})}{1-2e^{-2t}\cos(2\theta_w)+e^{-4t}}\right] dt.$$

Similarly, we have

$$\sum_{\tau \in L_T^o} \overline{\xi_{w\tau}(t_0) I_f^+(\tau)} =$$

$$(\tfrac{1}{2})\overline{\xi_{w\tau_2}(t_0)} \left\{\sum_{n=-\infty}^{-1} \overline{\xi_{n\tau_1}(t_2(w))} \int_0^\infty e^{-|(2n+1)t|} F(t_1(w);h_t;\tau_2)\,dt - \sum_{n=0}^{\infty} \overline{\xi_{n\tau_1}(t_2(w))} \int_0^\infty e^{-|(2n+1)t|} F(t_1(w);h_t;\tau_2)\,dt\right\}$$

$$= (\tfrac{1}{2})\overline{\xi_{w\tau_2}(t_0)} \int_0^\infty F(t_1(w);h_t;\tau_2) \left[\frac{e^{-t}e^{-2\sqrt{-1}\theta_w}}{1-e^{-2t}e^{-2\sqrt{-1}\theta_w}} - \frac{e^{-t}}{1-e^{-2t}e^{2\sqrt{-1}\theta_w}}\right] dt.$$

Combining the sums over $L_T^e$ and $L_T^o$ and using the fact that $\xi_{\tau_2}(\gamma) = -1$, we obtain

$$2\sum_{\tau \in L_T} \overline{\xi_{w\tau}(t_0)}\, I_f^+(\tau) =$$

(2.31)
$$\int_0^\infty F_f^A(t_1(w)h_t)\left[\frac{e^t\left(e^{-\sqrt{-1}\theta_w} - e^{\sqrt{-1}\theta_w}\right)}{1-2e^t\cos(\theta_w)+e^{2t}}\right] dt$$

$$+ \int_0^\infty F_f^A(\gamma t_1(w)h_t)\left[\frac{e^t\left(e^{\sqrt{-1}\theta_w} - e^{-\sqrt{-1}\theta_w}\right)}{1+2e^t\cos(\theta_w)+e^{2t}}\right] dt.$$

<u>Step IV.</u> Define $\hat{F}_f^A(\chi, \nu) = (-1)^{r_I} \varepsilon(\chi)(\sqrt{2\pi})^{-1} T^{(\chi,\nu)}(f)$ where $T^{(\chi,\nu)}(f)$ is given by (2.18). Then, by standard theorems of classical Fourier analysis, we have, for $t_1 \in T_1$, $h_t \in A_{\underline{p}}$,

(2.32)
$$F_f^A(t_1 h_t) = (\sqrt{2\pi})^{-1} \sum_{\chi \in \hat{T}_1} \overline{\chi(t_1)} \int_{-\infty}^{\infty} e^{-\sqrt{-1}\nu t} \hat{F}_f^A(\chi, \nu)\, d\nu,$$

and, since $\gamma \in T_1$, the same reasoning yields

(2.33)
$$F_f^A(\gamma t_1 h_t) = (\sqrt{2\pi})^{-1} \sum_{\chi \in \hat{T}_1} \overline{\chi(\gamma t_1)} \int_{-\infty}^{\infty} e^{-\sqrt{-1}\nu t} \hat{F}_f^A(\chi, \nu)\, d\nu.$$

In (2.32) and (2.33), $d\nu$ is just Lebesque measure on $\mathbb{R}$.

Substituting (2.32) and (2.33) into (2.31), we see that

$$2\sum_{\tau \in L_T} \overline{\xi_{w\tau}(t_0)}\, I_f^+(\tau) = (\sqrt{2\pi})^{-1}\left(e^{-\sqrt{-1}\theta_w} - e^{\sqrt{-1}\theta_w}\right)$$

$$\times \left\{ \sum_{\chi \in \hat{T}_1} \overline{\chi(t_1(w))} \int_{-\infty}^{\infty} \hat{F}_f^A(\chi, \nu) \int_0^\infty \frac{e^{-\sqrt{-1}\nu t}\, e^t}{1-2e^t\cos(\theta_w)+e^{2t}}\, dt\, d\nu \right.$$

$$\left. - \sum_{\chi \in \hat{T}_1} \overline{\chi(\gamma t_1(w))} \int_{-\infty}^{\infty} \hat{F}_f^A(\chi, \nu) \int_0^\infty \frac{e^{-\sqrt{-1}\nu t}\, e^t}{1+2e^t\cos(\theta_w)+e^{2t}}\, dt\, d\nu \right\}.$$

Since $\hat{F}_f^A(\chi, \nu) = \hat{F}_f^A(\chi, -\nu)$, we can write

$$\int_{-\infty}^{\infty} \hat{F}_f^A(\chi, \nu) \int_0^{\infty} \frac{e^{-\sqrt{-1}\nu t} e^t}{1 \mp 2e^t \cos(\theta_w) + e^t} \, dt \, d\nu$$

$$= \frac{1}{2} \int_{-\infty}^{\infty} \hat{F}_f^A(\chi, \nu) \int_0^{\infty} \frac{\lambda^{\sqrt{-1}\nu}}{1 \mp 2\lambda \cos(\theta_w) + \lambda^2} \, d\lambda \, d\nu .$$

Using the evaluation of the inner integrals given in [4], p. 297, we then have

$$2 \sum_{\tau \in L_T} \overline{\xi_{w\tau}(t_0)} I_f^+(\tau) = (\sqrt{2\pi})^{-1} (e^{-\sqrt{-1}\theta_w} - e^{\sqrt{-1}\theta_w})$$

(2.34) $$\times \left\{ -\frac{\pi}{2} \sum_{\chi \in \hat{T}_1} \overline{\chi(t_1(w))} \int_{-\infty}^{\infty} \hat{F}_f^A(\chi, \nu) \left[ \frac{\sinh(\nu(\theta_w - \pi))}{\sinh(\nu\pi) \sin(\theta_w)} \right] d\nu \right.$$

$$\left. - \frac{\pi}{2} \sum_{\chi \in \hat{T}_1} \overline{\chi(\gamma t_1(w))} \int_{-\infty}^{\infty} \hat{F}_f^A(\chi, \nu) \left[ \frac{\sinh(\nu\theta_w)}{\sinh(\nu\pi)\sin(\theta_w)} \right] d\nu \right\} .$$

Step V (The final formula). We now use everything in sight. Eliminate $\sin(\theta_w)$ from (2.34) and plug into (2.23) to obtain $I_f(t_0)$. The substitute into (2.22).

Theorem 2.35. $F_f^T(t_0) = (-1)^r \sum_{\tau \in L_T} \Theta_\tau(f) \overline{\xi_\tau(t_0)}$

$$+ \left(\frac{\sqrt{-1}}{2}\right) (-1)^{r_I} [W(G, A)]^{-1} \sum_{w \in W(G,T)} \det(w) \sum_{\chi \in \hat{T}_1} \varepsilon(\chi) \times$$

$$\left\{ \overline{\chi(t_1(w))} \int_{-\infty}^{\infty} T^{(\chi, \nu)}(f) \left[ \frac{\sinh(\nu(\theta_w - \pi))}{\sinh(\nu\pi)} \right] d\nu + \overline{\chi(\gamma t_1(w))} \int_{-\infty}^{\infty} T^{(\chi, \nu)}(f) \left[ \frac{\sinh(\nu\theta_w)}{\sinh(\nu\pi)} \right] d\nu \right\} .$$

To formulate this result in a more representation theoretic way, we observe that

$$\sum_{\tau \in L_T'} \Theta_\tau(f)\overline{\xi_\tau(t_0)} = \sum_{\{\tau\} \in L_T'/W(G,T)} \sum_{w \in W(G,T)} (-1)^n \varepsilon(w\tau)\Theta_{\omega(w\tau)}(f)\overline{\xi_{w\tau}(t_0)}$$

$$= \sum_{\{\tau\} \in L_T'/W(G,T)} \Theta_{\omega(\tau)}(f)(-1)^n\varepsilon(\tau) \sum_{w \in W(G,T)} \det(w)\overline{\xi_{w\tau}(t_0)}$$

$$= \overline{\Delta_T(t_0)} \sum_{\omega \in E_2(G)} \Theta_\omega(f)\overline{\Theta_\omega(t_0)} .$$

Thus,

$$F_f^T(t_0) = (-1)^r \sum_{\tau \in L_T^s} \Theta_\tau(f)\overline{\xi_\tau(t_0)} + (-1)^r\overline{\Delta_T(t_0)} \sum_{\omega \in E_2(G)} \hat{f}(\omega)\overline{\Theta_\omega(t_0)}$$

$$+ \left(\frac{\sqrt{-1}}{2}\right)(-1)^{r_I}[W(G,A)]^{-1} \sum_{w \in W(G,T)} \det(w) \sum_{\chi \in \hat{T}_1} \varepsilon(\chi)$$

$$\times \left\{ \overline{\chi(t_1(w))}\int_{-\infty}^{\infty} T^{(\chi,\nu)}(f)\left[\frac{\sinh(\nu(\theta_w - \pi))}{\sinh(\nu\pi)}\right] d\nu + \overline{\chi(\gamma t_1(w))}\int_{-\infty}^{\infty} T^{(\chi,\nu)}(f)\left[\frac{\sinh(\nu\theta_w)}{\sinh(\nu\pi)}\right] d\nu \right\}.$$

Now suppose that $G = SU(1,1)$. Then $r = 1$, $r_I = 0$, $W(G,T) = \{1\}$, $[W(G,A)] = 2$ and $T_1 = \{\pm 1\}$. We shall write $T^{(+,\nu)}(f)$ for $T^{(\chi,\nu)}(f)$ when $\chi$ is the trivial character on $T_1$ and $T^{(-,\nu)}(f)$ when $\chi$ is the non-trivial character on $T_1$. For $t_0 \in T'$, write $t_0 = \begin{pmatrix} e^{\sqrt{-1}\theta} & 0 \\ 0 & e^{-\sqrt{-1}\theta} \end{pmatrix}$, $0 < \theta < \pi$. Here, $L_T$ may be identified with $\mathbb{Z}$ and $L_T'$ with $\mathbb{Z}\setminus\{0\}$ and we have

$$F_f^T(t_0) = -\sum_{n=-\infty}^{\infty} \Theta_n(f)e^{-\sqrt{-1}n\theta} + \frac{\sqrt{-1}}{4}\left\{\int_{-\infty}^{\infty} T^{(+,\nu)}(f)\left[\frac{\sinh(\nu(\theta-\pi)) + \sinh(\nu\theta)}{\sinh(\nu\pi)}\right] d\nu \right.$$

$$\left. + \int_{-\infty}^{\infty} T^{(-,\nu)}(f)\left[\frac{\sinh(\nu(\theta-\pi)) - \sinh(\nu\theta)}{\sinh(\nu\pi)}\right] d\nu \right\}.$$

If $n \neq 0$, then the discrete series character corresponding to $n$ is $\Theta_{\omega(n)}(f) = \mathrm{sgn}(n)\Theta_n(f)$. Moreover, the character $T^{(+,0)}$ in the principal series may be written as the sum of two irreducible characters which we shall denote

$T^{(+,+)}$ and $T^{(+,-)}$ ([3]), and $\Theta_0(f) = \frac{1}{2}(T^{(+,+)}(f) - T^{(+,-)}(f))$. Employing elementary addition formulas, we obtain

$$F_f^T(t_0) = -\frac{1}{2}(T^{(+,+)}(f) - T^{(+,-)}(f)) - \sum_{\substack{n=-\infty \\ n\neq 0}}^{\infty} \operatorname{sgn}(n)\Theta_{\omega(n)}(f)e^{-\sqrt{-1}n\theta}$$

$$(2.36) \qquad + \frac{\sqrt{-1}}{4}\left\{\int_{-\infty}^{\infty} T^{(+,\nu)}(f)\left[\frac{\sinh(\nu(\theta - \frac{\pi}{2}))}{\sinh(\frac{\nu\pi}{2})}\right]d\nu - \int_{-\infty}^{\infty} T^{(-,\nu)}(f)\left[\frac{\cosh(\nu(\theta - \frac{\pi}{2}))}{\cosh(\frac{\nu\pi}{2})}\right]d\nu\right\}.$$

From a theorem of Harish-Chandra [7(b)], it follows that

$$\lim_{\theta \to 0} \frac{1}{\sqrt{-1}}\frac{d}{d\theta}(F_f^T(t_0)) = 8\pi f(1).$$

We thus have

$$(2.37) \qquad 8\pi f(1) = \sum_{n=-\infty}^{\infty} |n|\,\Theta_{\omega(n)}(f) + \int_0^{\infty} T^{(+,\nu)}(f)\left[\frac{\nu}{2}\coth\left(\frac{\nu\pi}{2}\right)\right]d\nu + \int_0^{\infty} T^{(-,\nu)}(f)\left[\frac{\nu}{2}\tanh\left(\frac{\nu\pi}{2}\right)\right]d\nu.$$

This is the Plancherel formula for $G$ ([7(a)], [7(c)]). We note that, in our parameterization, the so-called class-one principal series correspond to the non-trivial characters on $T_1$. This results from the form of the characters of the principal series given by (2.18) (see [7(g)], §24).

Finally, we remark that in [11(b)], we take $T_1$ to be the connected component of the identity in $A_K$. In our present development for $SU(n,1)$, $A_K$ is connected for $n \geq 2$, and we have made the definition (2.26) simply to include $SU(1,1)$.

## REFERENCES

[1] F. Adams, Lectures on Lie Groups, Benjamin, New York, 1969.

[2] S. Gelbart, "Fourier Analysis on Matrix Space", Memoirs Amer. Math. Soc., No. 108, 1971.

[3] I. M. Gel'fand, M.I. Graev and I.I. Pyatetskii-Shapiro, Representation Theory and Automorphic Functions, Saunders, Philadelphia, 1969.

[4] Gradshteyn and Ryzhik, Tables of Integrals, Series and Products, Academic Press, New York, 1965.

[5] E. Hewitt and K. A. Ross, Abstract Harmonic Analysis, v. II, Springer, Berlin, 1970.

[6] A. W. Knapp and K. Okamoto, "Limits of Holomorphic Discrete Series", to appear in J. Fcnl. Analysis.

[7] Harish-Chandra, (a)"Plancherel formula for the 2 × 2 real unimodular group", Proc. Nat. Acad. Sci. 38(1952), 337-342.

(b) "A formula for semisimple Lie groups", Amer. J. Math. 79(1957), 733-760.

(c) "Harmonic analysis on semisimple Lie groups", in Annual Science Conference Proceedings, Belfer Graduate School of Science, Yeshiva U., New York, 1966.

(d) "Some results on an invariant integral on a semisimple Lie algebra", Annals of Math. 80(1964), 551-593.

(e) "Invariant eigendistributions on a semisimple Lie group", Trans. Amer. Math. Soc. 119(1965), 457-508.

(f) "Discrete series for semisimple Lie groups, I", Acta Math. 113(1965), 241-318.

(g) "Two theorems on semisimple Lie groups", Annals of Math. 83 (1966), 74-128.

(h) "Discrete series for semisimple Lie groups, II", Acta Math. 116 (1966), 1-111.

(i) "Harmonic analysis on semisimple Lie groups", Bull. Amer. Math. Soc. 76 (1970), 529-551.

[8] T. Hirai, (a) "The characters of irreducible representations of the Lorentz group of n-th order", Proc. Japan Acad. 41 (1965), 526-531.

(b) "Classification and the characters of irreducible representations of SU(p,1), Proc. Japan Acad. 42 (1966), 907-912.

[9] K. Okamoto, "On the Plancherel formulas for some types of simple Lie groups", Osaka J. Math. 2 (1965), 247-282.

[10] P. J. Sally, Jr., and J. A. Shalika, (a) "The Plancherel formula for SL(2) over a local field", Proc. Nat. Acad. Sci. 63 (1969), 661-667.

(b) "The Fourier transform on $SL_2$ over a non-archimedean local field", to appear.

[11] P. J. Sally, Jr., and G. Warner, (a) "Fourier inversion for semisimple Lie groups of real rank one", Bull. Amer. Math. Soc. 78 (1972),

(b) "The Fourier transform on semisimple Lie groups of real rank one", to appear.

[12] J. A. Shalika, "A theorem on semisimple p-adic groups", to appear in Annals of Math.

[13] G. Warner, Harmonic Analysis on Semi-Simple Lie Groups, 2 vols., Springer, Berlin, 1972.

[14] A. Weil, L' intégration dans les groupes topologiques et ses applications, Hermann, Paris, 1941.

# CHARACTERS OF THE GROUP GL(∞, q)

by

Elmar Thoma
Technical University of Munich

Let $G$ be a group and $A(G)$ be the set of all functions $f\colon G \longrightarrow \mathbb{C}$ such that $f(x) = 0$ for all $x \in G$ except for a finite number of $x \in G$. A function $\alpha\colon G \longrightarrow \mathbb{C}$ is called a character of $G$ if $\alpha(xy) = \alpha(yx)$ for all $x, y \in G$ and $\sum_{x,y \in G} \alpha(y^{-1}x) f(x) \overline{f(y)} \geq 0$ for all $f \in A(G)$. By using pointwise addition and pointwise multiplication with non-negative real numbers, it is obvious that the set of all characters $L(G)$ is a convex cone. By using the weakest topology such that all mappings $\alpha \longrightarrow \alpha(x)$ are continuous for all $x \in G$, the set $\{\alpha \in L(G) \mid \alpha(E) = 1\} = K(G)$ is a convex and compact cut of $L(G)$. Here $E$ is the identity element of the group. Since it is known that the natural order of $L(G)$ induces a lattice, we can use Choquet's existence and uniqueness theorem. Therefore it is interesting to determine all characters of the set $E(G)$ of extremal points of $K(G)$ for special choices of $G$. This problem is also the direct generalization of that of finding all characters of irreducible representations of a finite group. Furthermore it is interesting to see the usefulness of certain general theorems in solving concrete problems.

Let $q$ be a power of a prime number and $GF(q)$ the field of $q$ elements. Let $GL(n, q)$ be the group of regular $n \times n$-matrices with coefficients from $GF(q)$, $n = 1, 2, \ldots$. Let $GL(\infty, q)$ be the group of all infinite matrices of the form $\begin{pmatrix} A & & 0 \\ & 1 & \\ 0 & & \ddots \end{pmatrix}$ where $A \in GL(n, q)$, $n = 1, 2, \ldots$.

The problem is now to find all characters of $E(GL(\infty, q))$. If $A \in GL(n, q)$, we consider $A$ also as an element of $GL(\infty, q)$ by adding 1's on the diagonal and 0's outside. I denote this element of $GL(\infty, q)$ also by $A$. In this sense $GL(n, q)$ is a subgroup of $GL(\infty, q)$.

Theorem 1. If $\alpha \in K(GL(\infty, q))$, then $\alpha \in E(GL(\infty, q))$ if and only if $\alpha \begin{pmatrix} A & 0 \\ 0 & B \end{pmatrix} = \alpha(A)\,\alpha(B)$ for all $A \in GL(n, q)$, $B \in GL(m, q)$, and all $n, m \in \{1, 2, \ldots\}$. (Recall the above notation!)

This theorem does have interesting consequences.

Corollary 1. $E(GL(\infty, q))$ is closed in $E(GL(\infty, q))$.

Proof: The equation of Theorem 1 remains true by taking pointwise limits.

Corollary 2. If $\alpha, \beta \in E(GL(\infty, q))$ then the pointwise product $\alpha\beta$ belongs to $E(GL(\infty, q))$.

Let me now give a description of the characters I found. Let $V$ be the linear space of all sequences $x = (x_1, x_2, \ldots)$, where $x_i \in GF(q)$ and $x_i = 0$ for all $i \in \{1, 2, \ldots\}$ except for a finite number of $i$'s. If $A \in GL(\infty, q)$, we can consider $A$ as a linear mapping of $V$ into $V$ by $x \longrightarrow xA$. Denote by $V_A$ the set $\{x \in V \mid xA = x\}$. Because $A \in GL(\infty, q)$, $\operatorname{codim} V_A = \dim {}^{V}\!/V_A$ is a finite number. Denote this number by $s(A)$. For $k = 0, 1, 2, \ldots$ the following functions $\phi_k\colon G \longrightarrow \mathbb{C}$, defined by $\phi_k(A) = \frac{1}{q^{ks(A)}}$, are characters from $E(GL(\infty, q))$. Let $M(G)$ be the multiplicative group of $GF(q)$. This is a cyclic group of order $q-1$. Let $\theta\colon M(G) \longrightarrow \mathbb{C}$ be an injective linear character; then the functions $\psi_\ell(A) = \theta^\ell(\det A)$ are linear characters of $GL(\infty, q)$ for $\ell = 0, \ldots, q-2$.

Theorem 2. $\phi_k\psi_\ell$ are characters of $E(GL(\infty, q))$ for $k = 0, 1, 2, \ldots$ and $\ell = 0, 1, \ldots, q-2$.

Of course $\phi_{reg}$, defined by $\phi_{reg}(E) = 1$ and $\phi_{reg}(A) = 0$ if $A \neq E$, is a character of $E(GL(\infty, q))$. The characters of Theorem 2 can be characterized in the following manner.

If $\alpha$ is a character of $GL(\infty, q)$, denote by $\alpha_n$ the restriction of $\alpha$ to $GL(n, q)$ (recall $GL(n, q)$ is a subgroup of $GL(\infty, q)$). $\alpha_n$ is a character of $GL(n, q)$. Therefore it is a unique linear combination with non-negative coefficients of the characters of the irreducible representations of this finite group $GL(n, q)$. Let us say $\alpha_n$ is "slim" if at least one of these coefficients is zero.

Theorem 2'. *If* $\alpha \in E(GL(\infty, q))$, *then* $\alpha$ *is a character of Theorem 2 if and only if there exists at least one natural number* $n$ *such that* $\alpha_n$ *is slim.*
I do have strong indications that the characters of Theorem 2, together with the regular character $\phi_{reg}$, are all characters of $E(GL(\infty, q))$.

It is obvious what I understand by $SL(\infty, q)$.

Theorem 3. $\phi_k$ *for* $k = 0, 1, 2, \ldots$ *and* $\phi_{reg}$ *are characters of* $E(SL(\infty, q))$.
If my conjecture about $E(GL(\infty, q))$ is true, then the characters of Theorem 3 are all characters of $E(SL(\infty, q))$.

I cannot give any proofs of these theorems during this lecture, but I would like to mention that, beside the general theory, I very strongly use the results of a paper of J.A. Green's (Trans. Am. Math. Soc. 80 (1955)) about the irreducible characters of $GL(n, q)$, and a paper of mine (Math. Zeitsch. 119 (1971)).

Please turn over

Vol. 146: A. B. Altman and S. Kleiman, Introduction to Grothendieck Duality Theory. II, 192 pages. 1970. DM 18,–

Vol. 147: D. E. Dobbs, Cech Cohomological Dimensions for Commutative Rings. VI, 176 pages. 1970. DM 16,–

Vol. 148: R. Azencott, Espaces de Poisson des Groupes Localement Compacts. IX, 141 pages. 1970. DM 16,–

Vol. 149: R. G. Swan and E. G. Evans, K-Theory of Finite Groups and Orders. IV, 237 pages. 1970. DM 20,–

Vol. 150: Heyer, Dualität lokalkompakter Gruppen. XIII, 372 Seiten. 1970. DM 20,–

Vol. 151: M. Demazure et A. Grothendieck, Schémas en Groupes I. (SGA 3). XV, 562 pages. 1970. DM 24,–

Vol. 152: M. Demazure et A. Grothendieck, Schémas en Groupes II. (SGA 3). IX, 654 pages. 1970. DM 24,–

Vol. 153: M. Demazure et A. Grothendieck, Schémas en Groupes III. (SGA 3). VIII, 529 pages. 1970. DM 24,–

Vol. 154: A. Lascoux et M. Berger, Variétés Kähleriennes Compactes. VII, 83 pages. 1970. DM 16,–

Vol. 155: Several Complex Variables I, Maryland 1970. Edited by J. Horváth. IV, 214 pages. 1970. DM 18,–

Vol. 156: R. Hartshorne, Ample Subvarieties of Algebraic Varieties. XIV, 256 pages. 1970. DM 20,–

Vol. 157: T. tom Dieck, K. H. Kamps und D. Puppe, Homotopietheorie. VI, 265 Seiten. 1970. DM 20,–

Vol. 158: T. G. Ostrom, Finite Translation Planes. IV. 112 pages. 1970. DM 16,–

Vol. 159: R. Ansorge und R. Hass. Konvergenz von Differenzenverfahren für lineare und nichtlineare Anfangswertaufgaben. VIII, 145 Seiten. 1970. DM 16,–

Vol. 160: L. Sucheston, Constributions to Ergodic Theory and Probability. VII, 277 pages. 1970. DM 20,–

Vol. 161: J. Stasheff, H-Spaces from a Homotopy Point of View. VI, 95 pages. 1970. DM 16,–

Vol. 162: Harish-Chandra and van Dijk, Harmonic Analysis on Reductive p-adic Groups. IV, 125 pages. 1970. DM 16,–

Vol. 163: P. Deligne, Equations Différentielles à Points Singuliers Reguliers. III, 133 pages. 1970. DM 16,–

Vol. 164: J. P. Ferrier, Seminaire sur les Algebres Complètes. II, 69 pages. 1970. DM 16,–

Vol. 165: J. M. Cohen, Stable Homotopy. V, 194 pages. 1970. DM 16,–

Vol. 166: A. J. Silberger, $PGL_2$ over the p-adics: its Representations, Spherical Functions, and Fourier Analysis. VII, 202 pages. 1970. DM 18,–

Vol. 167: Lavrentiev, Romanov and Vasiliev, Multidimensional Inverse Problems for Differential Equations. V, 59 pages. 1970. DM 16,–

Vol. 168: F. P. Peterson, The Steenrod Algebra and its Applications: A conference to Celebrate N. E. Steenrod's Sixtieth Birthday. VII, 317 pages. 1970. DM 22,–

Vol. 169: M. Raynaud, Anneaux Locaux Henséliens. V, 129 pages. 1970. DM 16,–

Vol. 170: Lectures in Modern Analysis and Applications III. Edited by C. T. Taam. VI, 213 pages. 1970. DM 18,–

Vol. 171: Set-Valued Mappings, Selections and Topological Properties of $2^X$. Edited by W. M. Fleischman. X, 110 pages. 1970. DM 16,–

Vol. 172: Y.-T. Siu and G. Trautmann, Gap-Sheaves and Extension of Coherent Analytic Subsheaves. V, 172 pages. 1971. DM 16,–

Vol. 173: J. N. Mordeson and B. Vinograde, Structure of Arbitrary Purely Inseparable Extension Fields. IV, 138 pages. 1970. DM 16,–

Vol. 174: B. Iversen, Linear Determinants with Applications to the Picard Scheme of a Family of Algebraic Curves. VI, 69 pages. 1970. DM 16,–

Vol. 175: M. Brelot, On Topologies and Boundaries in Potential Theory. VI, 176 pages. 1971. DM 18,–

Vol. 176: H. Popp, Fundamentalgruppen algebraischer Mannigfaltigkeiten. IV, 154 Seiten. 1970. DM 16,–

Vol. 177: J. Lambek, Torsion Theories, Additive Semantics and Rings of Quotients. VI, 94 pages. 1971. DM 16,–

Vol. 178: Th. Bröcker und T. tom Dieck, Kobordismentheorie. XVI, 191 Seiten. 1970. DM 18,–



Vol. 179: Seminaire Bourbaki – vol. 1968/69. Exposés 347-363. IV. 295 pages. 1971. DM 22,–

Vol. 180: Séminaire Bourbaki – vol. 1969/70. Exposés 364-381. IV, 310 pages. 1971. DM 22,–

Vol. 181: F. DeMeyer and E. Ingraham, Separable Algebras over Commutative Rings. V, 157 pages. 1971. DM 16.–

Vol. 182: L. D. Baumert. Cyclic Difference Sets. VI, 166 pages. 1971. DM 16,–

Vol. 183: Analytic Theory of Differential Equations. Edited by P. F. Hsieh and A. W. J. Stoddart. VI, 225 pages. 1971. DM 20,–

Vol. 184: Symposium on Several Complex Variables, Park City, Utah, 1970. Edited by R. M. Brooks. V, 234 pages. 1971. DM 20,–

Vol. 185: Several Complex Variables II, Maryland 1970. Edited by J. Horváth. III, 287 pages. 1971. DM 24,–

Vol. 186: Recent Trends in Graph Theory. Edited by M. Capobianco/ J. B. Frechen/M. Krolik. VI, 219 pages. 1971. DM 18.–

Vol. 187: H. S. Shapiro, Topics in Approximation Theory. VIII, 275 pages. 1971. DM 22,–

Vol. 188: Symposium on Semantics of Algorithmic Languages. Edited by E. Engeler. VI, 372 pages. 1971. DM 26,–

Vol. 189: A. Weil, Dirichlet Series and Automorphic Forms. V, 164 pages. 1971. DM 16,–

Vol. 190: Martingales. A Report on a Meeting at Oberwolfach, May 17-23, 1970. Edited by H. Dinges. V, 75 pages. 1971. DM 16,–

Vol. 191: Séminaire de Probabilités V. Edited by P. A. Meyer. IV, 372 pages. 1971. DM 26,–

Vol. 192: Proceedings of Liverpool Singularities – Symposium I. Edited by C. T. C. Wall. V, 319 pages. 1971. DM 24,–

Vol. 193: Symposium on the Theory of Numerical Analysis. Edited by J. Ll. Morris. VI, 152 pages. 1971. DM 16,–

Vol. 194: M. Berger, P. Gauduchon et E. Mazet. Le Spectre d'une Variété Riemannienne. VII, 251 pages. 1971. DM 22,–

Vol. 195: Reports of the Midwest Category Seminar V. Edited by J.W. Gray and S. Mac Lane.III, 255 pages. 1971. DM 22,–

Vol. 196: H-spaces – Neuchâtel (Suisse)- Août 1970. Edited by F. Sigrist, V, 156 pages. 1971. DM 16,–

Vol. 197: Manifolds – Amsterdam 1970. Edited by N. H. Kuiper. V, 231 pages. 1971. DM 20,–

Vol. 198: M. Hervé, Analytic and Plurisubharmonic Functions in Finite and Infinite Dimensional Spaces. VI, 90 pages. 1971. DM 16.–

Vol. 199: Ch. J. Mozzochi, On the Pointwise Convergence of Fourier Series. VII, 87 pages. 1971. DM 16,–

Vol. 200: U. Neri, Singular Integrals. VII, 272 pages. 1971. DM 22,–

Vol. 201: J. H. van Lint, Coding Theory. VII, 136 pages. 1971. DM 16,–

Vol. 202: J. Benedetto, Harmonic Analysis on Totally Disconnected Sets. VIII, 261 pages. 1971. DM 22,–

Vol. 203: D. Knutson, Algebraic Spaces. VI, 261 pages. 1971. DM 22,–

Vol. 204: A. Zygmund, Intégrales Singulières. IV, 53 pages. 1971. DM 16,–

Vol. 205: Séminaire Pierre Lelong (Analyse) Année 1970. VI, 243 pages. 1971. DM 20,–

Vol. 206: Symposium on Differential Equations and Dynamical Systems. Edited by D. Chillingworth. XI, 173 pages. 1971. DM 16,–

Vol. 207: L. Bernstein, The Jacobi-Perron Algorithm – Its Theory and Application. IV, 161 pages. 1971. DM 16,–

Vol. 208: A. Grothendieck and J. P. Murre, The Tame Fundamental Group of a Formal Neighbourhood of a Divisor with Normal Crossings on a Scheme. VIII, 133 pages. 1971. DM 16,–

Vol. 209: Proceedings of Liverpool Singularities Symposium II. Edited by C. T. C. Wall. V, 280 pages. 1971. DM 22,–

Vol. 210: M. Eichler, Projective Varieties and Modular Forms. III, 118 pages. 1971. DM 16,–

Vol. 211: Théorie des Matroïdes. Edité par C. P. Bruter. III, 108 pages. 1971. DM 16,–

Vol. 212: B. Scarpellini, Proof Theory and Intuitionistic Systems. VII, 291 pages. 1971. DM 24,–

Vol. 213: H. Hogbe-Nlend, Théorie des Bornologies et Applications. V, 168 pages. 1971. DM 18,–

Vol. 214: M. Smorodinsky, Ergodic Theory, Entropy. V, 64 pages. 1971. DM 16,–